Locations of Knowledge in Dutch Contexts

Knowledge Infrastructure and Knowledge Economy

Edited by

Karel Davids (*VU University, Amsterdam*)
Larry Stewart (*University of Saskatchewan, Saskatoon*)

VOLUME 6

The titles published in this series are listed at *brill.com/knie*

Locations of Knowledge in Dutch Contexts

Edited by

Fokko Jan Dijksterhuis
Andreas Weber
Huib Zuidervaart

BRILL

LEIDEN | BOSTON

Cover illustration: Nineteenth-century depictions of the Old Leiden Observatory. Leiden University Library.

Library of Congress Cataloging-in-Publication Data

Names: Dijksterhuis, Fokko Jan, editor. | Weber, Andreas, 1979- editor. | Zuidervaart, H. J., editor.
Title: Locations of knowledge in Dutch contexts / edited by Fokko Jan Dijksterhuis, Andreas Weber, Huib J. Zuidervaart.
Description: Leiden ; Boston : Brill, [2019] | Series: Knowledge infrastructure and knowledge economy, 1877-2323 ; volume 6 | Includes bibliographical references and index.
Identifiers: LCCN 2019029295 (print) | LCCN 2019029296 (ebook) | ISBN 9789004264878 (hardback) | ISBN 9789004264885 (ebook)
Subjects: LCSH: Science—Study and teaching—Netherlands—History. | Science—Netherlands—History. | Learning and scholarship—Netherlands—History. | Geography—Social aspects—Netherlands—History. | Knowledge, Sociology of—History.
Classification: LCC Q183.4.N4 L63 2019 (print) | LCC Q183.4.N4 (ebook) | DDC 509.492—dc23
LC record available at https://lccn.loc.gov/2019029295
LC ebook record available at https://lccn.loc.gov/2019029296

Typeface for the Latin, Greek, and Cyrillic scripts: “Brill.” See and download: brill.com/brill-typeface.

ISSN 1877-2323
ISBN 978-90-04-26487-8 (hardback)
ISBN 978-90-04-26488-5 (e-book)

This book is printed on acid-free paper and produced in a sustainable manner.

Contents

Figures and Tables

Figures

Tables

Notes on Contributors

Azadeh Achbari

received her PhD from the Faculty of Science, VU University of Amsterdam, the Netherlands. She is currently revising her dissertation, *Rulers of the Winds, How academics came to dominate the science of the weather, 1830–1870*, for publication as a monograph. Her dissertation and a number of related publications focus on discipline formation and the history of meteorology and marine studies during the long nineteenth century.

Fokko Jan Dijksterhuis

studies the cultural history of early modern knowledge practices. His early research focused on the history of early modern optics, investigating the intersections between mathematics, physics and instruments in the seventeenth century. He has an ongoing specific interest in matters of light, exploring new perspectives on early modern optics from the arts, instruments, philosophy, and economy.

Alette Fleischer

holds a MA degree as Art Historian and obtained in 2010 a PhD degree in the History of Science and Technology at the University of Twente. Her dissertation was titled: *Rooted in fertile soil: Seventeenth-century Dutch gardens and the hybrid history of material and knowledge production*. Since then, as an independent scholar who earns a living as a self-employed museum guide, she has published several articles on the topic of gardening and the production and consumption of knowledge in the Dutch Republic.

Floor Haalboom

teaches history of (veterinary) medicine and environmental history at Erasmus University Medical Center Rotterdam and Utrecht University. She is a member of the Utrecht University Descartes Centre for the History and Philosophy of the Sciences and the Humanities. As a post-doctoral historian, she writes about the modern history of environmental and health problems associated with modern livestock farming, including infectious diseases shared by humans and livestock. She is currently working on a global environmental history of livestock feed, started during her fellowship at the Rachel Carson Center for Environment and Society in Munich in 2018–2019.

Marijn J. Hollestelle
works as a project leader for QANU, conducting activities in the field of external quality assurance in higher education and research. He has worked as a researcher at the Foundation for the History of Technology at Eindhoven University of Technology, where he has done research on the history of polymer science. He published parts of this research in *The plastics revolution: how the Netherlands became a global player in plastics* (2017), together with H.W. Lintsen and H.N.M. Hölsgens, and in this present volume.

Dirk van Miert
is an associate professor of Early Modern Cultural History at Utrecht University and a member of the Descartes Centre Utrecht. He publishes studies on the history of early modern knowledge, in particular the Republic of Letters, the history of universities, and the history of scholarship.

Ilja Nieuwland
is a historian of science working for the Huygens Institute for the History of the Netherlands and the Vrije Universiteit Amsterdam. His research focuses on the cultural history of paleontology and paleontological museum collections. Recently, he published *American Dinosaur Abroad. A Cultural History of Carnegie's Plaster Diplodocus* (University of Pittsburgh Press, 2019).

Abel Streefland
received his PhD from Leiden University. He wrote a dissertation on Jaap Kistemaker and uranium enrichment in The Netherlands, 1945–1962. He currently works as a university historian at the Library of the Technical University in Delft. His research focusses on the post-war development of science and technology in The Netherlands.

Andreas Weber
is an assistant professor in the department of Science, Technology and Policy Studies (STePS) at the University of Twente in the Netherlands. Most of his research examines the history of natural history and chemistry in the context of the former Dutch empire in insular Southeast Asia. He is also involved in a digital humanities project, "Making Sense of Illustrated Handwritten Archives," directed toward developing search capability for illustrated archival material.

Martin P.M. Weiss
is a historian of science at the German Maritime Museum / Leibniz Institute for Maritime History in Bremerhaven and an associate lecturer at the University of

Brunswick. He studied physics and history of science in Aachen and Utrecht and received his PhD from Leiden University in 2013.

Gerhard Wiesenfeldt
is a historian of science and a lecturer in the School of Historical and Philosophical Studies at the University of Melbourne. He has worked extensively on the history of early modern natural philosophy, with a focus on universities, especially the University of Leiden. His current research studies the role of practitioners in the *longue durée* of academic philosophy.

Huib J. Zuidervaart
is a retired senior researcher from the Huygens Institute for the History of the Netherlands in Amsterdam, an institution of the Royal Netherlands Academy of Arts and Sciences – KNAW. He was the Editor-in-chief of *Studium*, the Belgian-Dutch journal for the History of Science, Medicine and Universities (2008–2017). Currently, with a colleague, he is finishing the last two volumes of the correspondence of the microscopist Antoni van Leeuwenhoek (1632–1723), a project started in 1939.

CHAPTER 1

The Netherlands as a Laboratory of Knowing: Introduction to Locations of Knowledge in Dutch Contexts

Fokko Jan Dijksterhuis and Andreas Weber

No, Netherland does not bow! in the kingdom of science,
she maintains her glory, stands on the highest stage.

HELMERS, 1812[1]

Between 1824 and 1844 the Dutch baron and parliamentarian Hendrik Collot d'Escury van Heinenoord (1773–1845) wrote a multi-volume book on the sciences and arts in the Netherlands: *Holland's Roem in Kunsten en Wetenschappen* ('The Glory of Holland in the Arts and Sciences').[2] He used the above lines as epitaph, from an ode on the Dutch nation. Collot erected a literary monument for a new state, the Kingdom of the Netherlands, that had just been established in 1815, after the defeat of Napoleon. It was the successor of the Dutch Republic, the idiosyncratic federation of provinces in the Low Countries that existed from 1588 to 1795. With the book Collot situated the new state in the global theatre of nation states as a prominent contributor to scientific and artistic progress.

The essays in this volume address knowledge practices in the Netherlands. They are not, however, exclusively about 'Dutch' knowledge production. They are particularly about locations: space- and time-specific configurations of actors, artefacts, and activities. The Dutch setting is what binds together the various locations discussed in this volume. The authors discuss the ways in

1 J.F. Helmers, *De Hollandsche Natie, in zes zangen* (Den Haag: J. Allart, 1812), v^{e} zang. "Neen Neerland zwicht niet! in het rijk der wetenschap,/Handhaaft zij haren roem, staat op den hoogsten trap."

2 H. Collot d'Escury, *Holland's Roem in Kunsten en Wetenschappen*, 7 volumes (Den Haag: Van Cleef, 1825–1844).

 | DOI:10.1163/9789004264885_002

which knowledge practices were shaped by the places of knowledge production. The cases are situated in the Netherlands, at different kinds of locations and at different moments between 1600 and 1980. We use 'the Netherlands' as shorthand for a historically fluid category referring to a more or less coherent geographical area in the Low Countries that had porous borders and an evolving sociocultural make-up. The Netherlands encompasses a range of locations with specific features. In some cases these features are fairly idiosyncratic – such as the decentral place of noble courts in the early modern Republic – in other respects the locations are comparable to other places around the globe.[3] Together the chapters present a variety of locations that enables comparisons and provides further insight into the way locations shaped knowledge practices. They were part of the broader history of the Dutch Republic and the Kingdom of the Netherlands that shaped the Netherlands and its relationships with the world. Some readers may be disappointed that we do not follow in the trails of Collot and praising the glory of the Dutch.

1 'Lab' – Staging Knowing in the Netherlands

The common designation 'the Netherlands' refers to a state with changing make-up whose borders were changeable and often permeable (see figure 1.1).[4] 'The Netherlands' also indicates a fluid spatiality and identity as a geographical node and sociocultural space. At the same time 'the Netherlands' is defined as much by its internal make-up as by its relationships with the outside world: from the early modern networks of urban 'nations' and world trade to the global fabric of modern colonies and post-colonies.

In the Low Countries, a new and rather idiosyncratic state emerged from the sixteenth century which included the northern provinces, the Republic of Seven United Netherlands (or Provinces; 1588–1795). The Republic was a multilayered society where power and wealth were distributed at all levels: between

3 Cf. Paul A. Elliott, *Enlightenment, Modernity and Science: Geographies of Scientific Culture and Improvement in Georgian England* (London / New York: I.B. Tauris, 2010).

4 James C. Kennedy, *A Concise History of the Netherlands* (Cambridge: Cambridge University Press, 2017); *A Miracle Mirrored. The Dutch Republic in European Perspective*, eds. Karel Davids and Jan Lucassen (Cambridge: Cambridge University Press, 1995); *Nederlandse cultuur in Europese context 1650–2000. Rekenschap*, eds. D. Fokkema and F. Grijzenhout (Den Haag: Sdu, 2001); *A History of Science in the Netherlands*, eds. K. van Berkel, A. van Helden and L. Palm (Leiden: Brill, 1999); *Technology and the Making of the Netherlands. The Age of Contested Modernization, 1890–1970*, eds. Johan Schot, Harry Lintsen and Arie Rip (Boston, MA: MIT Press, 2010).

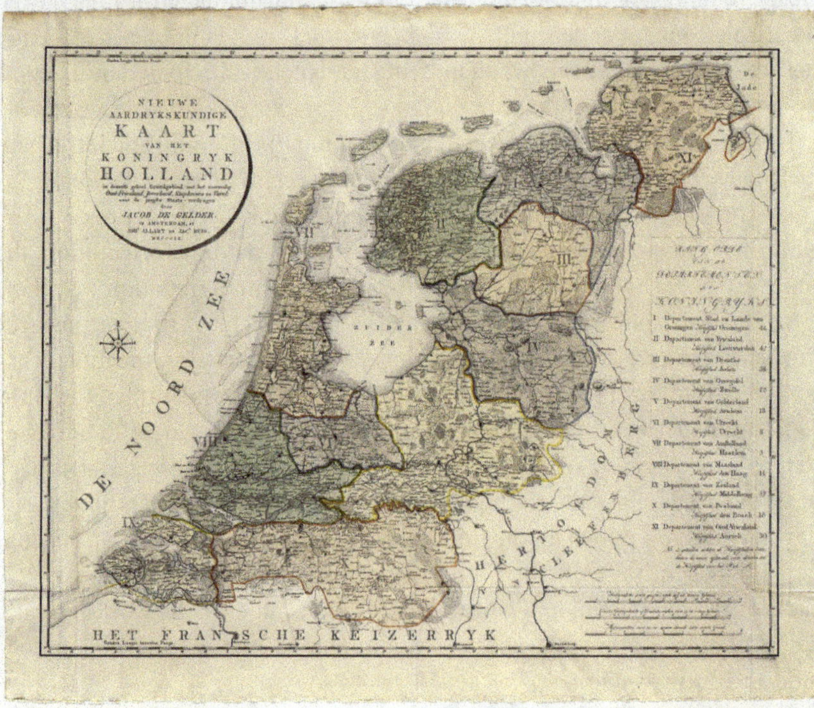

FIGURE 1.1 "Nieuwe Aardrykskundige Kaart van het Koningryk Holland" by Jacob de Gelder, 1810

This map of the Low Countries is rather unfamiliar to Dutch eyes. Not only are the provincial borders unrecognizable, the whole contours of the state are different. Still, this is quite characteristic of the Netherlands. Large parts of the south and east regions were not demarcated definitively and spheres of influence and affinity stretched beyond political borders. Moreover, overseas networks and terrorities defined 'the Netherlands' throughout the early modern and modern periods. The map is from the period of French domination (1795–1813) when large parts of the structure of the modern Kingdom of the Netherlands were created – including the royal foundation. In that sense it is also typical of the Netherlands, being a state that has coevolved with European and global histories.

SOURCE: SPECIAL COLLECTIONS, UNIVERSITY OF AMSTERDAM

the towns and regions making up the provinces; between the patrician families and institutions in the urban centres. In principle, every Dutch province was its own sovereign. And in every province the power was distributed over the various cities and its magistrates, as well as over the landed classes. However, the provinces were bound together by a union, ruled by the States-General with representatives from all seven provinces, who arranged foremost matters of foreign policy and national defence. On the levels of both the state and the province, Stadtholders (noblemen from the Dillenburg Nassau family) were chosen to represent military interests, but with only limited governing powers. The whole concept of centrality was alien to the Dutch Republic, at least until the late eighteenth century when unifying tendencies anticipated the patriotic aspirations for a nation state.

The Kingdom of Holland, erected in 1806 upon the groundwork of the patriotic and French reorganization of the state, was a centralised Napoleonic state, where the foundations were laid for the Kingdom of the Netherlands, a constitutional monarchy with central government and national orientation, that came into being in 1815. Paradoxically, this turn to a modern nation went hand in hand with a shift in power from the urban trade centres to the rural structures of agricultural and industrial regions.[5] More importantly, the nineteenth-century nation was built upon overseas territories that transformed the Dutch global trade network into a colonial empire.[6] Where the Dutch Republic existed as a transit zone of both European and global commercial networks of trade, the Kingdom of the Netherlands capitalized on the exploitation of overseas territories – especially in the Malay Archipelago and the Caribbean. If there is any continuity in 'the Netherlands,' it is the position at the coast as an entrepôt in commercial and cultural connections all over the world. Up until today Dutch economy and politics are dominated by the interests of global trade and entrepreneurship.[7]

Collot's ode to Dutch glory fitted the new conceptions of what knowledge was and what constituted a state. Enlightenment thinkers increasingly interpreted knowledge as a universal entity that progressed over time. With this conception they had begun writing a history of knowledge as a narrative of growth and improvement and the gradual uncovering of truth and proper method.[8] Besides disciplinary divisions and an increasing separation of formal and tacit knowledge practices, The Enlightenment combined the idea of universal knowledge with the emergent concept of the nation state. Specific nations would contribute to scientific progress and such contributions would reflect the glory and character of the nation. This paradoxical amalgam of

---

5 P. Brusse and W.W. Mijnhardt, *Towards a new template for Dutch history. De-urbanisation and the balance between city and countryside* (Zwolle: Waanders, 2011).

6 Catia Antunes and Jos Gommans, *Exploring the Dutch Empire. Agents, Networks and Institutions, 1600–2000* (London: Bloomsbury, 2015).

7 See also Harold Cook, "Amsterdam, entrepôt des savoirs au XVIIe siècle," *Revue d'histoire moderne & contemporaine* 55–2 (2008): 19–42.

8 As pioneered and exemplified by Jean-Étienne Montucla, *Histoire des Mathematiques* (Paris: Jombert, 1758; extended edition 1799–1801); Joseph Priestley, *The History and Present State of Discoveries Relating to Vision, Light and Colours* (London: Johnson, 1772) (preceded by *The History and Present State of Electricity* of 1768).

universality and individuality opened a competition between nations over inventions, discoveries and priorities. The rivalry over universal knowledge was part of the contest over global power and wealth. Politically, economically, and culturally, the nation state was defined by its overseas territories and its cultural reputation anchored in realms of universality.[9]

Collot propelled the Netherlands into the race by highlighting the unique character of the Dutch and the incomparable achievements of the Golden Age in particular. This volume turns his story inside out. It is not about the Dutchness of knowledge, let alone the glory of Dutch knowledge production. The Netherlands provides a geographical focus that enables comparison and cross-reference of different types of locations. In this way the Netherlands – from the early modern Republic to the modern Kingdom – is a laboratory to investigate the various properties and effects of locations on knowledge production. A laboratory not necessarily in the cliché sense of white coats isolating and testing phenomena, but rather as a multifarious site where phenomena are created and brought under control – or not, when an experiment fails.

In this respect, the volume takes a cue from an article written by Klaas van Berkel on the international relevance of Dutch history.[10] Van Berkel's 'The Dutch Republic. Laboratory of the Scientific Revolution' may be seen as a reflection upon his earlier history of science in the Netherlands. Originally published in Dutch in 1985, it was included in the 1999 handbook *A History of Science in the Netherlands*.[11] In his 2010 article, Van Berkel asked how a Dutch history relates to a history of science in general. He argues that the case of the Netherlands can be used as a historical laboratory to investigate the nature and causes of early modern transformations of science and society.

---

9 Jürgen Osterhammel, *Die Verwandlung der Welt. Eine Geschichte des 19. Jahrhunderts* (München: Beck, 2009); R. Drayton, *Nature's Government: Science, Imperial Britain, and the "Improvement" of the World* (New Haven, CT: Yale University Press, 2000); Michael Werner and Bénédicte Zimmermann, "Vergleich, Transfer, Verflechtung. Der Ansatz der Histoire croisée und die Herausforderung des Transnationalen," *Geschichte und Gesellschaft* 28–4 (2002) 607–636.

10 Klaas van Berkel, "The Dutch Republic. Laboratory of the Scientific Revolution," BMGN – *Low Countries Historical Review* 125–2,3 (2010): 81–105; Klaas van Berkel and Leonie de Goei, "The International Relevance of Dutch History: Introduction," BMGN – *Low Countries Historical Review* 125–2,3 (2010): 3–6.

11 Klaas van Berkel, *In het voetspoor van Stevin. Geschiedenis van de natuurwetenschap in Nederland 1580–1940* (Meppel: Boom, 1985). It was translated into English and enlarged with some essays as *A History of Science in the Netherlands*, eds. K. van Berkel, A. van Helden and L. Palm (Leiden and Boston: Brill, 1999). This book can be still considered the most up-to-date general history of science in the Netherlands.

Our volume takes up the idea of a laboratory, but proposes to broaden both the content and its application. Understanding locations as places of knowledge transformation extends the concept of lab beyond the confines of a secluded research and development laboratory. Van Berkel maintains a rather positivistic conception of a laboratory as a demarcated and institutionalized setting where phenomena are isolated and brought under control (see figure 1.2a). In our view it is problematic to transpose this concept to earlier periods. Moreover, such a narrow conception precludes analytical opportunities to study knowledge practices. By broadening the concept to places like the field, the workshop, the construction site, and even the city, the laboratory becomes a place where historical cases can be studied about the question of how locally situated socio-cultural configurations produce specific knowledge practices.[12] Locations are laboratories where new configurations of actors and ideals transform processes of knowing.

Van Berkel's reflections focus upon the relationship between the Netherlands and the Scientific Revolution in the seventeenth century. This can be extended towards the eighteenth-century and beyond to include developments as the Enlightenment, industrialization, colonization and globalization. However, such a move also raises the historiographical question of how to approach the relationship between local history and such 'universal' events. Van Berkel argues that the small and demarcated context of the Dutch Republic can help to identify dynamics and crucial variables of general developments. He identifies three elements of society in the Dutch Republic that in his view were conducive for the rise of modern science: (1) an open and sustained interaction between scholars and craftsmen; (2) a prominent role for academic institutions as centres of both teaching and research; and (3) a congruence of scientific and mercantile values.[13] The Republic reflected general trends like the artisanal involvement, but also had more singular features, for example in higher education. Newly established universities in the Republic were not burdened by the scholastic heritage of old established universities elsewhere in Europe.

---

12 D. Turnbull, *Masons, Tricksters and Cartographers: Comparative Studies in the Sociology of Scientific and Indigenous Knowledge* (London and New York: Psychology Press, 2000); Pamela O. Long, *Artisan/practitioners and the Rise of the New Sciences, 1400–1600* (Corvallis, OR: Oregon State University Press, 2011); D.N. Livingstone, *Putting Science in Its Place. Geographies of Scientific Knowledge* (Chicago: University of Chicago Press, 2003).

13 In a review article of BMGN Benjamin Schmidt argues that such efforts to identify specific Dutch features is rather common in Dutch historiography. Benjamin Schmidt, "Dikes and Dunes: On Dutch History and Dutchness," BMGN – *Low Countries Historical Review* 133–1 (2018): 82–99.

FIGURES 1.2A–1.2B Two nineteenth-century depictions of the Leiden Observatory. The contrasting images nicely illustrate different views of 'the location of science.' The engraving of the 'Hoogeschool' emphasizes the seclusion of research and the autonomy of science. The pictoresque environments of the bastion park offer a tranquil place for observation and contemplation. This is one of the aspects of locations illustrated by the essays in this volume: knowledge is cultivated in specific places where expertise and interests meet. Another aspect is emphasized in the photograph of the town and its academic buildings: knowledge is embedded in wider social and cultural settings. The photograph from 1861 not only shows the messy appearance of a city but also signal the dispersed and often ad hoc character of scientific practice.
SOURCE: LEIDEN UNIVERSITY LIBRARY AND ERFGOED LEIDEN

However, such an approach considers the Scientific Revolution to be some universal development that affected local circumstances and took specific shape in regional contexts. Such an approach often leads to histories of pioneers and followers, centres and peripheries, boiling down to questions of how the Dutch Republic contributed to 'the' Scientific Revolution. Historiography of the Industrial Revolution instead emphasizes the multiplicity of processes of industrialisation from early modern period onwards. Rather than *one* Industrial Revolution of coal, textile and private entrepreneurship, originating in the English Midlands, different pathways can be recognized, such as the typical interactions between state, gentry and learned elites in the German lands.[14] For the Netherlands, Karel Davids has put the early modern rise and decline of technological leadership in comparative perspective.[15] Taken together, such scholarship shows that *explicanda* like the 'Scientific Revolution,' 'Enlightenment,' 'Industrial Revolution' tend to evaporate when put in the Petri dish of local history.[16] Instead of investigating universal history on a local scale, a reverse approach considers a location as a setting that gives shape to situated values and practices (including perceptions of the natural world).

2 'Location' – Place and Knowing

The previous section has made clear that what defines a location is far less interesting than what it does. A location is not just a place where things happen to happen. It is a stage that gives shape and meaning to a performance, a performance of knowing in our case. A location creates the setting for knowledge production by bringing people and practices, objects and interests together, and by structuring their interactions. Space is a necessary condition for science; where else could it take place, the historian and sociologist of science

14 Ursula Klein, *Nützliches Wissen. Die Erfindung der Technikwissenschaften* (Göttingen: Wallstein Verlag, 2016); *Landschaften agrarisch-ökonomischen Wissens. Strategien innovativer Ressourcennutzung in Zeitschriften und Sozietäten des 18. Jahrhunderts*, ed. Marcus Popplow [Cottbuser Studien zur Geschichte von Technik, Arbeit und Umwelt, vol. 30] (Münster: Waxmann Verlag, 2010). Thomas Misa shows how heterogeneous industrialization in different places in England was: Thomas Misa, *Leonardo to the Internet. Technology and Culture from the Renaissance to the Present* (Baltimore: Johns Hopkins University Press, 2004).

15 Karel Davids, *The Rise and Decline of Dutch Technological Leadership. Technology, Economy and Culture in the Netherlands, 1350–1800* (Leiden: Brill, 2008).

16 See above: Withers, "Place and the Spatial Turn"; *Negotiating Knowledge in Early Modern Empires. A Decentered View*, eds. L. Kontler, A. Romano, S. Sebastiani and B.Z. Török (New York: Palgrave, 2014); Annapurna Mamidipudi, *Towards a theory of innovation in handloom weaving in India* (Maastricht: Maastricht University, 2016).

Steven Shapin once rhetorically asked.[17] Knowing situated in time and place and the outcomes of inquiry cannot be abstracted from where they are produced. As Ludwik Fleck detailed in his monumental 1935 study, 'Erkennen' ('Understanding') is a social and cultural phenomenon rooted in the interactions and evolutions of concrete communities.[18] Fleck's emphasis on the local and material circumstances of research has been taken up again in the last quarter century. In this introduction we draw upon some concepts that have been developed to detail and problematize the situatedness of knowledge production.[19]

Besides providing spatial configurations for ways of knowing and making, locations also mediate relationships. They connect different actors, confronting and aligning interests and ambitions, and they trigger the development of specific knowledge practices. The idea of location as a junction of social, cultural, political, and economic relationships employs the concept of trading zone as developed by Peter Galison and Pamela Long.[20] Discussing twentieth-century particle physics, Galison showed how seemingly incommensurable world readings can become connected when people act together, irrespective of the different interpretations of the materials shared and exchanged. Pamela Long has transposed this idea to the early modern period in an effort to understand interactions between scholars and artisans. The early modern period witnessed a transition of knowledge cultures in which roles of actors were in flux and not institutionally fixed. In Long's discussion of a number of trading zones, the rearrangement of such roles is illustrated and analysed.[21] The

17 Steven Shapin, "Review of David Livingstone: Science, Space and Hermeneutics," *British Journal for the History of Science* 36 (2003): 89–90.

18 Ludwig Fleck, *Entstehung und Entwicklung einer wissenschaftlichen Tatsache. Einführung in die Lehre vom Denkstil und Denkkollektiv* (Frankfurt am Main: Suhrkamp, 1980) 56–60.

19 Hans-Jörg Rheinberger, *Toward a History of Epistemic Things. Synthesizing Proteins in the Test Tube* (Redwood City: Stanford, 1997) ('Experimentalsystem'); see next footnote for 'trading zones'; see also *Sites of Mediation. Connected Histories of Places, Processes, and Objects in Europe and Beyond, 1450–1650*, eds. Susanna Burgharz, Lucas Burkart and Christine Göttler (Leiden and Boston: Brill, 2016); Kocku von Stuckrad, *Locations of Knowledge in Medieval and Early Modern Europe: Esoteric Discourse and Western Identities* (Leiden and Boston: Brill, 2010); *Science as Cultural Practice: Vol. 1: Cultures and Politics of Research from the Early Modern Period to the Age of Extremes*, eds. Moritz Epple and Claus Zittel (Berlin: De Gruyter, 2014).

20 Peter Galison, *Image & Logic. A material culture of microphysics* (Chicago: The University of Chicago Press, 1997); Pamela O. Long, "Trading Zones in Early Modern Europe," *Isis* 106–4 (2015): 840–847.

21 Long, *Artisan/practitioners* responds to Edgar Zilsel's 1942 argument that the influence of artisanal knowledge practices on scholars was crucial to the Scientific Revolution. She emphasizes the interaction and mutual transformation of knowledge practices into new ones, precipitating into modern institutionalized domains from the eighteenth century.

concept of trading zone emphasizes the fluidity of roles of actors, as they are shaped by interactions rather than the other way around. The contributions in this volume build upon such notions of localizing knowledge practices and showcase the variety of configurations and their outcomes.

What also binds the cases in this volume together is the evolving scene of the Netherlands over the early modern and modern periods. It thus ties in with attempts by David Livingstone and others to write a history of knowledge from a geographical perspective.[22] A geography of knowledge perspective emphasizes the ways in which specific knowledge practices emerge from a variety of sites and how circulation between different sites creates translocal qualities of knowledge. On a regional scale this means that different sites interact in cultural contexts that both assign meaning and value to knowledge practices and are shaped by them. Building upon Livingstone, Simon Naylor argues that such contexts are nested at different levels: city, region, nation, as well as global networks of power, commerce and knowledge.[23] The Netherlands presents various instances of such multi-layered contexts. For instance, the early networks of cities were intertwined with their global linkages. Or, the rise of modern disciplinary structures in the sciences was entangled with the construction of the Dutch national and colonial system. A long-term view enables us to illustrate and understand that such contexts of identification and valuation were often only temporarily anchored. Naylor identifies a third aspect of the geography of knowledge, which he describes as the way in which science reciprocally creates geographies. This is not central to this volume, although it is touched upon in some of the essays, such as Fokko Jan Dijksterhuis's account of Golden Age Amsterdam and Azadeh Achbari's analysis of imperial networks of meteorology in the long nineteenth century. In a more general sense, one of the underlying ideas of this volume is that such knowledge practices also contributed to the shaping of 'the Netherlands.'

Long identifies three features of early modern trading zones: driven by 'technological' projects, substantial interest of elites, and involving persons with markedly fluid careers.

22 Livingstone, *Putting Science in Its Place*; A. Ophir and S. Shapin, "The place of Knowledge: The Spatial Setting and Its Relation to the Production of Knowledge," *Science in Context* 4:1 (1991): 3–21; Charles W.J. Withers, "Place and the 'Spatial Turn' in Geography and in History," *Journal of the History of Ideas* 70–4 (2009): 637–658; Diarmid A. Finnegan, "The Spatial Turn: Geographical Approaches in the History of Science," *Journal of the History of Biology* 41 (2008): 369–388; *Geographies of Nineteenth-Century Science*, eds. David N. Livingstone and Charles W.J. Withers (Chicago: University of Chicago Press, 2011).

23 Simon Naylor, "Introduction: historical geographies of science – places, contexts, cartographies," *The British Journal for the History of Science* 38 (2005): 1–12. For a general reflection see also L.L. Roberts, "Situating Science in Global History: Local Exchanges and Networks of Circulation," *Itinerario* 33 (2009): 9–30.

3 Structure of This Volume: Three Aspects of Locations

We divided the volume in three sections that highlight different aspects of locations: a *kind* of location, the city; a specific *feature* of locations, their connections over time and place; and an *effect* of locations, the transformation of knowing. The essays also interconnect these three aspects, for example the way exchanges over networks transform the matters that are moved. Likewise, global connections have been defining for cities and national institutes alike. Recent historiography has made clear that circulation is a main agent of knowledge transformation. This volume draws upon this perspective, but focuses on the transformative working of locations rather than circulation between locations.[24]

The **City** is a specific kind of location. It can be considered as a location itself, but it also contains specific urban places. In his sketch of city improvement in Golden Age Amsterdam, Dijksterhuis explores how the city as such gave rise to particular forms of knowledge production. Gerhard Wiesenfeldt shows how the characteristics of Leiden as an industrial town affected the organization and ambitions of the university. In his essay he presents the university as an urban institution and the 'engineering school' as a product of Leiden's industrial character in that time. Huib Zuidervaart uses the case of Middelburg to discuss anatomical theatres as a product of an ambitious urban culture that was the cornerstone of the Dutch Republic. Anatomical theatres created a new kind of node that gave rise to a knowledge practices in which the interests of professionals and urban bureaucratic elites were integrated. Zuidervaart describes a dynamics that is usually associated with learned societies of the eighteenth century. The three cases in this section present a varied picture of the urban cultures and remind us of the central role cities played in the Dutch Republic. Historiography of the city as a place of knowing has been developing only recently, most notably in Deborah Harkness's account of Early Modern London and Sven Dierig's study of nineteenth-century Berlin.[25] Trendsetting

24 *Silent Messengers. The circulation of Material Objects of Knowledge in the Early Modern Low Countries*, eds. Sven Dupré and Christoph Luthy (Berlin: LIT Verlag, 2011); *Cultural Transfer Through Translation. The Circulation of Enlightened Thought in Europe by Means of Translation*, ed. Stefanie Stockhorst (Amsterdam: Rodopi, 2010).

25 D.E. Harkness, *The Jewel House. Elizabethan London and the Scientific Revolution* (New Haven, CT and London: Yale University Press, 2007); Sven Dierig, *Wissenschaft in der Maschinenstadt. Emil Du Bois-Reymond und seine Laboratorien in Berlin* (Göttingen: Wallstein, 2006). The topic is addressed in a more general way in *Science and the City*, eds. Sven Dierig, Jens Lachmund and Andrew Mendelsohn [Osiris 18] (Chicago: Chicago University Press, 2003). See also H. Cook, *Matters of Exchange. Commerce, Medicine, and Science in the Dutch Golden Age* (New Haven, CT: Yale University Press, 2007).

in this respect has been the late 1990s project on urban achievements in early modern Europe.[26] It presented an interdisciplinary, comparative account of Antwerp, Amsterdam, and London, along different themes aimed at understanding the dynamics of flourishing and the relationship with urban settings. In urban history, issues of knowledge production are also addressed, querying how specific features of cities stimulate innovation as well as the nature and agency of the city as such.[27] Alette Fleischer draws upon these debates by highlighting how 'the' city is defined by mycelial structures of early modern towns, rather than as a locally delimited entity.

The section **Networks** exemplifies how locations establish connections over time and place and how these connections produce specific ways of knowing. In her study on the Danzig naturalist and collector Jacob Breyne, Fleischer approaches the early modern city as a node of mercantile routes. Her reconstruction of the activities of the Breyne family showcases how personal and commercial ties fostered knowledge transfer. In a similar vein, but then with a focus on the modern period, Achbari argues how European aspirations in global trade and empire laid the foundation for a new scientific discipline of meteorology. By describing the gradual development of international conferences in medicine and natural history, Ilja Nieuwland presents a parallel case, but then with a special focus on the nationalization and professionalization of other scientific disciplines as they emerged in the nineteenth century. Both essays tellingly show how networks that enabled the coordination of a large number of individuals at different places appear as the very location of knowledge production. Abel Streefland extends this line of reasoning to the twentieth century by showing how the efforts of the Netherlands to position itself in international politics shaped Dutch nuclear science.

The final section on **Transformations** shows how locations process knowledge. Instead of departing from the notion that transformations always lead to new knowledge, this section rather focusses on processes of combining, recombining and reinterpreting already available knowledge through

---

26 *Urban Achievement in Early Modern Europe. Golden Ages in Antwerp, Amsterdam and London*, eds. Patrick O'Brien, Derek Keene, Marjolein 't Hart and Herman van der Wee (Cambridge: Cambridge University Press, 2001).

27 Bert De Munck, "Disassembling the City: A Historical and an Epistemological View on the Agency of Cities," *Journal of Urban History* 43–5 (2016): 811–829; Gert-Jan Hospers, "Creative Cities: Breeding Places in the Knowledge Economy," *Knowledge, Technology, & Policy* 16–3 (2003): 143–162; Ali Madanipour, *Knowledge Economy and the City: Spaces of Knowledge* (Oxon: Routledge, 2013). An urban history perspective on early knowledge societies is developed in *Knowledge and the Early Modern City. A History of Entanglements*, eds. Bert De Munck and Antonella Romano (Oxon: Routledge, 2019).

different methods of test, proof and rejection. It presents widely different sorts of locations – stable, lectern, museum, R&D lab – to be able to distinguish and compare different settings of knowledge production. Dirk van Miert illustrates how the oratory processing of knowledge in the form of disputation produced new textual knowledge. Seen from this perspective, the lectern can be read as a Petri dish and the lecture hall as the laboratory where the research took place. Martin Weiss shows how Teylers Museum in Haarlem gradually transformed from a laboratory where new scientific knowledge was made into a display of what science was considered to be. Floor Haalboom shifts the focus to cowsheds in the countryside in which academic and industrial attitudes were juxtaposed. In doing so, all involved actors brought forward new conceptions of hygiene. Marijn Hollestelle, in turn, contrasts academic and industrial laboratories as different sites of R&D where polymer chemistry was processed.

The essays in this volume also display links between the sections and thus the different aspects of the way locations mediate knowledge production. Zuidervaart offers an exquisite example of the way the urban setting of Middelburg gave a specific shape to medical and anatomical knowledge, thus showing how the city also transformed a body of knowledge. Conversely, the transformations that Weiss sketches as occurring in the 19th-century Teylers Museum were also embedded in the setting of urban culture. Fleischer shows how the early modern merchant town can be understood from a network perspective. The exemplary laboratories presented by Hollestelle were essentially networked in the fabric of industry and academia. Achbari discusses the rise of a new location, the sea, that both transformed meteorological inquiry and was transformed by the mercantile, imperial, and scientific interests underlying the rise of a new discipline. Nieuwland shows how scientific societies established new locations of knowing precisely by capitalizing on dislocation, whereas Streefland argues that the Netherlands could engineer its place in Cold War nuclear research by carefully managing an international network of contacts.

The long term perspective of this volume also allows for highlighting the transformations that locations underwent themselves. Their nature and the role co-evolved with knowledge practices, as is exemplified by the case of Teylers Museum, as discussed by Weiss in this volume. In a more general sense the emergence of modern universities is closely tied to concrete knowledge practices like disputations, lectures, and collecting, as well as the national interests of scientific research. The transformation of the city from the early modern period to the late twentieth century reminds us of the question raised by the mythic king Theseus: to what extent is the ship still the ship after all the parts have been gradually replaced over time. Likewise, the global networks of the Dutch continued to give shape to 'the Netherlands,' even when the nature of such networks changed from mercantile voyages to colonial governance to international collaboration.[28] Whatever its parts, the Netherlands continues to navigate across the seven seas. The editors of this volume hope to have produced a useful compass to allow readers to take up the challenges presented in these chapters.

By conceptualizing locations of knowing as time-specific configurations of actors, artefacts and activities, the essays of this volume speak a clear language: over the last four hundred years, the Netherlands have witnessed the emergence of a variety of different locations at which scientific and other forms of knowledge were produced, disputed and often also rejected. Many of them are still visible until the present day. Taken together the essays in this volume show that understanding the histories of the Dutch Republic and the Kingdom of the Netherlands through its 'locations of knowing' offer historians a fruitful heuristics to bring different scales of analysis together: on one hand, our approach allows to study the complexities and often experimental character of knowledge production on a local scale. On the other hand, a locations perspective also serves as reminder that local dynamics of knowledge production cannot be analyzed disconnected from developments on a regional and global scale. Many of the locations we are discussing in this volume form part of wider geographies of communication, travel and exchange in which actors and artefacts not only from the Netherlands but also other parts of the world played an important role. By approaching the history of the Dutch Republic and the Kingdom of the Netherlands through some of its localities, we invite readers to work towards a spatialized understanding of knowledge production in the Netherlands in which local dynamics as well as regional and global entanglements remain never far out of sight.

---

28 The problem of 'transformation' is touched upon in the introduction to *Transformations of Knowledge in Dutch Expansion*, eds. Susanne Friedrich, Arndt Brendecke and Stefan Ehrenpreis (Berlin: De Gruyter, 2015).

PART 1

Cities

CHAPTER 2

Science in the Theatre of Towns

Fokko Jan Dijksterhuis

1 Introduction

In 1649 the Amsterdam publisher Joan Blaeu published *Tonneel der Steden* ('Theatre of Towns and Cities'), a pinnacle of Dutch Golden Age cartographical printing.[1] It is a lavishly produced collection of maps and descriptions of all the towns and cities of the Dutch Republic, including sections on the many and diverse curiosities and achievements in the Low Countries. It is also a remarkable book, because it is an atlas of *cities* rather than lands, seas or skies. Traditionally the focus of atlases had been domains; from principalities up to the world and the cosmos. With the increasing dominance of urban centers in the Low Countries atlases of cities had begun to appear.[2] In the *Tonneel der Steden* the building blocks of the Dutch Republic stood center stage. The fact that this atlas too, like the *Atlas Maior*, was a collectible treasure for the wealthy citizens of the Republic reflects the confidence Dutch towns and cities had gained over the previous century. By the first half of the seventeenth century the Low Country cities had established their international economic and political powers and cultural prestige. These cities knew that they were the ultimate foundation of the Republic's position and it was only fitting to have this reflected in iconological production.

The self-consciousness of the Dutch cities manifested itself in a new genre in chorography: city descriptions.[3] Chorography was the part of geography

1 The book was first published in Latin: *Novum ac Magnum Theatrum Urbium Belgica* (Amsterdam: Joan Blaeu, 1649). The Dutch translation was published the same year: *Toonneel der Steden van 's Konings Nederlanden, met hare Beschrijvingen* (Amsterdam: Joan Blaeu, 1649). Note the designation in the Dutch of the Netherlands as 'royal.'

2 Georg Braun (1541–1622) and Frans Hogenberg (1535–1590) issued an atlas of cities in the 16th century (with engravings by Joris Hoefnagel and others): the highly popular *Civitates orbis terrarum* (Cologne: Petrum a Brachel, 1572–1618) that went through many editions and expansions. Hogenberg introduced the urban identification from Flanders into the German lands.

3 Eßer Raingard, "Städtische Geschichtsschreibung in den Niederlanden im 17. Jahrhundert: Chorographie und Erinnerungskultur," in: *Bild und Wahrnehmung der Stadt*, ed. P. Johanek, [Städteforschung. Reihe A, Darstellungen; vol. 63] (Wien-Köln-Weimar: Böhlau Verlag, 2012), 105–120; Eddy Verbaan, *De woonplaats van de faam. Grondslagen van de stadsbeschrijving in*

 | DOI:10.1163/9789004264885_003

that gave historical-geographic descriptions of lands, traditionally focusing on domains and dynasties.[4] The Republic did not fit this regional and personal identification, being constituted in a multi-level structure of cities, provinces and the States-General, in which wealth and power were distributed. As the cornerstone of political and economic power the city became the point of reference for cultural identity. In 1611 the Harderwijk historian Johannes Isacus Pontanus (1571–1639) published *Rerum et urbis Amstelodamensium historia*. It was the first Dutch city description, making the city the analytic unity of chorography.[5] Pontanus built the reputation of Amsterdam by situating it historically – working back to Batavian times, describing its impressive achievements in trade and the high level of urban amenities, and, not lastly, the glorious expeditions that had set sail from the harbor. Other Dutch cities would follow suit and during the 17th and 18th centuries numerous city descriptions were published to express urban pride.[6]

Urban chorography is an example of the way the early modern city was a site of knowledge production. The cities were junctions of trade, power and culture that functioned as nodal points of exchange, terminals of processing and financers of projects and education. The city was a stage for inquisitive and inventive pursuits that produced new ways of knowing. Yet, a stage is not a place where things just happen to be done; it gives meaning and shape to the performance.[7] The city too sets the stage for urban affairs and this spatial dimension is constitutive to what is made and done in town.[8] In this paper I want to explore the way the urban theatre gave specific shape and meaning to early modern knowledge production.

---

de zeventiende-eeuwse Republiek (Hilversum: Verloren, 2011). An inspiration may have been Lodovico Guicciardini's highly popular ode to the Low Countries from 1567, in which he emphasized the role of the towns.

4 See for example Lesley Cormack, *Charting an empire: geography at the English universities, 1580–1620* (Chicago: University of Chicago Press, 1997).

5 E.O.G. Haitsma Mulier, "De eerste Hollandse stadsbeschrijvingen uit de zeventiende eeuw," *De Zeventiende Eeuw* 9–2 (1993): 97–111. The Dutch translation was published in 1614 as *Historische beschrijvinghe der seer wijt beroemde coop-stadt Amsterdam*.

6 Geert Medema, *Achter de façade van de Hollandse stad. Het stedelijk bouwbedrijf in de achttiende eeuw* (Nijmegen: Vantilt, 2011), 14–20.

7 See David Livingstone, *Putting science in its place. Geographies of scientific knowledge* (Chicago: University of Chicago Press, 2003), 5–12.

8 Medema, *Achter de façade*, 20–23; *Urban Achievement in Early Modern Europe. Golden Ages in Antwerp, Amsterdam and London*, eds. Patrick O'Brien, Derek Keene, Marjolein 't Hart, and Herman van der Wee (Cambridge: Cambridge University Press, 2001), in particular the introduction: Patrick O'Brien, "Reflections and mediations on Antwerp, Amsterdam and London in their golden ages," in: O'Brian et al., *Urban Achievement*, 3–35.

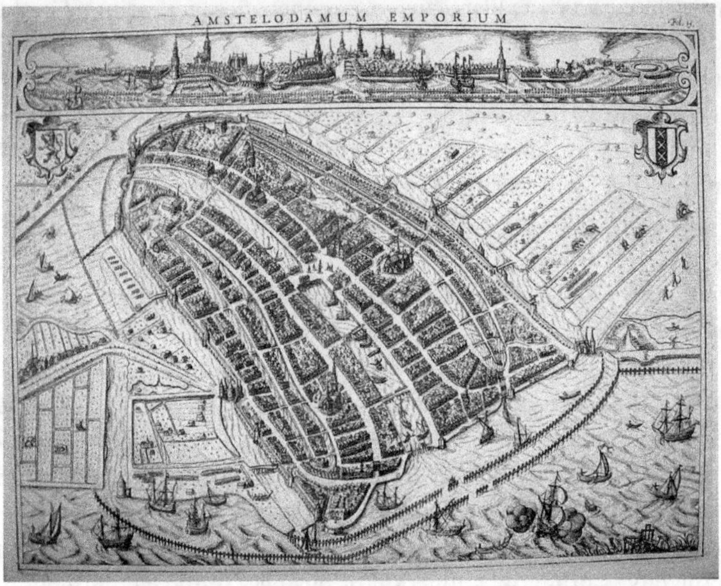

FIGURE 2.1 Birds-eye view of Amsterdam drawn in 1544 by Cornelis Anthonisz. (ca. 1505–1553), as reproduced in Joh. Pontanus, *Rerum et urbis Amstelodamensium historia* (Amsterdam: Hondius, 1611)
SOURCE: RIJKSMUSEUM, AMSTERDAM

I will discuss a rather specific case: Amsterdam, the most prominent city of the Dutch Republic. In comparison to the other European states the Dutch Republic was a rather idiosyncratic society. It was a highly urbanized country of trade and industry without a true center. The power was with the cities, but this was a distributed power. The Republic essentially was a conglomerate of cities that cooperated at the level of provinces and via the provinces also on a national level. The power within the cities was distributed in the same way, among a class of patricians, urban associations, and so on. This patriciate may have been gradually developing into an urban aristocracy; it retained its bourgeois and de-centralized character. This affected the way knowledge was cultivated. The quality of urban space and services was central to the reputation of an early modern city and thus figured prominently in city depictions.[9] Accordingly, the organization and management of public works by the city government was key to a city's existence. This paper will focus on this dimension of early modern Amsterdam, leaving such spheres as the free

9 Ibidem, 14–23.

market of knowledge workers like teachers, apothecaries, and painters in the background.[10]

Together with the metropolises elsewhere in Europe, the cities of the Dutch Republic played a decisive role in the development of the early modern sciences. In comparison to the crown and the gown, in history of science the historiography of the town as site of early modern knowledge production is relatively underdeveloped. The court, including gardens and societies as well as amateurs and gentlemen, has risen high on the agenda of history of early modern science. Recent historiography has shown the universities as particularly early-modern places of scholarship where, amongst other things, the pursuits of courtly learning were aligned with academic traditions. As regards the urban theatre, in the history of science the market and the coffeehouse have been studied as sites of knowing; the city *as such* only sporadically, however.[11] Harkness's magnificent study of sixteenth century London stands out as an inspiring example.[12] Cities – Amsterdam in particular – do appear prominently in studies of global trade as sites of knowledge production.[13] Still, in these cases the city is primarily a junction of business traffic, an entrepôt of knowledge.[14] In urban history knowledge production has received growing interest, as regards the early modern period most notably in the project on urban

---

10 Fokko Jan Dijksterhuis, "Fit to Measure. 'Bequamheit' in Mathematics in the Dutch Republic," in: *Public Offices, Personal Demands. Capability in Governance in the seventeenth-century Dutch Republic*, eds. J. Hartman and J. Nieuwstraten (Newcastle upon Tyne: Cambridge Scholars Publishing, 2009), 80–100; Tim Nicolaije, "Dwaasheid of retoriek? Cornelis van Leeuwen en de 'Belachelijke Geometristen,'" *Studium* 5–1 (2012): 1–14.

11 *The Cambridge History of Science. Volume 3. Early Modern Science*, eds. Katherine Park and Lorraine Daston (Cambridge: Cambridge University Press, 2006) gives a good overview of recent historiography of science, including the role of courts and universities in early modern knowledge cultures. The geography of knowledge is acknowledged in the introduction, and while it contains chapters on specific sites such as markets, homes, libraries, courts, gardens and laboratories and networks, the city is not spotlighted.

12 Deborah Harkness, *The Jewel House. Elizabethan London and the Scientific Revolution* (New Haven, CT: Yale University Press, 2007).

13 Harold Cook, *Matters of Exchange. Commerce, Medicine, and Science in the Dutch Golden Age* (New Haven, CT: Yale University Press, 2007); Thomas J. Misa, *Leonardo to the Internet. Technology and Culture from the Renaissance to the Present* (Baltimore: Johns Hopkins University Press, 2004).

14 Harold Cook, "Amsterdam, entrepôt des savoirs au XVIIe siècle," *Revue D'histoire Moderne et Contemporaine* 552–2 (2008): 19–42; Patrick O'Brien, *Urban Achievement in Early Modern Europe: Golden Ages in Antwerp, Amsterdam and London* (Cambridge: Cambridge University Press, 2001) focusses on the comparison of economic, social and cultural achievements of three main towns, and less about the particular urban dynamics that produced these.

achievements in Antwerp, Amsterdam and London.[15] In this paper I want to see how the way the city's social and cultural structure and dynamics contributed to the processing of knowledge. Besides urban history perspectives on knowledge cultures, I build upon the work done in recent history of technology where the city is presented as a knowledge junction, but that tends to link this exclusively to modern, industrial cities.[16] I also build upon the wonderful work that has been done in recent history of Dutch cities and city planning that has developed a socio-material perspective on cities and show how cities too have stories to tell.[17] By linking up these with questions of knowledge production, I will try and explore the city as a site of knowing.

2 Securing Cities

The success of the Dutch Republic was the success of its cities. The Dutch towns and cities had to be won over for the Dutch cause of political and religious liberty, and often literally conquered from the Spanish. Amsterdam had a special place in the Dutch Republic, being the most wealthy and powerful of all its Dutch competitors. At the same time this city had not sympathized with the Revolt and had to be forced to take sides in 1578. It profited tremendously from the blockade of Antwerp in 1585 that resulted in an accelerated growth of trade, industry, and population. Eventually the Revolt became successful and from the late 1580s the front definitively fanned out from the cities in the west to ever-expanding circles of besieged cities in the north, east and south.

The success of the Revolt was to a large degree the result of the innovative warfare methods developed by the two Stadtholders Maurits and Willem Lodewijk, the Nassau counts and cousins who had taken over the leadership of the Revolt after Willem of Orange's assassination in 1584. They reformed organization and training of their armies and introduced powerful techniques

15 O'Brien et al., *Urban Achievement*. See also *Knowledge and the Early Modern City. A History of Entanglements*, eds. Bert De Munck and Antonella Romano (Oxon: Routledge, 2019).

16 Misa, *Leonardo to the Internet*, 59–96 (Chapter 3: Geographies of Industry, 1740–1851); Adrienne van den Bogaard, "Site-Specific Innovation," in: *Technology and the Making of the Netherlands. The Age of Contested Modernization, 1890–1970*, eds. Johan Schot, Harry Lintsen and Arie Rip (Cambridge, MA: MIT Press, 2010), 125–178.

17 Medema, *Achter de façade*; Ed Taverne, *In 't land van belofte: in de nieue stadt. Ideaal en werkelijkheid van de stadsuitleg in de Republiek, 1580–1680* (Maarsen, 1978); Frank Ankersmit, "Over stad, stadsgeschiedenis en het kunstwerk," in: *De Stad*, ed. Mieke Dings (Rotterdam: Uitgeverij 010, 2006), 12–20; Karl Schlögel, *Im Raume lesen wir die Zeit. Über Zivilizationsgeschichte und Geopolitiek* (München: Fischer-Taschenbuch-Verlag, 2003).

of siege. A full-scale program of fortifying cities was a cornerstone of their campaign. As the frontline advanced the conquered cities were protected from recapture by innovative fortification works. These works were supervised by the chief engineer Adriaen Anthonisz. (1527–1607) assisted by three engineers for the northern, eastern and southern regions.[18] In this way the Revolt transformed the cities of the Low Countries. Besides new fortifications that transformed the appearance of the cities, standing garrisons were established that transformed the make-up of the city by introducing an autonomous and influential element to urban life.[19]

The transformation of the cities was not a unilateral operation prescribed by the Stadtholders and their engineers. The fortifications were the outcome of negotiations over the location, dimensions and design of hornworks, bastions, ravelins, moats and gates. The lay-out of fortifications required a profound survey of the terrain around a city to assess the possibilities for building robust constructions, optimal defense lines, and a labor- and material-efficient design. As the city itself had to pay for the new fortifications, it had a decisive say in the plans and required a detailed estimate of the costs involved.[20] In many cases several alternative plans were made before agreement was reached (see figs. 2.1–2.4).[21] In order to calculate the investment and infrastructural consequences, the cities demanded realistic and accurate plans that unambiguously explained the proposed works and made them controllable. In this process the engineer played a central role by designing, elaborating and adapting plans. Rather than as the creator of the fortress, the engineer can be seen as the intermediary that made transparent and aligned military and urban interests. They acquired their expertise in situ and made it mobile by bringing it to new construction sites.[22] In the course of the revolt a system of mapping and planning

---

18 Frans Westra, *Nederlandse ingenieurs en de fortificatiewerken in het eerste tijdperk van de Tachtigjarige Oorlog, 1573–1604* (Alphen aan den Rijn: Canaletto, 1992); Charles van den Heuvel, *"Papiere bolwercken." De introductie van de Italiaanse stede- en vestingbouw in de Nederlanden (1540–1609) en het gebruik van tekeningen* (Alphen aan den Rijn: Canaletto, 1991). Adriaen Anthonisz. was a cartographer and surveyor from Alkmaar. He had played a crucial part in the defense of his hometown against the Spanish attack in 1573, the victory that is said to have been the turning point in the Revolt. He was appointed 'master of fortifications' by the States of Holland in 1579, and 'superintendant of fortifications in Holland and Utrecht' by William of Orange in 1584.

19 Jonathan Israel, *The Dutch Republic. Its Rise, Greatness, and Fall, 1477–1806* (Oxford: Oxford University Press, 1995), 241–275.

20 Charles van den Heuvel, "De versterkte stad," in: *De Stad*, ed. Mieke Dings (Rotterdam: Uitgeverij 010, 2006), 23–36.

21 The various plans for fortification of Deventer are discussed in Westra, *Nederlandse Ingenieurs*, 41–47.

22 Compare David Turnbull, *Masons, Tricksters and Cartographers. Comparative Studies in the Sociology of Scientific and Indigenous Knowledge* (London: Psychology Press, 2000).

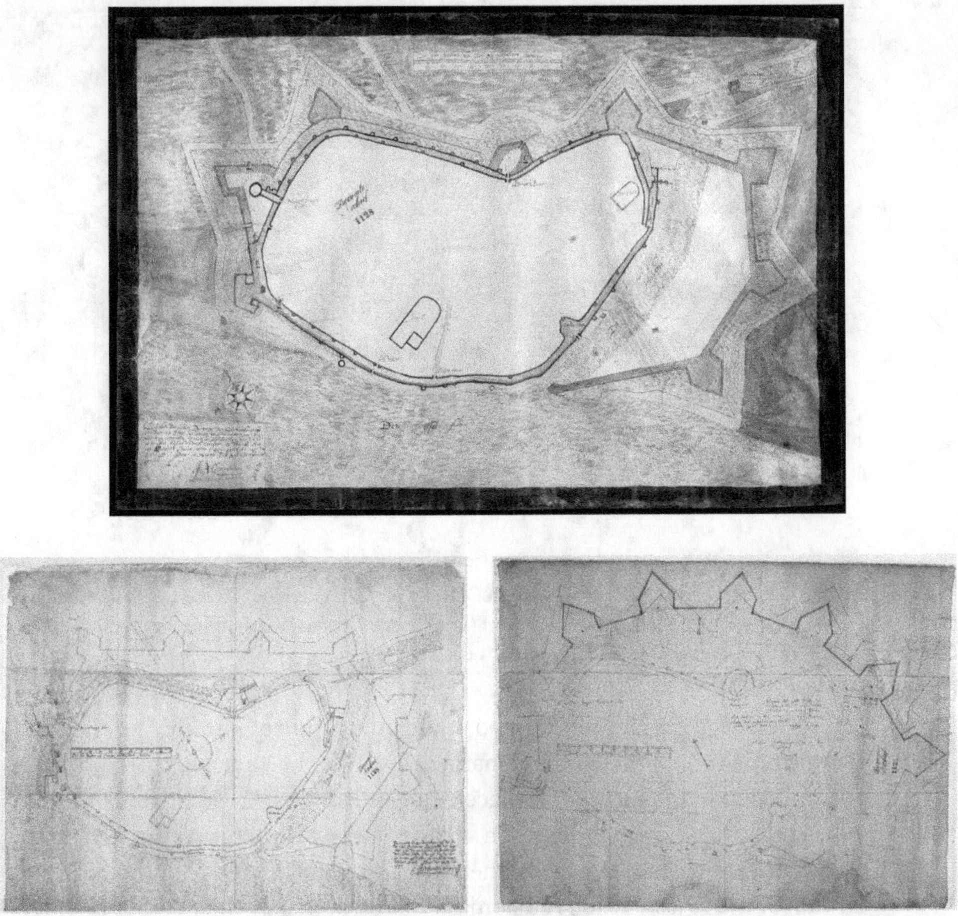

FIGURES 2.2–2.4 Plans for the defense works of Deventer from 1594–1595 by Adriaen Anthonisz. (2.2 and 2.3) and Jacob Kemp (2.4).
SOURCE: DEVENTER STADSARCHIEF AND ATHENAEUMBIBLIOTHEEK, 0691_808, 0691_805, 1691_806.

developed that had features of standardized design rules which made the fortification of Dutch cities manageable for the urban administrators, the famous 'Old-Dutch' fortification system (see fig. 2.5).[23] Thus the Dutch cities not only took shape by the new fortification system, but just as much gave shape to it.

23 On the Italian and Spanish context of the process of enveloping expertise and the nature of Dutch 'schools' of fortification, see: Van den Heuvel, *Papiere Bolwercke*; idem, "Les ingénieurs dans la boue. Adapter les styles internationaux aux pratiques néerlandaises en matière de fortifications," in: *Vauban, bâtisseur du Roi-Soleil. Fortifications, urbanisme, architecture*, eds. Isabelle Warmoes and Victoria Sanger (Paris: Somogy, 2007), 42–47.

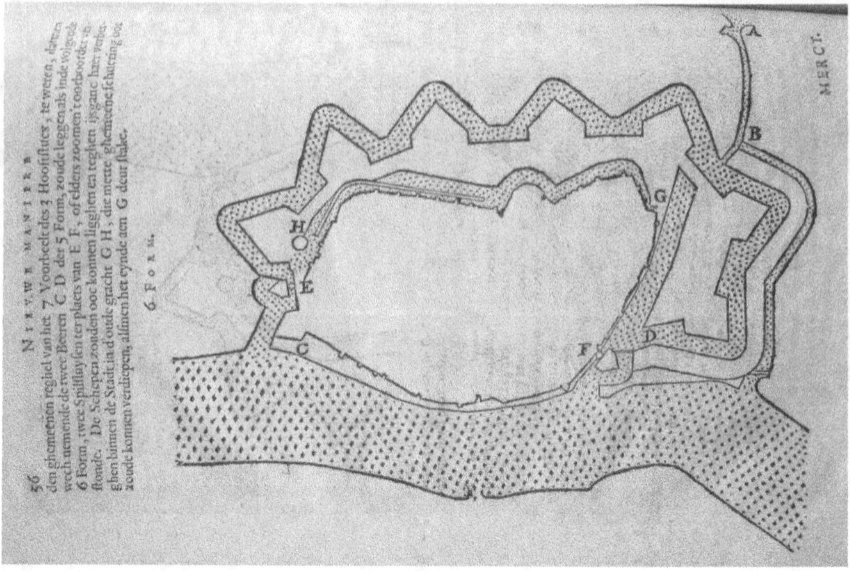

FIGURE 2.5 The defense works of Deventer as they appeared in Simon Stevin, *Nieuwe Maniere van Sterctebou door Spilsluysen* (Rotterdam: Jan van Waesberghe, 1617), 56
SOURCE: GENT UNIVERSITY LIBRARY

The structuring of a city contributed to the production of knowledge. The case of Amsterdam illuminates the dynamics involved. Once again the main city of the Republic took a special place as regards infrastructural works. It had only modest fortifications but was a site for drastic restructuring through three phases of extension in the 1580s, the 1610s and the 1660s. Only in the political turmoil of the 1650s – when Amsterdam was assaulted by the new Stadtholder Willem II – the city started new fortification works.[24] The extension of the city was a large technological project that presents an illuminating case of knowledge production. The civil setting emphasizes the urban features of the process, which is further illuminated by the subsequent infrastructural works on the water management of the city.

3 Working Mud

After the Stadtholder's assault on Amsterdam in 1650 the city carried out some ad hoc defense work. More structural plans were made in 1651 and 1652 as part

24 The assault was an attempt at manifesting his power in the dispute over authority in the Dutch Republic between the States and the Stadtholder. This is no place to discuss the complex political history of the Dutch Republic. For an introduction see Israel, *The Dutch Republic*.

of the plan for a large scale expansion of the city on the south-east side.[25] In 1660 the planning of the area began when a committee of the city council was installed to evaluate five concepts that had been proposed and make a report for a 'roeyinge,' a plan for the layout of streets and canals.[26] The city surveyor Cornelis Danckerts de Rij (1596–1662) was instructed to make twelve copies of a map of the existing buildings for the planners to work on.[27] In January 1662 the committee gave its report, starting from the points where the plans agreed, deciding on some of the differences, and leaving open a couple of issues.[28] After the plan was established, the city architect, the city surveyor and the city engineer were instructed to mark it out in the terrain. With an extra surveyor and a team of diggers, the structure of streets and canals was indicated by plotting and digging ditches.[29] Now the inhabitants of the area could 'regulate' themselves: prepare for expropriation and removal.[30] The terrain outside the city limits was after all not empty land but was inhabited by many people and enterprises. The city had the power to expropriate and clear rules existed for compensation. To prevent laborious taxations the city later decided on a fixed price of 16 guilders per rod, accepting the estimated extra costs of 45,000 guilders.[31]

The projected extension of the city was a large administrative challenge that, on the one hand, had to resolve the city's problems of accommodating a vastly growing population and, on the other hand, was to yield immense profits for the city. The new extension – the fourth since 1578 – went beyond mere administration, though. The city council wanted to prevent the problems

---

25 Data on planning, organization, and executing of the expansion come from the grandiose Jaap Evert Abrahamse, *De Grote Uitleg van Amsterdam. Stadsontwikkeling in de zeventiende eeuw* (Amsterdam: Thoth, 2010). On the basis of immense archival work he has created a narrative of the project from the earthworks up. See also Marjolein 't Hart, "The glorious city: monumentalism and public space in seventeenth-century Amsterdam," in: O'Brien et al., *Urban Achievement*, 128–150. 't Hart focuses on monumental aspects, but also draws comparisons with Antwerp.

26 Abrahamse, *Grote Uitleg*, 145–147. The committee contained two people involved in the fortification works: Joan Blaeu and the trader, counselor and mayor Cornelis Witsen. The concepts were made by city architect Daniel Stalpaert, city carpenter and 'onderfabrieksmeester' Gerrit Barentz Swanenburg (three plans), and burgomaster Joan Huydecoper. Boudewijn Bakker, "Het primaat van de orde. De Grote Uitleg als bestuurlijk kunstwerk," in: *Amsterdams Werelderfgod. De Grachtengordel na 400 jaar*, eds. Marloes Slooff and Jaap Evert Abrahamse (Amsterdam: Thoth, 2012), 28–47, reads the plans as an esthetic design along Stoic lines, without however historically substantiating his interpretation.

27 Abrahamse, *Grote Uitleg*, 145.

28 Ibidem, 148.

29 Ibidem, 158.

30 Ibidem, 150.

31 Ibidem, 160.

the previous extension of the 1610s had produced. That had been mainly an administrative operation of managing the issue of parcels while maintaining the existing pre-urban structure. As a result the city had tremendous difficulties regulating social and economic life and managing the infrastructure.[32] The extension of the 1660s was an integrated design for the new city in which all aspects of city structure were planned and controlled: business and industry, traffic and defense, allotment and sale of land, water management.[33] This included discarding existing informal structures and expropriating the people living and working in the 'empty' land. The design made manifest the architectural principles of order and proportion that were being codified by authors such as Simon Stevin and Nicolaus Goldmann, and that reflected the modular thinking also apparent in fortification and land reclamation.[34]

The plan was executed with the help of a professional government apparatus. Central was the *stadsfabriek* (city firm) that consisted of a group of officially appointed 'city masters,' such as the city carpenter and the city mason.[35] Although the precise organization differed from location to location, all Dutch cities had a 'city works' to develop and maintain public works. In Golden Age Amsterdam the task of the 'stadsfabriek' was incomparably larger, having to realize the huge expansion of the city – in a way, its very existence. At the time of the fourth extension the main figure was the city architect Daniël Stalpaert (1615–1676), who held this office from 1648 to his death in 1676.[36] He was appointed to supervise the building of the new town hall and other official buildings but acquired a central position in the city planning from the 1660s extending his work from the design and building of objects to the layout of the new city. The office of city architect was without precedent and was not continued after. The 'stadsfabriek' saw to the daily execution of the plans

32 Ibidem, 119.

33 Ibidem, 215–216.

34 Heuvel, "Ingénieurs dans la boue"; Jeroen Goudeau, *Nicolaus Goldmann (1611–1665) en de wiskundige architectuurwetenschap* (Groningen: Philip Elchers, 2005); Erik de Jong, C. Steenbergen and P. de Zeeuw, "De Beemster. Een arena van natuur, kunst en techniek," in: *Nederland als kunstwerk. Vijf eeuwen bouwen door ingenieurs*, ed. Toon Lauwen (Rotterdam: NAi Uitgevers, 1995), 53–167.

35 Stadsfabriek literally translates to 'city factory' or 'city works.' Because of the type of organization 'city firm' (or 'city farm') seems more appropriate, comparable to the later Ferme génerale in France and tax farming in general. Gea van Essen, *Het Stadsfabrieksambt. De organisatie van de publieke werken in de Noordelijke Nederlanden in de zeventiende eeuw* (Ph.D. dissertation, Utrecht University, 2011).

36 Gea van Essen, "Daniel Stalpaert (1615–1676), stadsarchitect van Amsterdam en de Amsterdamse stadsfabriek in de periode 1647 tot 1676," *Bulletin KNOB* 99–4 (2000): 101–121.

and the contracting-out of the work. By the time of the fourth extension, the 'stadsfabriek' was directly supervised by the city council. The treasurer was in charge and directed the 'onderfabrieksmeester' (vice master of the city works), as the office of master of the city works had been discarded in the 1640s.[37] The 'stadsfabriek' was organized in the way most governmental bodies in the Dutch Republic were: through offices that were contracted out.[38] Holders of the offices were thus not in service of the government. Likewise the competence of the masters was solely judged on the basis of their work and the government did not assess or stipulate their qualifications directly.

The fortification works were supervised by engineers who were recruited from outside: trained and practiced in the army. After the assault of 1650, the city had appointed a special engineer, *stadsingenieur* Jan Heijmansz. Coeck (after a long-time service to the Spanish army), to manage the fortification works.[39] From 1663, the work on the city extension was coordinated with the fortification works, under the supervision of Coeck in direct instruction of the treasurer. Because of the close connection between the fortification works and the city planning, engineers were also involved with the extension assisting with specific designs and constructions.

Surveyors played an important role in the layout of the new city, beginning with the implementation of the macro structure. The basic unit of city planning was the bloc; the building line was the most important structuring element. The real work of digging the canals and leveling up the city blocks came next. It was executed by contractors and involved a competitive trade in mud and waste that were scarce but indispensable for the job.[40] Further details of the alignment of parcels were made. The projected ground level was determined by measuring the water level and applying sounding rods to indicate the building surface and the slope of the streets. In this way prospective buyers could plan the construction of their houses.[41] In the meantime, the definitive allotment was made that ought to solve the puzzle of distributing the various urban functions in an optimal way, which would guarantee maximum profit for the city. Trade and industry were situated along the radial streets, consisting of shops with upstairs living floors, whereas the streets along the canals were reserved for private (wealthy) housing. The business lots along the side-streets should be big enough for the shops without taking up too much of the

37 Essen, *Stadsfabrieksambt*, 35–40.
38 See also Hartman and Nieuwstraten, *Public Offices, Personal Demands.*
39 Abrahamse, *Grote Uitleg*, 121.
40 Ibidem, 166.
41 Ibidem, 174.

valuable land along the canals.[42] The puzzle of optimal distribution of parcels was solved by surveying and calculation and eventually the city council agreed with a plan that promised maximum profits and minimum costs. The building lines were marked out accurately as their enforcement was the main instrument of city planning. The alignment should be observed carefully as violation could result in the demolition of a newly built house. After the granting of the lots, a special clerk was appointed for the inspection.[43]

In the case of the surveyors we also see that the city contracted out the work, and more importantly, the assessment of competence. As in the case of engineers and masters stipulation of qualifications was delegated to the structures and standards of the profession. The profession of surveyor was relatively open and formal qualifications were only required for contractors of surveying works with legal status. To acquire such a qualification a surveyor had to be admitted by the 'Hof van Holland.' This was the sole institutionalized proof of a surveyor's competence and there were no preconditions of mathematical education or training. An official surveyor could employ assistants on the basis of his own judgement. So, not all of the surveyors involved in the measuring out of the new city were formally qualified.[44] The expansion of Amsterdam was a huge site of knowledge production that required sustained levels of expertise. Characteristically the city did not provide this expertise itself, nor did it regulate its quality directly. As with the work to be done, quality of knowledge was contracted out to the city masters and the professional groups.[45]

4 Conducting Water

In the course of the 1670s, the extension project petered out following the crisis of the Republic. In 1672 the Dutch Republic was attacked by all major European states and the basis of its economic and political power was effectively swept away. In response, the office of Stadtholder, abandoned in 1651, was restored with the appointment in 1672 of the young Willem III (1650–1702)

42 Ibidem, 188–189.

43 Ibidem, 192–193.

44 E. Muller and Kees Zandvliet, *Admissies als landmeter in Nederland voor 1811. Bronnen voor de geschiedenis van de landmeetkunde en haar toepassing in administratie, architectuur, kartografie en vesting- en waterbouwkunde* (Alphen aan den Rijn: Canaletto, 1987); H.C. Pouls, *De landmeter. Inleiding in de geschiedenis van de Nederlandse landmeetkunde van de Romeinse tot de Franse tijd* (Alphen aan den Rijn: Canaletto, 1997).

45 See also Karel Davids, "Amsterdam as a centre of learning in the Dutch golden age, c. 1580–1700," in: O'Brien et al., *Urban Achievement*, 305–325.

and the cities more or less took a step down. The 'disaster year' 1672 was particularly disastrous for Amsterdam in economic terms (but not necessarily politically): growth subsided and expansion stalled.[46] The city was directly threatened by the advance of the French army and the land around the city was inundated. A group around Gillis Valckenier (1623–1680) seized upon the imminent disaster to dispose of his main rivals in the urban government, a transition that happened throughout the Dutch Republic upon the restauration of Stadtholderly power. The other principal figures of the group were Johannes Hudde (1628–1704), Nicolaas Witsen (1641–1717), and Coenraad van Beuningen (1622–1693). Besides serving as burgomasters regularly they held positions with the Admiralty, the Dutch East India Company (VOC), and so on, as well as diplomatic posts. The direct response to the military threat of Amsterdam consisted of renewed and intensive fortification work and meant an immediate halt to the extension work. After peace had been restored, the city faced tremendous obstacles for continuing the public works. The instantaneous repair and construction of the fortification had left the city with large debts. At the same time the urban economy went into a long recession and the extension came to a halt. Only in the 1680s the issue of parcels got going again, but the fourth extension was never fully completed.[47]

On top of all these tribulations, the city was struck by a flood in 1675. Being close to the sea and almost at sea level, high water was not uncommon for Amsterdam. The All Saints Flood that November was particularly devastating for the city as well as large areas in the provinces.[48] In response the city decided to improve the city's water defense by a new system of dikes and locks along the sea front. A central figure in these plans was Hudde, a long-time governor in Amsterdam boards and from 1672 prominent in the city council. Between 1672 and 1703 he held the office of Mayor as often as rules allowed, two of every three years. Several interests had to be taken into account: the seawall ought not to hinder the seaport nor obstruct the water traffic through the canals, while at the same time the residents in the area should not lose out too much.[49]

---

46 The Dutch Republic was attacked by England, France, Münster, Cologne, and Bavaria. The Dutch States Army was quickly defeated, a large part of the Republic conquered and Amsterdam seriously threatened. In response Orangists took over power in cities and provinces and the Stadtholder was re-installed in office. See Israel, *Dutch Republic*.

47 Abrahamse, *Grote Uitleg*, 180, 210.

48 M.K.E. Gottschalk, *Stormvloeden en rivieroverstromingen in Nederland. Deel III: de periode 1600–1700* (Assen: Van Gorcum, 1977), 272.

49 Fokko Jan Dijksterhuis, "Hudde, de wiskunde en de stad," *Studium* 11–1 (2018): 62–77.

In March 1681 the city council took the final decision. The quay along the city front would be reinforced and a series of eight locks built between the sea and the main canals. The 'aquatic fortifications' should be operative within two years. The organization of the works differed greatly from the extension works. For the execution of the plan a member of the council, close ally and cousin of Hudde, Witsen, was appointed to manage the project. The project thus was not handed to the 'stadsfabriek'; the city masters were directly instructed by the city council. The plan succeeded: in January 1683 the burgomasters and treasurers could inspect the new water defense. Until it began to fall in disrepair in the 1700s it protected the city from further floods.[50]

In each lock a sounding stone was placed that indicated the mean high tide of the river IJ and the nearby Zuiderzee.[51] They are known as the 'Hudde stones.' One stone still survives in the 'Eenhoornsluis' and reads: "zee dyks hooghte zynde negen voet vyf duym boven stadtspeyl" (sea dike's height being nine feet five inch above city level). The sounding stone established the height of the sea dike with respect to the level of the water in the Amsterdam canals. The level was based on precise measurements of the heights of the Zuiderzee and a determination of the mean high tide. This was combined with data of the levels in the canals in the various neighborhoods. To implement the level, an intricate geodetic web was established throughout the city, using surveying measurements along the canals and in the narrow streets. As we have seen, expertise to do this work was available and had been crucial to the management and extension of the city. The new level was, however, not local to one canal or neighborhood. It fixed a water level that applied to the whole city, creating an imaginary plane that transformed the city into a unified whole. The stones in the sea locks were used to monitor the water level and take appropriate action to prevent flooding of the city.

The city's high water defense may have been successful; floods were not the main challenge to the city's water management. As the city lay low in a wet area, the water flow through the canals could hardly be controlled. It was extremely difficult to get clean water into city and the dirty water out. The canals in the densely populated and industrious metropolis were highly polluted. Amsterdam was probably the filthiest city of Europe; the smell of the air was often unbearable. The management of the waters was a complex problem that involved various interests that often conflicted: the city had to be defended

---

50 Abrahamse, *Grote Uitleg*, 328–330.

51 In the following centuries the city level became the 'Standard Amsterdam Level' (NAP), the national reference that has been adopted by several international boards. P.I. van der Weele, *De geschiedenis van het N.A.P.* (Delft: Nederlandse Commissie voor Geodesie, 1971).

from the sea, traffic through the canals should not be obstructed, the levels in the canals should not damage to houses and workshops along the canals, etc. etc. The locks in the canals were required to maintain the appropriate water levels in each neighborhood and were controlled by lockkeepers with specific instructions to respond to changes in the water flow. In addition to the urban interests, the city had to take into account the interests of surrounding regions. It depended on the neighboring water boards for its water supply and could not manipulate the water flow at will. The city extensions only increased the problems as the system of canals grew larger and more complex. Throughout the (shiny but smelly) Golden Age solutions to the problem were tried, like a plan to use mills to pump the water through the canals.[52] It was all in vain.

As burgomaster, Hudde had taken up the problem of water management with renewed energy.[53] He revived the plan to make locks in the dam that blocked the river Amstel (after which Amsterdam is named). He sent for the city masters to discuss the plan. Soon after it was decided that this waterworks would also be contracted out rather than realized by the 'stadsfabriek.'[54] The idea was to flush out the canals in a controlled fashion when enough water had accumulated. In combination with basins to collect waste water that would be pumped in with mills, this should enhance the quality of the water in the canals. To no avail, the Amstel locks did not solve any of the problems they were built for.[55] As regards the efforts to clean up the canals, administrative measures were more successful than infrastructural projects. For example, regulations for the disposal of waste by brewers and a system of barrels to collect and convey garbage were of some effect.[56]

As is reflected by the virtual high water plane that ran through the city, the management of the levels and flow of the water were becoming a responsibility of the city government instead of the individual neighborhoods. The regulations for operating the locks in the canals were geared to the requirements in other parts of the city, prescribing standard procedures for the lockkeepers and increased demands regarding their discipline and skills, including requirements in literacy.[57] The increased centralization and rationalization of water management did not, however, preclude private initiatives. Throughout the seventeenth century, men of all trades proposed plans to the city government

52 Abrahamse, *Grote Uitleg*, 307.

53 L. Jansen, "De oorsprong van het Stadswaterkantoor," *Werk in uitvoering* 4 (1953–54): 142–143.

54 Abrahamse, *Grote Uitleg*, 314.

55 Ibidem, 315.

56 Ibidem, 300–305.

57 Ibidem, 319–321.

to solve (once and for all) the water challenges.[58] These could be officials and contractors of the city, builders and engineers, entrepreneurs and schemers, as well as learned amateurs. In 1682 one Elias Sandra came to Hudde to present a plan to direct clean water from the river Vecht – about 15 kilometers to the east – to the city by means of a system of canals and sluices.[59] Two years later he published the plan that consisted of three alternatives for the transport of freshwater, proposals for fountains, and detailed estimates.[60] The booklet also included a report of the discussion Sandra had had with Hudde. The latter had raised administrative objections, such as the fact that the plan required permission on a provincial level and the consent of affected landholders and water boards. Hudde also discussed financial and technical problems, explaining that the city's pockets were empty and that the soil was not suited for constructing a canal.[61] Sandra replied to these all objections in detail, explaining how the investment – that was large indeed – could be recovered by marketing the water.[62] Like many plans for the improvement of the city's waters, Sandra's proposal never materialized.

5 New Philosophy in the City

The success of the water defense – as regards the management of the works as well as its effectiveness – is often attributed to a particular character of the group of governors that dominated the Amsterdam administration in the last quarter of the seventeenth century. They are an indication of what Karel Davids has called the late-seventeenth-century peak in Amsterdam's role in advancement of knowledge.[63] The group around Hudde and Witsen that had risen to power in 1672 was particularly interested in the new philosophy, mathematics and natural history and stood in close contact with prominent savants like

58 Ibidem, 321–325.

59 Marijke Donkersloot-De Vrij, *De Vechtstreek. Oude kaarten en geschiedenis van het landschap* (Weesp: Heureka, 1985), 78–79; J.A. Groen, *Een Cent per Emmer. Het Amsterdamse drinkwater door de eeuwen heen* (Amsterdam: Amsterdamse Waterleiding maatschappij, 1978), 29–33.

60 Elias Sandra, *Ontwerp en beschryvinge om het soet water uyt de riviere de Vegt op driederley wyse te brengen binnen de stad Amsterdam; en vervolgens door de geheele stad, om yder huys met soet en vars water te voorsien. Mitsgaders fonteynen te maken, daer men 't begeerd....* Gepractiseerd door E.S. t' Amsterdam [gedrukt voor den Auteur], 1684.

61 Sandra, *Ontwerp*, 45: "Tegenwerpingen van de Heer Burgemeester Johan Hudde, gedaen in October 1682."

62 Sandra, *Ontwerp*, 46–47.

63 Davids, "Amsterdam as a centre of learning," 320–322.

Christiaan Huygens, Isaac Vossius, Frederick Ruysch and Jan Swammerdam. They were trailblazers of Cartesianism and were probably involved in the circles around Spinoza.[64] Van Beuningen was interested in philosophy, theology and philology, combining Cartesian ideas with a mystical attitude. Hudde had studied mathematics with Frans van Schooten and contributed to the latter's edition of Descartes's *Geometry*. 'Hudde's rule' for minima and maxima of polynomials was highly valued by seventeenth-century mathematicians. He collaborated with Spinoza on lens grinding. In 1671 he and Christiaan Huygens made a survey of the Lower Rhine and the Holland IJssel to advice the States-General on the management and amelioration.[65] As a student, Witsen gave a disputation at the Amsterdam Athenaeum on comets, that also discussed optical topics. It was dedicated to his cousin, the 'incomparable philosopher and mathematician' Johannes Hudde. In later life Witsen became a grand collector, and built a network of informants for geographical and natural historical information from around the globe. These men mobilized their idiosyncratic learning for the well-being of the city, presenting their savant activities in mathematics, cartography and geography as contributions to the urban interests in trade and administration.

It is often assumed that their predilection for mathematics and philosophy affected their governmental work, producing a kind of 'scientific management.' Traces of a 'scientific approach' may be discerned in the water defense works, like the integral character that transcended the existing local organization of water management to the level of the city as a whole. I am rather skeptical, however, about such suggestions, as they tend to privilege top-down processes and formal knowledge. The waterworks project built on robust traditions of measurement and surveying. Furthermore, it remains to be seen whether the project was markedly different in terms of design and management. Hudde did indeed bring a large body of unique learning and versatility to Amsterdam and in the 1660s he had been pivotal in the stimulation of the new philosophy.[66] Yet, despite Hudde's mathematical proficiency I do not see that he was more rational and calculating than his predecessors. The whole planning and executing of the city expansion had been a carefully managed process of measuring, proportioning, and calculating. The only notable difference was the

64 Cook, *Matters of Exchange*, 286; See also Israel, *Dutch Republic*.

65 Christiaan Huygens, *Oeuvres Complètes de Christiaan Huygens* (Den Haag, 1888–1950), tome 6, 60–71.

66 Eric Jorink and Huib Zuidervaart, "Inleiding: Johannes Hudde (1628–1704), 'groot ende onvergelykelyk wiskunstenaar'," *Studium* 11–1 (2018): 1–14; Theo de Jong and Huib Zuidervaart, "'Philosophus et Mathematicus Incomparabilis': Johannes Hudde (1628–1704), de Verlichte burgemeester van Amsterdam," *Studium* 11–1 (2018), 26–54.

FIGURE 2.6 Map of Amsterdam with the extension drawn in by Daniel Stalpaert, 1662. Map published on a later date by Nicolaes Visscher.
SOURCE: AMSTERDAM UNIVERSITY LIBRARY

way the city sidestepped the *Stadsfabriek* in the planning and realization, but this is a governmental innovation that may reflect new developments in Dutch governance in the later Golden Age and the centralizing tendencies after 1672. Moreover, it seems that after 1670 his interest in philosophical issues of comets, generation, and so on, shifted towards the fields of novel practical mathematics. Even if Hudde had a rationalistic approach to urban projects, one should not overestimate the significance of the new philosophy and interpret it as some application of 'géométrie' to administrative challenges. As regards the organization of knowledge, the 'scientific attitude' of the Hudde faction does not seem to have made much of a difference.

6 Governing Expertise

The extensions of Amsterdam and the management of its waters were large infrastructural projects that posed immense organizational and financial challenges to the city. The city's flourishing and growth depended on these projects and the financial risks that had to be taken literally affected its survival. These were complex projects that involved large numbers of workers over long periods of time, requiring reliable expertise and sound planning. This raises the question how the city administrators knew which professionals were competent to carry out the works, and how they organized, selected and assessed the expertise necessary for the materialization of the city. To reflect on this issue, the situation in Amsterdam is compared with the Stadtholders' fortification campaign of the early seventeenth century.

As infrastructural projects the city's extensions and its water management were no less substantial than the fortifications half a century earlier. In both cases high-rated professionals were involved in design and realization. In the course of the war, forms of rationalization developed in the deliberations between the army and the cities. The cadre of engineers that supervised the fortifications was well organized and trained. Around 1600 the Stadtholders initiated adjustments to their training, supplementing the field practice with formal schooling. Willem Lodewijk established a chair in mathematics at Franeker University that also would provide lessons in practical mathematics in the vernacular.[67] In 1600 Maurits established an engineering school at the university of Leiden where his engineers were taught the same (when the campaign was suspended in the winter). Simon Stevin drew up a curriculum of formal mathematics tailored to the abilities and needs of fortification engineers, making a specific selection of geometrical and arithmetical topics.[68] The *Duytsche Mathematique*, as it was called, is generally hailed as pioneering effort in the modernization of engineering and as an indication of the extraordinary advancement of learning in the Dutch Republic.

The *Duytsche Mathematique* reflected the particular views of Stevin and Maurits on the mathematical sciences and on the value of formal learning for practical pursuits. Similar ideas about learned skills required and valued in the arts were voiced in 17th-century debates about the training of draughtsmen, architects, and also surgeons and apothecaries.[69] It was aimed at the demands of the military as Maurits and Willem Lodewijk perceived them to have

---

67 Arjen Dijkstra, *Between Academics and Idiots. A Cultural History of Mathematics in the Dutch Province of Friesland (1600–1700)* (Ph.D. dissertation, University Twente, 2012).

68 P.J. van Winter, *Hoger beroepsonderwijs avant-la-lettre. Bemoeiingen met de vorming van landmeters en ingenieurs bij de Nederlandse universiteiten van de 17e en 18e eeuw* (Amsterdam: Noord-Hollandsche Uitgevers Maatschappij, 1988).

69 See Elske Gerritsen, *Zeventiende-eeuwse architectuurtekeningen. De tekening in de ontwerp- en bouwpraktijk in de Nederlandse Republiek* (Zwolle: Wbooks, 2006), 17–31; Bert De Munck, "Corpses, live models, and nature: assessing skills and knowledge before the industrial revolution (case: Antwerp)," *Technology and Culture* 51–2 (2010), 332–356.

developed in the course of the campaign. The *Duytsche Mathematique* was primarily a military institution; the designation 'engineer' was strictly military in the seventeenth century. Although any aspiring mathematician could attend the classes and many surveyors participated, it did not give any formal qualification except for those who shared the military ranks. This does not mean that other mathematical practitioners did not see possibilities for the *Duytsche Mathematique*. At the time of its establishment, two prominent surveyors of the Republic seemingly undertook an effort to have civil engineering also established at the engineering school. In 1600 Jan Pieterszoon Dou from Holland and Johannes Sems from Friesland published *Practijck des Lantmetens en Van het gebruyck der geometrische instrumenten*, expounding similar conceptions about theory and practice in surveying as Stevin and Maurits held. Their books sold well and were standard repertoire for surveyors, but the *Duytsche Mathematique* never became an official training for 'civil engineers' (as these practitioners can be called anachronistically). To be admitted as a surveyor one had to pass exams at a water board, a province, or – later on – the VOC, but there were no requirements to enter the exam. Neither did it make a difference to have finished the *Duytsche Mathematique*: in Holland a surveyor had to pass the exam for admission all the same. In some other provinces the situation was different, though; in Friesland for example a certificate from the mathematics course at the university gave access to admission.[70]

As contrasted to the military mathematics of the engineers, urban mathematics was mainly left to a free market where aspirants competed over education, positions and reputation. There was no guild system such as existed for surgeons and other professions for deciding who was allowed to practice a trade and no institutionalized training existed to acquire qualifications as was the case with medical doctors.[71] Anybody could start a mathematics school, become a private mathematics teacher, and give mathematical advice. As a result, mathematicians had to find out for themselves who was competent and who was not. They had to do so publicly by challenging each other with mathematical problems. This gave rise to a particular mathematical genre in the Golden Age: pamphlets in which disputes between mathematicians were battled out.[72] In Amsterdam a mass battle was waged in the early 1660s, when a shrinking market induced fierce competition between the 'Rekenmeesters'

---

70 Winter, *Beroepsonderwijs*.

71 Dijksterhuis, "Fit to measure"; Pouls, *Landmeter*; Muller and Zandvliet, *Admissions*; Davids, "Amsterdam as a centre of learning," 306–308.

72 Renowned are the disputes between Ludolf van Ceulen and Simon van der Eycke over the quadrature of the circle around 1585, between Jan Stampioen and Jacob van Wassenaer (i.e. René Descartes) around 1640. See Dijksterhuis, "Fit to Measure."

who offered mathematics education for navigators, bookkeepers and so on.[73] In these challenges and disputes mathematicians showed their competence in mathematics as well as their civility.[74]

In the case of mathematics education this open system seems reasonable. However, in the case of large infrastructural projects were much was at stake, one would expect some kind of institutional system to develop as had been the case with fortification. No such thing happened, not even in Amsterdam and not even when the Hudde faction rose to power. Despite the large scale and sophisticated nature of urban infrastructural projects, the training of mathematical practitioners was not institutionalized by the urban administration. Whether this was deliberate policy or a matter of tradition remains to be seen. Many who were contracted for surveying and engineering work were not admitted or formally trained; in some case military engineers were involved for reason of their official competence.[75] In the case of the professionals of the city works, the government relied on the guilds that regulated the construction sector.[76] For specific projects, applications sometimes had to do a test of competence by designing a construction or solution, but this can also be interpreted as a form of putting out to tender.[77] Amsterdam never established an institution to train the 'officers' of the city.[78]

In the second half of the seventeenth century the *Duytsche Mathematique* was in decline. When the professor died in 1676 no successor was appointed. Criticisms had already been voiced against the lessons of the professors, claiming that they had lost track of the developments in fortification: they carried on teaching the Old-Dutch System of 1600 rather than adopting the new French methods.[79] At the same time the demand for formally educated fortification engineers was decreasing in the middle of the century, thus taking away the *raison d'être* of the school. The growth of civil engineering projects had not resulted in a comparable institution where city planners and polder designers were trained. The urban aristocracy, including the group of Hudde, did not ra-

---

73 Nicolaije, "Dwaasheid of retoriek?"

74 Willem Frijhoff and Marijke Spies, *1650. Bevochten Eendracht* (Den Haag: Sdu Uitgevers, 1999), 186. Dijksterhuis, "Fit to measure."

75 Abrahamse, *Grote Uitleg*, 119–123.

76 Medema, *Achter de façade*, 123–134. Essen, *Stadsfabrieksambt*.

77 Ibidem, 137–140.

78 Ibidem, 140–148.

79 These criticisms should be set in context: they came from Henrick Ruse who was looking for patronage and would eventually go to Brandenburg and Denmark. Taverne, *Land van Belofte*. The same Ruse had been appointed extraordinary engineer in Amsterdam in 1651. Abrahamse, *Grote Uitleg*, 121. See also Van den Heuvel, "Ingénieurs dans la boue." In particular French methods innovated fortification, these were adopted in the Low Countries in de later 17th century.

tionalize the training of city officials and professionals by institutionalizing some formal education. The Illustrious School that was established by the city was aimed at academia instead of administrative or professional training. The Amsterdam Athenaeum was a classical institution of higher learning for the urban elite and primarily intended to promote the *learned* reputation of the city.[80] In this sense the Athenaeum fits in with the self-assurance of the Dutch cities that was expressed by the development of an urban chorography.

7 Conclusion

Without coming anywhere near to completeness or analytical substance, I have tried to make clear that the city was a site of knowledge as well. The urban theatre had its own dynamics and make-up, giving shape to specific kinds of inquiry in specific forms. In this contribution I have focused on the construction site rather than the learned circles. The sociable activities of Hudde and the academic formalities of the Athenaeum had a distinct Amsterdam flavor that warrants discussion.[81] Besides these learned fashionings, the city had to be built and shaped physically as well and this was done in the planning and management of mud and sewage. In the interaction between local demands and practices and expertise brought from outside, the large infrastructural works were designed and realized. Besides mobilizing and producing knowledge these sites also constituted and created specific ways of assessing and valuing knowledge. The organization of urban knowledge in this context seems to reflect the structure of the Dutch city: rather than a hierarchical system like the military engineers, it was a networked system of distributed authority and delegated responsibility. What strikes me is that the 'philosophical' administrators who rose to power after 1672 did not change this: the training of urban officers was not institutionalized by the city.

The urban patricians may not have rationalized the management of the city by means of institutionalized education; they mobilized their learning for the city's interests in their own way. Hudde, for example, employed his mathematical proficiency to advance Amsterdam's international trade interests. Amsterdam was the key center of transfer and processing of goods between

---

80 Dirk van Miert, *Illuster Onderwijs. Het Amsterdamse Athenaeum in de Gouden Eeuw, 1632–1704* (Amsterdam: Bert Bakker, 2005). Compare Davids, "Amsterdam as a centre of learning," 317–319. O'Brien, "Reflections," 31–33 points out that none of three urban centres established a university, but relied on neighbouring towns for academic resources.

81 See Cook, *Matters of Exchange; Isaac Vossius (1618–1689), between science and scholarship*, eds. Eric Jorink and Dirk van Miert (Leiden: Brill, 2012).

the Baltic and the Atlantic. Access to the Baltic was controlled at the very tight passage between Denmark and Sweden, the Sound (Øresund). The Dutch were continually negotiating with the Danish about the toll to be paid by the passing cargo (besides securing the free passage of Dutch warships). In 1666 Hudde proposed a new way of calculating the cargo of a ship by determining its displacement. On the instruction of the States-General, the Dutch mission to København presented the supposedly rational, and fair, method to the Danish negotiators, but it did not convince them. Hudde's method is preserved in a voluminous book on ships, *Aeloude en hedendaegsche scheeps-bouw en bestier* (Ancient and Modern Shipbuilding and Handling, 1671), written and published by no other than Nicolaes Witsen.[82]

82 Dijksterhuis, "Hudde, de wiskunde en de stad."

CHAPTER 3

The '*Duytsche Mathematique*' and Leiden Family Networks, 1600–1620

Gerhard Wiesenfeldt

In 1600, the recently established University of Leiden converted the fencing hall that was operated as a private business by Ludolf van Ceulen into a place of scholarly learning.[1] Located below the university library and next to the anatomy theatre in a former church (the 'Faliede Begijnkerk'), it subsequently housed the new academic discipline of *Duytsche Mathematique* (Dutch Mathematics). This discipline was established following a request of the Dutch Stadtholder Maurits of Nassau, who wished to have classes in practical mathematics taught "for the service of the country and the advancement of those who will give themselves to the exercise of engineering."[2] This teaching was to be directed at young craftsmen and artisans without a scholarly background and therefore to be given in the vernacular.[3] After finishing their studies, these men were

---

1 The research for this contribution has been made possible by a Faculty Research Grant of the University of Melbourne. Over the past years, I've discussed the role of the *Duytsche Mathematique* with various colleagues, including Aaron Mitchell, Christoph Meinel and Fokko Jan Dijksterhuis. Eric Jorink and Huib Zuidervaart have been extremely helpful in their editorial role providing much valuable extra information.

2 "… tot dienst van den lande ende bevorderinge der geenre, die hun tot oeffeninge van het ingenieurscap sullen begeven." Request by Maurits of Nassau to the University of Leiden, in: *Bronnen tot de geschiedenis der Leidsche Universiteit*, ed. Philip C. Molhuysen (Den Haag: Martinus Nijhoff, 1913–1924), vol. I, 122 and 389*. Online on https://www.historici.nl/resources/bronnen-tot-de-geschiedenis-der-leidsche-universiteit-1574-1811. For the original printed instruction of the *Duytsche Mathematique*, see: *Maniere ende Ordre die sijn Extie. verstaen heeft dat sal worden achtervolght int doen van de Nederduytsche Lessen in de Vniversiteyt tot Leyden tot onderrichtinghe ende bevorderinghe der ghene die hun ten dienste van den Lande sullen begheven tot oeffeninghe van het Ingenieurschap ende andere Mathematische consten* (Leyden: Jan Paedts Jacobsz., 1600). Only known preserved copy Scheepvaartmuseum Amsterdam, sign. B-I-0073 (II, 37).

3 On the introduction of the *Duytsche Mathematique* in general, see: P.J. van Winter, *Hoger beroepsonderwijs avant-la-lettre. Bemoeiingen met de vorming van landmeters en ingenieurs bij de Nederlandse universiteiten van de 17e en 18e eeuw* (Amsterdam: Noord-Hollandsche Uitgevers Maatschappij, 1988); Jan A. van Maanen, *Facets of Seventeenth Century Mathematics in the Netherlands* (PhD thesis, Utrecht University, 1987); Charles van den Heuvel, "Wiscontighe Ghedachtenissen: Maurits over de kunsten en wetenschappen in het werk

 | DOI:10.1163/9789004264885_004

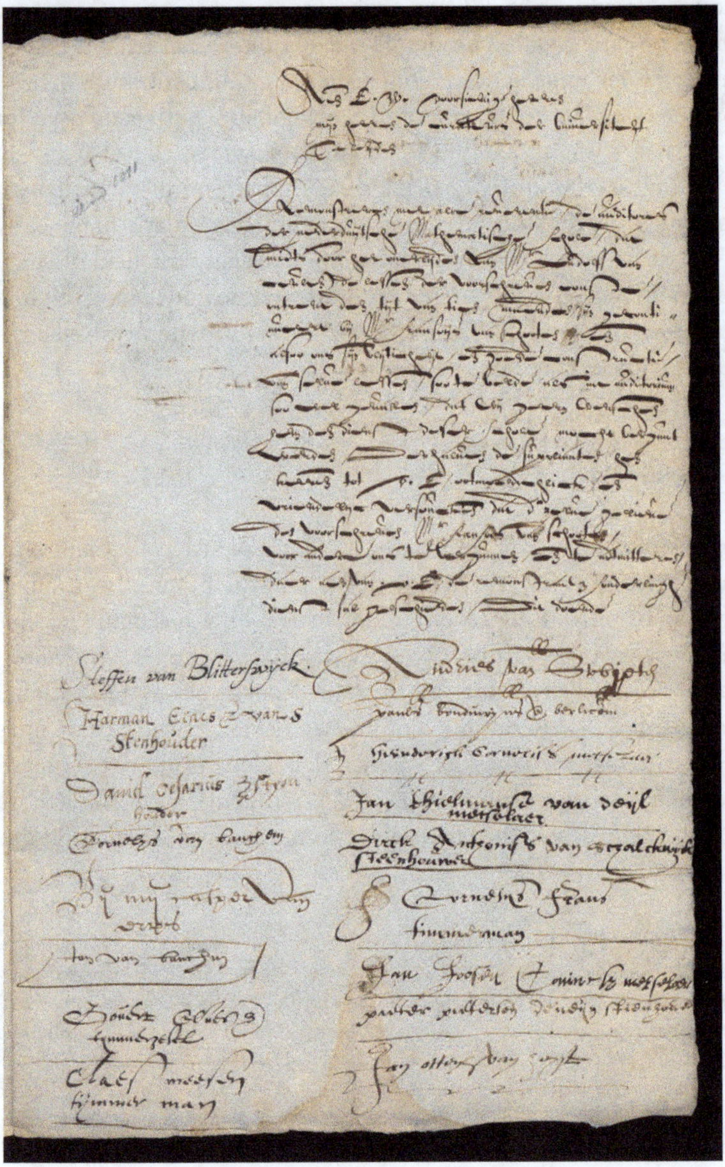

FIGURE 3.1 Request by students of the *Duytsche Mathematique* to the curators of the University of Leiden
SOURCE: UNIVERSITEITSARCHIEF LEIDEN, ARCHIEF VAN CURATOREN, 1574–1815, MS. AC 42 (BIJLAGEN TOT DE RESOLUTIËN, 1607–1618)

expected to pursue careers as military engineers or as surveyors. Compared to the customs of early modern academia outside the Dutch Republic, such teaching seemed to be unorthodox, to say the least. In this contribution, I will analyse how the teaching in the converted fencing hall – partly still used for its former purpose – was situated in the context of Leiden University. I will argue that the *Duytsche Mathematique* played an important role in mediating between the university and the city it was located in, particularly in the early decades of the seventeenth century. Within the university the field of study thus played a more central role than frequently argued. To understand this role it is important to understand the social fabric of the early modern university as well as its intellectual orientations.

1 The Family University

Universities, it is often argued, have always been conservative places, and early modern universities especially so. Their institutional structure has been aimed at preserving and transmitting received knowledge, only hesitantly accepting new knowledge claims, particularly those that have emerged *extra muros*. Sometimes this conservative structure is valued positively as institutional wisdom that guarantees the continuity of intellectual endeavour,[4] sometimes it is derided as the pigheadedness of staunch defenders of out-dated ideas.[5] The close relation between academic teachers and students, particularly those that aspire to become academics themselves, forms a central element in this

---

van Stevin," in: *Maurits Prins van Oranje*, ed. Kees Zandvliet (Amsterdam: Rijksmuseum / Zwolle: Waanders [2000]), 107–121. Charles van den Heuvel, "Simon Stevin, De *Crychconst* en de *Duytsche Mathematique*," in: *Simon Stevin (1548–1620) De geboorte van de nieuwe wetenschap* (Turnhout: Brepols, 2004), 101–111; Klaas van Berkel, "Simon Stevin et la fondation de l'école militaire de Leyde en 1600," in: *Simon Stevin, De la vie civile, 1590. Présentation et traduction de Catherine Secretan*, eds. Catherine Secretan and Pim den Boer (Lyon: ENS Éditions, 2005), 93–108. On the architectural situation of the *Duytsche Mathematique*, see: Gregory Gramiger, *Verortungen von Wissen: Die Räume und Sammlungen der Universität Leiden 1575–1700* (PhD thesis, ETH Zürich, 2014), 201–219.

4 Classical expositions of this perspective are: Clark Kerr, *The Uses of the University*, 5th ed. (Cambridge, MA: Harvard University Press, 2001), esp. 71–81; John H. Newman, *Discourses on the Scope and Nature of University Education* (Dublin: Duffy, 1852). Cf. *The Idea of a University*, eds. David C. Smith and Anne Karin Langslow (London / Philadelphia: Jessica Kingsley Publishers, 1999).

5 Cf. Martha Ornstein, *The Role of Scientific Societies in the Seventeenth Century* (Chicago: University of Chicago Press, 1913), 213–257; Richard Westfall, *The Construction of Modern Science: Mechanism and Mechanics* (Cambridge: Cambridge University Press, 1977), 205.

structure; the genealogy of knowledge is often superimposed on genealogies of academic progression. Most clearly expressed in the German term *Doktorvater*, such a construction of academic families relates the conservatism of universities to their social cohesion. Throughout most of their history European universities educated students whose fathers had been to university themselves or at least came from a social stratum occupied by academics.[6] Even the most significant opening of European academia prior to the middle of the twentieth century – the admission of women to universities around 1900 – did not change the social cohesiveness of the academic sphere, as the women entering academia predominantly came from the families that had previously sent only their sons.[7] Until the widespread reforms after World War II, the academic cohort of each generation tended to include only a small portion of students that did not come from an academic family background.

So, while preserving academic traditions appears as second nature to universities, historians and sociologists have often pondered how innovations can happen in such conservative structures in the first place.[8] There has been a tendency in early modern historiography to regard actual changes predominantly as the result of political pressure from outside the university that provided the momentum to overcome the institutional resistance dominating the university culture.[9] It needs to be noted, however, that while institutional conservatism had been dominant in the history of European universities, it did not apply to all universities at all times. Particularly in periods of radical social change, newly founded universities could have a social formation different from the established pattern. As disciplinary structures had not yet emerged within scholarship, the order of knowledge within a learned institution was to

6 For an overview see Maria Rosa di Simone, "Admission," in: *A History of Universities in Europe: Volume 2, Universities in Early Modern Europe (1500–1800)*, ed. Hilde de Ridder-Symoens (Cambridge: Cambridge University Press, 1996), 285–325, esp. 311–323; Fritz Ringer, "Admission," in: *A History of Universities in Europe: Volume 3, Universities in the Nineteenth and Early Twentieth Centuries (1800–1945)*, ed. Walter Rüegg (Cambridge: Cambridge University Press, 2004), 233–267, esp. 257–266.

7 Cf. Patricia M. Mazón, *Gender and the Modern Research University: The Admission of Women to German Higher Education, 1865–1914* (Stanford, CA: Stanford University Press, 2003).

8 Lewis Elton, "Can universities change?" *Studies in Higher Education* 6 (1981): 23–33; Claudia Meister-Scheytt and Tobias Scheytt, "The Complexity of Change in Universities," *Higher Education Quarterly* 59 (2005): 76–99; Niklas Luhmann, *Universität als Milieu* (Bielefeld: Haux, 1992).

9 Cf. Charles E. McClelland, *State, Society and University in Germany, 1700–1914* (Cambridge: Cambridge University Press, 1980).

a large extent contingent on the local context, in which the social formation played a significant role.[10]

In this perspective, the establishment of a new academic subject can sometimes be interpreted in terms of social change as well as in terms of intellectual needs. New forms of scholarship emerged with new social types, whether it was the Renaissance humanist, the naturalist, the savant or the engineer.[11] In some cases, this could affect existing social stratification. So, in order to understand the role of an academic discipline at an early modern university, it can be useful to look at it in terms of its social structure. This might sound reasonable and more than thirty years after the formulation of the Edinburgh School's strong programme almost trivial. Yet, among historians of science at early modern universities, the social background of their actors has not always been high on their agenda.[12] A frequent assumption seems to be that university professors – once appointed – belong to a specific social stratum, as long as nobody in their lifetimes questioned their status. So, if the social background of academics is studied, it is usually for individual cases where an unusual background – say, for someone coming from a family of craftsmen – is seen as the cause for exclusion from an academic career or at least as a strong obstacle against inclusion.[13]

The situation is somewhat different among historians of universities, but even in this field, there is substantially more research on the social background of students, whereas the social background of academic teachers is rarely the object of research.[14] The existing research on academic teachers often stresses the dynastic aspects of early modern universities: "The majority of the seventeenth- and eighteenth-century universities had indeed become diseased by inbreeding."[15] While partially agreeing with this assessment, the historian Peter Moraw has introduced the notion of the *Familienuniversität*

---

10 Cf. Gerhard Wiesenfeldt, *Leerer Raum in Minervas Haus: Experimentelle Naturlehre an der Universität Leiden, 1675–1715* (Amsterdam: KNAW, 2002), 278–286.

11 Cf. Gadi Algazi, "Scholars in Households: Refiguring the Learned Habitus, 1480–1550," *Science in Context* 16 (2003): 9–42.

12 This is not to argue that there is not substantial work on the academic habitus and its exclusionary function (e.g. William Clark, *Academic Charisma and the Origins of the Research University* [Chicago: University of Chicago Press, 2006]), but very little systematic work how this exclusion worked for specific groups.

13 Cf. Pamela H. Smith, *The Body of the Artisan: Art and Experience in the Scientific Revolution* (Chicago: University of Chicago Press, 2004).

14 For a discussion of the Dutch student population see: Willem Frijhoff, *La société néerlandaise et ses gradués, 1575–1814. Une recherche sérielle sur le statut des intellectuels à partir des registres universitaires*, (Amsterdam / Maarssen: Rodopi, 1981); Martine Zoeteman, *De studentenpopulatie van de Leidse universiteit, 1575–1812: 'Een volk op zyn Siams gekleet eenige mylen van Den Haag woonende'* (Leiden: Leiden University Press, 2011).

15 Peter A. Vandermeersch, "Teachers," in: *A History of Universities in Europe: Volume 2, Universities in Early Modern Europe (1500–1800)*, ed. De Ridder-Symoens, 210–255, esp. 227.

(family university) to give a better understanding of the role of families in early modern universities.[16] With this concept, Moraw referred to the role that family networks had in effective governance of early modern universities up to the possession of individual chairs or even whole faculties. Often enough, family hierarchies determined changes in knowledge hierarchies and provided the social cohesion that the post-Renaissance university lacked in intellectual coherence. Moraw is somewhat critical of the power that families could exercise and points out the difference to the nineteenth century universities, in which academic merit was the dominating factor, even if family relations still played a role. Yet, at least the more functional of these family networks produced quite fruitful researches. Probably best known is the Bernoulli family in mathematics, but there are many other examples, both in the Dutch Republic and elsewhere.[17]

An impressive case for the power of family networks – in university governance and in knowledge production – had taken place in Copenhagen, where a historical overview of the university has been written as a genealogy of two families.[18] The families Winstrup/Rosen and Bartholin/Fincke had monopolised three of the four faculties; only Theology was granted some independence. The family Bartholin/Fincke included 49 university professors during the seventeenth century – six of whom have their own entry in the *Dictionary of Scientific Biography*. The family was more or less in possession of the Medical and the better part of the Arts faculty. It included not only a Thomas (1616–1680) and Rasmus (1625–1698) Bartholin and a few other notable physicians of that name, but also Ole Roemer (1644–1710) – Rasmus Bartholin's son-in-law – and Ole Worm (1588–1655), who shared his father-in-law Thomas Fincke (1561–1656) with Caspar Bartholin. While this is evidence of the family's intellectual status, its political power can best be seen in an appointment affair in 1648. In that year, Simon Paulli had been appointed as a new professor of anatomy as well as the personal physician of the Danish King. Unfortunately for Paulli, Thomas Bartholin wanted to move from his professorship in philosophy to anatomy. Within a few months, Paulli was dismissed from the chair, despite apparent support from the King, as he remained Royal Physician until

---

16 Peter Moraw, "Aspekte und Dimensionen älterer deutscher Universitätsgeschichte," in: *Academia Gissensis. Beiträge zur älteren Gießener Universitätsgeschichte*, ed. Peter Moraw and Volker Press (Marburg: Elwert, 1982), 1–43.

17 For the role of the Bernoulli family at Basel University see: Edgar Bonjour, *Die Universität Basel von den Anfängen bis zur Gegenwart, 1460–1960.* 2nd ed. (Basel: Helbing & Lichtenhahn, 1971), 307–319.

18 Ditlef Tamm, "Københavns Universitet 1621–1732," in: *Københavns Universitet 1479–1979. Bind 1: Almindelig Historie 1479–1788*, eds. Svend Ellehøj, Leif Grane and Kai Hørby (Copenhagen: Gad, 1991), 199–314.

his death in 1680.[19] If this is an indication that strong family networks could even resist monarchical power, it also evidence of the central role of families as expressions of a particular social formation at early modern universities.

2 Leiden and the *Duytsche Mathematique*

When Leiden University was founded in 1575, it was placed in an unusual historic situation. As 'bastion of liberty' it represented the intellectual conception of the Dutch revolt against the Spanish rule.[20] Hence, the university founders did not want to rely on imports from other countries to set up the new university, a few notable exceptions notwithstanding. In consequence, the academic traditions of the *Familienuniversität* did not yet exist. These would establish themselves during the seventeenth century, and the family dynasty of Frans van Schooten the Elder, the Younger, and Petrus van Schooten as professors of *Duytsche Mathematique* from 1615 until 1679 is a particularly strong example. Yet, the first fifty years of the university provided opportunities for aspiring academics not found elsewhere. This does not imply that family ties did not play a major role in the university, but these family relations were of a different structure than elsewhere.

The role the *Duytsche Mathematique* came to play in this context is partly a result of these differences.[21] I want to show that far from being a marginal and unorthodox annex, the field of study had an important place within the university, which provided a way to negotiate the social diversity of early Dutch Academia and the city of Leiden in particular. This role might not have been intended and possibly not even welcomed by Simon Stevin, when he conceptualised the education of engineers at the university. The result of these negotiations between city, university and Dutch Stadtholder paved a way for empirical learning in mathematical and philosophical sciences.

Already the original appointments for the teaching of *Duytsche Mathematique* reflected the complicated relationships between the three parties, as the university followed Maurits's advice in 1600 and charged two

19 Cf. Vilhelm Møller-Christensen and Albert Gjedde, "Det medicinske Fakultet 1479–1842," in: *København Universitet 1479–1979. Bind VII: Det lægevidenskablige Fakultet,"* eds. Johannes C. Melchior et al. (Copenhagen: Gad, 1979), 1–90.

20 Willem Otterspeer, *Groepsportret met Dame I. Het bolwerk van de vrijheid. De Leidse universiteit, 1575–1672* (Amsterdam: Bert Bakker, 2000). Henrike L. Clotz, *Hochschule für Holland: die Universität Leiden im Spannungsfeld zwischen Provinz, Stadt und Kirche, 1575–1619* (Stuttgart: Franz Steiner Verlag, 1998).

21 Cf. Otterspeer, *Groepsportret*, 198–202.

mathematicians with the task: Ludolf van Ceulen and Simon van der Merwen.[22] Prior to his appointment, Van Ceulen had worked as mathematical practitioner as well as a fencing master.[23] While this brought him into the well-known controversy with Joseph Scaliger over quadratures, it also brought considerable fame through a range of publications including his famous *Vanden Circkel* (1596) and an appointment to a commission on navigation alongside Simon Stevin.[24] In contrast to this, Van der Merwen had much more local credentials as a mathematician. He had been a member of the Leiden *Vroedschap* (town council) since 1576 and had functioned as fortifications officer after the siege of 1574 and as city treasurer from 1580 to 1593, as well as burgomaster in 1598.[25] The double appointment of Van Ceulen and Van der Merwen thus gave representation to both the mathematical practitioners around Stevin and the technical expertise in the city.

The full extent to which the social formation of the university and its city was involved in the establishment of *Duytsche Mathematicque* became evident with the appointment of Frans van Schooten the Elder as professor of *Duytsche Mathematicque* at Leiden University in 1615 as successor of Van Ceulen and Van der Merwen, who both had died in 1610.[26] I use the term 'professor' because I could not find a single piece of evidence that the men who taught Dutch mathematics at Leiden were called anything but professor. In the case of Van Schooten this was explicitly mentioned in the appointment resolution.[27]

---

22 Cf. *Bronnen*, I, 122 (January 1600).

23 Cf. Huib Zuidervaart and Marlise Rijks, "'Most rare workmen': optical practitioners in early seventeenth-century Delft," *British Journal for the History of Science* 48 (2015): 53–85.

24 Friedrich Katscher, *Einige Entdeckungen über die Geschichte der Zahl Pi sowie Leben und Werk von Christoffer Dybvad und Ludolph van Ceulen*, Österreichische Akademie der Wissenschaften; Mathematisch- naturwissenschaftliche Klasse; Denkschriften; vol. 116, no. 7 (Wien, 1979). Jan P. Hogendijk. "The Scholar and the Fencing Master: The Exchanges between Joseph Justus Scaliger and Ludolph van Ceulen on the Circle Quadrature (1594–1596)," *Historia Mathematica* 37 (2010): 345–375.

25 Sterling A. Lamet, *Men in Government: The Patriciate of Leiden, 1550–1600* (PhD thesis, University of Massachusetts, 1979), 404.

26 Van Schooten had already taken over the lectures in 1611, apparently out of his own initiative. He enrolled as 'Lector Mat.' on 15 February 1612. Cf. *Bronnen*, II, 42 (1612) and 47 (Febr. 1613).

27 Universiteitsarchief Leiden, Archief van Curatoren, Ms. AC 20, Resolutiën van Curatoren ende Burgermeesters, 1595–1617, f. 354. "Ende de voorst. van Schooten verzogte dat hy mogte worden vereerd metten titule van Professeur [...] ende Burgemn. gehoord hebbende het goede rapport hen gedaan van de diensten ende cragten des voorst. van Scohoten [...] denzelven aangenoomen tot Professeur der Duytsche Mathematyke." In a similar ilk, I couldn't find any use of the term "engineering school" for the teaching of Dutch mathematics in the university context. The term was probably introduced by Bierens de Haan in the nineteenth century (cf. Daniel Bierens de Haan, *Bouwstoffen voor de geschiedenis der wis- en natuurkundige wetenschappen in de Nederlanden* (Amsterdam,

His predecessors were classified as "professories extraordinaris" in a list compiled by the Academic Senate in 1602.[28] This meant that they were not part of the academic elite of ordinary professors, but so were the bulk of teaching staff in other disciplines including luminaries such as Carolus Clusius and Justus Scaliger, whose names appeared next to those of Van Ceulen and Van der Merwen on the Senate's list. What is puzzling about the appointment of Frans van Schooten, however, is not that he was appointed a professor, but that he was appointed at all. After all, he had considerable competition. The other candidate for the job was Samuel Marolois, who was not only a respected mathematician, but also had the support of the Stadtholder. In 1612 and 1614 Maurits of Nassau had sent two letters to the curators of the university, asking to appoint Marolois.[29] His father, Nicolaas Marolois, was a Huguenot refugee that had been in the service of Maurits since the 1570s. Samuel himself was already serving on various senior administrative bodies of the States of Holland, including the board to examine candidates to be admitted as surveyors, a position previously held by Van Ceulen and Van der Merwen.[30] He was also an eminent expert on fortification, practical mathematics and perspective. Between 1614 and 1616, he published a five-volume *Opera mathematica*, which included an edition of Hans Vredeman de Vries' *Perspective*.[31]

Frans van Schooten, on the other hand, descended from a baker's family in Leiden and was in the process of establishing himself as a surveyor.[32] He, too, sent two letters to the curators, in which he pointed out that he had followed

---

KNAW, 1878), 263). For the university, the Duytsche Mathematique appears to have been an ordinary discipline, if one that was taught under extraordinary circumstances.

28 Universiteitsarchief Leiden, Archieven van Senaat en Faculteiten, 1575–1877, Ms. 286, Codices Actorum Academicorum VIb, 1595–1604, f. 189.

29 *Bronnen*, II, 43* and 67*. Marolois had requested Maurits intercession in 1611. Idem, 68*.

30 E. Muller and K. Zandvliet, *Admissies als landmeter in Nederland voor 1811* (Alphen aan de Rijn: Canaletto, 1987), 157. According to the lemma by Cornelis de Waard in: *NNBW*, 2, kol. 873–874, Van der Merwen was also serving on a longitude committee of the States General.

31 Samuel Marolois, *Opera mathematica. Ou oeuvres mathematicques. Traictans de geometrie, perspective, architecture, et fortification* (The Hague: Hondius, 1614–1616). Cf. Kirsti Andersen, *The Geometry of an Art: The History of the Mathematical Theory of Perspective from Alberti to Monge* (New York: Springer, 2007), 298–308. Marolois had received an honorarium from the States General for the edition of Vredeman de Vries (cf. J.J. Dodt, *Archief voor kerkelijke en wereldsche geschiedenissen*, vol. VI (Utrecht: Van der Monde, 1836–1846), 367).

32 While Van Schooten's profession is given as surveyor in register of his wedding in 1609, no official record of his admission as a surveyor has been found. On his background see Jantien Gea Dopper, *A life of learning in Leiden. The mathematician Frans van Schooten (1615–1660)* (PhD thesis, Utrecht University, 2014), 18–19. According to Dopper, he had tried to set up a mathematical school in Gouda and changed plans after the death of Van Ceulen.

the lectures of Ludolf van Ceulen, assisted him subsequently and that he possessed all the necessary instruments to give the lectures.[33] From this it may be concluded that the outcome had been a foregone conclusion. So, how is it possible that he was appointed and not Marolois, who by any standard appears as the candidate with better qualifications and strong political support? An answer can be found in three undated letters written in support of Van Schooten, making reference to the appointment process. These letters therefore must have been written between 1611 and 1614.[34] It is not so much about the content, which is fairly standard in requesting his appointment and providing a good reference for his teaching abilities, but in the signatures of the letters. The signatories comprise a group of in total 63 people, some of them stating their professions. Among those were five men who declared to be surveyors, a few stonemasons, bricklayers and carpenters, a schoolmaster and a couple of students.[35] This seems to suggest that the support for Van Schooten was mainly coming from local craftsmen and some enthusiastic students. Yet, following up the actual names of the signers reveals a different picture, especially for those that did not regard it as necessary to accompany their name with a profession. One of them was Franciscus Junius, junior, son of the theology professor of the same name, another one was Jan Otten van Zeyst, brother of Nicolaas van Zeyst. The latter was secretary of the university as well as son in law of one of the university curators. In 1600 he had been responsible for the furnishing of the classroom of the *Duytsche Mathematique* in the Faliede Begijn church.[36] And last but not least Nicolaas van Zeyst was also pensionary of the city of Leiden. Finally, there was the signature of Hendrick Cornelisz. Bilderbeeck, who stated to be a sworn surveyor in one letter and to be a bricklayer in the other. While both statements were true, he also happened to be the father-in-law of Johannes Meursius, professor of history.[37]

33 Universiteitsarchief Leiden, Archief van Curatoren, Ms. AC 42, Bijlagen tot de Resolutiën, 1607–1618.

34 Ibidem.

35 The names of the signatories of the first letter have been listed in: *Bronnen*, I, 392*. The full list is reproduced in the appendix to this contribution, which includes additional biographical information from other sources.

36 "Hebben voorts die voorsz. Curateurs ende Burgemeesteren tot een Leesplaetse geordonneert een gedeelte van de Faliebagynen Kercke ende de Bibliotheeque om met deelen afgeschooten en voorts soo met Banken ende stoelen als anders gemaeckt te werden bequaem tot een gemeene Leesplaetse, om 't welck te bevorderen gecommitteert zijn de voorn. Mr. Simon Fransz. van der Merwen ende Mr. Nicolaes van Zeyst, Pensionaris der stad Leyden ende Secretaris van de voorn. Curateurs ende Burmeesteren." *Bronnen*, I, 122/3.

37 Archive Leiden, DTB. Marriage of Johannes Meursius and Anna Henricksdr van Bilderbeeck, 20 April 1612.

So, while the support for Van Schooten appeared to come from local craftsmen and enthusiastic students, it also came from family members of senior figures within the university. The local support apparently was strong enough to withstand Maurits's repeated requests and appoint Van Schooten instead of Marolois. In terms of the concept of family university, this is to some extent a well-known phenomenon and, compared to Simon Paulli's dismissal at Copenhagen, even on a moderate scale of resistance to government intervention by a university in pursuit of increasing the autonomy of the university. Yet, it was not the university alone that acted against Maurits here. After all, Bilderbeeck and his fellow subscribers Jan Otten van Zeyst and Harman Claesz. were the three most senior craftsmen in service of the city of Leiden.[38] To understand why city and university joined forces in this affair, it is instructive to look closer at the family network surrounding Leiden University.

3 The Family Networks in Leiden

At first glance, the appointment of Frans van Schooten as professor of *Duytsche Mathematique* seems an unlikely case for a joint political action of city and university against the Stadtholder. The fact that it happened on the matter of *Duytsche Mathematique* is evidence that the university establishment cared much more about the subject than is sometimes assumed. On the other hand, it is understandable that important parts of the city elite, such as Bilderbeeck, Zeyst and Claesz. – broadly speaking the technical expertise in city governance – had a strong interest in the subject. They were all surveyors who most likely had taken classes under Van Ceulen and Van der Merwen. However, they were able to muster resources within the university. Looking more broadly on the situation reveals that the family ties between Bilderbeeck and Meursius were not a coincidence. The family network of Leiden University was deeply tied to other family networks within the city, including networks of craftsmen and artisans.

A prosopographic study of the family background of academics at Leiden until 1725 has shown that over the first 50 years a remarkably high number of professors without an academic family record were appointed.[39] Of the 65 ap-

38 E. Pelinck, "De functionarissen belast met de zorg voor de stadsbouwwerken te Leiden (1575–1850)," *Jaarboekje voor de geschiedenis en oudheidkunde van Leiden en omstreken* 59 (1967): 59–76, esp. 61–64.

39 Aaron Mitchell and Gerhard Wiesenfeldt, "Craftsmen, merchants and scholars: Hiring practices at the Universities of Leiden and Edinburgh, 1575–1750," *Yearbook for the European Culture of Science* 8 (2015): 163–187.

pointments that were made until 1624 and of which the relevant information could be retrieved, 12 appointments (more than 1 in 6) went to men, whose fathers were craftsmen and artisans, or had another occupation of relative low social status (such as boatsmen and organ player), or were characterised as being poor.[40] After 1625 the number of academic appointments to candidates with a non-traditional family background dropped significantly to 5 out of 63,[41] which shows a process of social closure within the university. Professorial hierarchies had been established; the purely academic family networks were evolving, settling and then proved strong enough to sustain themselves. In contrast, the first 50 years of Leiden University show family networks that extended beyond the university. After all, academics were not the only social group in early modern Europe who organised their professional existence in family relations; craft guilds and city regents were not much different in this respect. What is apparent in the letter of support for Van Schooten and significant for an understanding of the social formation of Leiden University is that in their early period the family networks between academics, regents and artisans overlapped to allow for a transition between the different social spheres.

A comparison between Leiden and two related university foundations – Edinburgh in 1583 and Franeker in 1585 – shows that Leiden was indeed an unusual case.[42] In Franeker, a few sons of farmers were initial appointees, which reflects the special position of the farmer's state in the Frisian constitution.[43] Thereafter, the only appointment to someone from a non-traditional background until 1650 was that of Adriaan Metius (1571–1635) to the professorship of mathematics in 1598. Yet, Metius' previous collaboration with Tycho Brahe at Uraniborg signifies a special case of a university that at least until 1700 was closed off in terms of social mobility. At Edinburgh, the appointments were usually made through the town council and dominated by the merchant guilds, which guarded carefully that no family member of the craft guilds could enter academic careers. The Leiden *Vroedschap*, on the other hand, had had a strong representation of arts and crafts long before the university was set up.[44]

---

40 The number does not include the much larger group of academics, whose father did not have an academic background, but was characterised as a merchant, a land-owner or a 'burgher' without further qualification.

41 Mitchell and Wiesenfeldt, *Craftsmen, merchants and scholars*. On the professorial families at Leiden University also see Ronald Sluijter, *"Tot ciraet, vermeerderinge ende heerlyckmaeckinge der universiteyt.' Bestuur, instellingen, personeel en financiën van de Leidse universiteit, 1572–1812* (Hilversum: Verloren, 2004), 152–155.

42 Ibidem.

43 Cf. Hotso Spanninga, *Gulden vrijheid? Politieke cultuur en staatsvorming in Friesland, 1600–1640* (Hilversum: Verloren, 2012).

44 Lamet, *Men in Government*.

The distinction between producing and trading guilds that dominated many other places in early modern Europe and frequently led to a formal denigration of manual labour was not prevalent in the city of Leiden. Given the political influence of the city government in the governance of the university, this led to a certain, if limited openness of the university to sons of craftsmen and artisans that would not be found elsewhere.

4 Mathematics and Leiden – City and University

The *Duytsche Mathematique* thus appeared in the Leiden context as a less unusual subject than it might have been elsewhere.[45] Both in terms of social embedding and disciplinary content, it was fairly well integrated in *academia* right from its start. It followed a practical ideal that would appeal to many influential men in the city who would in turn have close relations – family or otherwise – to university professors. When the field of study was introduced in 1600, it happened at a time when the relations between university and city were still emerging, enabling various actors to use the initiative of Simon Stevin to improve their positions. One important development coinciding and most probably related to the introduction of Dutch mathematics improved the position of mathematics within the university, which over time blurred the line between Dutch and Latin mathematics. Before 1600, mathematics was represented at Leiden, but not in a particularly strong way. While the philological arts of the trivium received prominent attention, the mathematical quadrivium remained in a low propaedeutic position. So, whereas Rudolph Snellius (1546–1613), the only professor of mathematics, was a prominent scholar within the world of learning, his position at the university was rather marginal, an attempt to give him a more prominent position by extending his portfolio to physics was rejected in 1598.[46] Yet, shortly after the introduction of teaching of *Duytsche Mathematique* and the appointment of Van der Merwen and Van Ceulen, Snellius was charged with the examinations of students of Dutch

---

45 This should not imply that the success of *Duytsche Mathematique* depended solely on the urban environment, as its significance at Franeker University related to different circumstances. Cf. Arjen Dijkstra. *Between Academics and Idiots: A Cultural History of Mathematics in the Dutch Province of Friesland (1600–1700)* (PhD thesis, Twente University, 2012).

46 Klaas van Berkel, "De geschriften van Rudolf Snellius. Een bijdrage tot de geschiedenis van het Ramisme in Nederland," *Tijdschrift voor der Geschiedenis van de Geneeskunde, Natuurwetenschap, Wiskunde en Techniek* 6 (1983): 185–194; *Bronnen tot de geschiedenis der Leidsche Universiteit*, ed. Philip C. Molhuysen (Den Haag: Martinus Nijhoff, 1913–1924), vol. I, 112.

mathematics in collaboration with the two other mathematicians.[47] He also had been promoted to ordinary professor of mathematics filling the ordinary professorship that had been reserved for philosophy. The bizarre wording of the appointment resolution, which stressed the fundamental role of philosophy as the "foundation of all sciences and scholarship" although a mathematician was appointed, suggests the implicit rationale that mathematics – and the Ramist philosophy associated with it through the person of Snellius – should be regarded more highly than the formal scholasticism taught by the unfortunate scholars who held the actual professorships of philosophy.[48]

The Ramist philosophy advocated by Snellius with its dialectic between theory and practice corresponded to the *Duytsche Mathematique* as practiced in Leiden. Its teaching was comprised of theoretical lectures, or demonstrations on paper "in't cleyn," and practical exercises, demonstrations in the field "in't groot," both of which were to be kept separate.[49] In the course of time, this dialectic between theory and practice appears to have become more significant for the teaching of mathematics than the difference in language between Latin and Dutch. As far as it can be established from the sources, there was a fair bit of overlap in content between the two. In both kinds of mathematics, you could hear an exposition of Euclid's elements or of the works of Archimedes. Of course, there were still considerable differences between the fields; status still was important, but there was the claim that you could translate one into the other. This claim was particularly strong in the work of Willebrord Snellius, who in 1613 succeeded his father as professor of Latin mathematics and translated the two forms of mathematics, literally, in both ways – Stevin and

---

47 Nationaal Archief, Den Haag, Archief van de Staten van Holland na 1572, Ms. 1369A, Stukken rakende de Universiteit te Leiden.

48 Universiteitsarchief Leiden, Archief van Curatoren, Ms. AC 20, Resolutiën van Curatoren ende Burgermeesters, 1595–1617, f. 84v.: "dat selve [de curatoren] om te doen bevorderen ende vereeren die studien der Philosophien (als wesende 't fondament aller wetenschap ende geleertheid) ende ten einde 't voorst. collegie bestaan mogte uit alle scientien, de welke in de Universiteyt worden geleert daer toe vercooren worde ymand van de Prof. der Philosophie voorst. Soo is 't dat curateurs ende Burgemeesteren voorst. hen volcomentlyc betrouwende op de geleertheid, ervarentheid ende bequaamheid van Rudolphus Snellius prof. Mathemaseos den zelven hebben verkozen [...] als ordinarum Professorem." Cf. Wiep van Bunge, *From Stevin to Spinoza: An Essay on Philosophy in the Seventeenth-Century Dutch Republic* (Leiden: Brill, 2001), 1–9; idem, "Before philosophy: Theory and practice in the emerging Dutch republic, 1580–1620," *The European Legacy: Towards New Paradigms* 4 (1999): 1–22.

49 Universiteitsarchief Leiden, Archief van Curatoren, Ms. AC 20, Resolutiën van Curatoren ende Burgermeesters, 1595–1617, f. 68. For a comparable case, see Klaas van Berkel, "Het Collegium Mechanicum van Isaac Beeckman," *Spiegel historiael* 15 (1980): 336–341.

Van Ceulen into Latin, the *Geometria* of Petrus Ramus into Dutch.[50] He also claimed an intimate connection between venerable ancient knowledge and the practical art of Dutch mathematics in his book series *Apollonius Batavus* (1608), *Eratosthenes Batavus* (1617) and *Typhis Batavus* (1624) on different parts of mixed mathematics.

In a certain way, the difference between Dutch and Latin mathematics seemed to have become more of an issue of language or of family ownership of teaching portfolios between the Van Schooten and the Snellius dynasties in Leiden, the former lasting until 1679. In another way the matter needed much more negotiation. Even if the social setting allowed for mediation, the traditional university hierarchies and privileges and Dutch mathematics did not necessarily go well with each other – not with respect to academic teachers, but with respect to students. While part of the audience of Van Ceulen and Van der Merwe consisted of ordinary students who took classes in practical mathematics as part of their ordinary studies, the official target audience, of course, was present there, too. So, the university had to decide whether and how far young craftsmen studying *Duytsche Mathematique* with no knowledge of Latin should be considered as regular students. On two occasions, the university took a decision that displayed an ambivalent position towards the role of craftsmen as students.

The first referred to the previously mentioned examinations. Some of these students had requested an official document from the university certifying their achievements. This did not only seem reasonable, but also in the interest of the university that wanted to distinguish itself from competing instructors for surveying.[51] So, examinations were set up, but it was unclear how these examinations related to regular graduations. Simon Stevin had requested the academic senate to state whether the university wished to award "a promotion or a certificate of their abilities" to students passing the examinations and which title should be associated with it.[52] Promotion seemed to go too far; at least this is implicit in the response of the senate, who in principle agreed with Stevin, but only spoke of the "Acte van bequaemheyt" ('deed of competence'), which the university would supply under the condition that the candidate would

50 Cf. Liesbeth de Wreede, *Willebrord Snellius (1580–1626). A Humanist Reshaping the Mathematical Sciences* (PhD thesis, Utrecht University, 2007).

51 Cf. H.C. Pouls, *De landmeter: inleiding in de geschiedenis van de Nederlandse landmeetkunde van de Romeinse tot de Franse tijd* (Alphen aan de Rijn: Canaletto, 1997).

52 "Enige promotie off acte van haer bequaemheyt," Nationaal Archief, Den Haag, Archief van de Staten van Holland na 1572, Ms. 1369A, stukken rakende de Universiteit te Leiden. Cf. *Bronnen*, I, 131 (August 1600).

provide a gratuity of 20 guilders to the senate.[53] Franeker went further here and indeed awarded promotions in practical mathematics.[54] An important note here is that Leiden claimed to hold the examinations under the auspices of the Senate, which meant that any influence from outside the university, i.e. Stevin, was foreclosed. Also, it implied that an ordinary professor as potential member of the senate would oversee the exams, which thus required the promotion of Rudolph Snellius in 1601.[55] In this light, the appointment of Van Schooten might indeed appear as part of larger strategy of the university to establish its autonomy against the influence of Maurits and Stevin.

The other question that needed to be settled was that of the status of students of Dutch mathematics in terms of the students' statutory rights. During the first two decades of the seventeenth century the *Album studiosorum* shows 25 students who had officially matriculated for "Mathematics," two even for "Geometry."[56] One of them was one of the signatories of both the letters in support of Van Schooten and thus may well have been a student of *Duytsche Mathematique*.[57] Yet, the majority of students of surveying or military engineering probably had not enrolled, as was the case with some ordinary university students. That there seems to be some debate about their status is apparent by a letter sent by 19 men to the curators. This letter was signed by nine regular students, four surveyors, four craftsmen and two persons with an unidentified quality.[58] They proposed that participants in courses in Dutch mathematics should enjoy the normal student privileges if they passed an examination, proving their mathematical knowledge.[59] It is unclear to what

53 Cf. *Bronnen*, I, 412*: "1602 Nov. 11. Diploma voor een "Meester in de Duytsche Mathematicque."

54 Cf. Van Winter, *Hoger beroepsonderwijs*, 51–63.

55 Cf. *Bronnen*, I, 411*: "1602, Nov. 11: Advies van den Senaat betr. de "examens in de Duytsche Mathematicque."

56 Cf. Zoeterman, *De studentenpopulatie*, 64–67.

57 This concerned Steffen van Blitterswyck, student in Geometry, from Leiden, who matriculated on 25 July 1609 at the age of 22. The other student in Geometry (the non-signatory) was Johannes à Pasquily from Dusseldorf, who enrolled at 2 September 1613, also at age 22.

58 Universiteitsarchief Leiden, Archief van Curatoren, Ms. AC 42, Bijlagen tot de Resolutiën, 1607–1618. See the appendix for the list of those who have signed this letter.

59 The content of the proposed examinations is published in: *Bronnen*, I, 391*/92*. The content to a large extent covers the content given by Stevin's original instructions for the teaching of *Duytsche Mathematique* (idem, 389*/90*), which implies that the proponents argued that this should apply to advanced candidates only. This would correspond to a phrase in Stevin's instruction stating that those "who have come thus far, should they have the desire to study the profound things more fundamental, should be allowed to do this." ("Die soe verre gecomen syn, hebben se alsdan lust die diepsinnige dingen grondelycker t'ondersoucken dat sullen sy mogen doen").

extent the proposal was put into place.[60] It is interesting for at least two reasons. By supplanting the traditional entry requirement – mastery of Latin – by mathematical ability, the proposal opened the student body to men with a non-traditional background. Given that this would not benefit them in gaining knowledge from attending lectures, since they could already attend all lectures given in Dutch, the proposal was aimed as a – mostly symbolic – act of inclusion of students of Dutch mathematics into the university.

Another interesting thing about the proposal is that one of the signatories was the young Jacob Golius, who later became professor of Arabic and subsequently also of Mathematics at Leiden University.[61] In the primacy of humanist learning for mathematics, Golius was not only an apt successor of Willebrord Snellius, but also a representative of the intellectual mainstream that dominated the university establishment. Golius's support for students of Dutch mathematics shows that the Leiden humanists around 1620 were not necessarily hostile to artisans entering the academic arena. The connection of Golius to practical mathematics was further strengthened when Frans van Schooten married his sister Maria in 1625.

5 Conclusion

The converted fencing hall of Ludolf van Ceulen was one of the more unusual teaching spaces in early modern academia – both by subject and by language of instruction. While fencing and dancing schools were an important part of early modern university education, they were usually kept separate from the university proper, at least in the institutional setting of the academy.[62] In Leiden this was different. The fencing school that had become a lecture theatre became a space in which the interests of city officials with a craft background were mediated with the established principles of academic scholarship in a place that reflected a new form of learning. While it reflected a broader ideal of

60 I have not been able to find any evidence for the claim by Dopper, *Life of Learning*, 171, that the proposal has been rejected. Winter, *Hoger beroepsonderwijs*, 19 and 23, who is referenced by Dopper in this context, claims that some were indeed admitted as students (without claiming that the discussed proposal had been influential in this).

61 Cf. Chikara Sasaki, *Descartes's mathematical thought* (Dordrecht: Kluwer, 2003), 203–224.

62 Maija Kallinen, "Lectures and Practices. The Variety of Mathematical and Mechanical Teaching at the University of Uppsala in the 17th Century," in: *Universities and Science in the Early Modern Period*, eds. Mordechai Feingold and Victor Navarro-Brotons (Dordrecht: Kluwer, 2006), 111–125, here 119; Rainer A. Müller, "Aristokratisierung des Studiums? Bemerkungen zur Adelsfrequenz an süddeutschen Universitäten im 17. Jahrhundert," *Geschichte und Gesellschaft* 10 (1984): 31–46.

combining theory and practice in scholarship, it also became a focus point for family networks in and around the university.[63]

Sixty years after the introduction of *Duytsche Mathematique*, the translator of Jean de Parival's *Les delices de la Hollande* added what he probably considered an amusing anecdote to de Parival's account in Leiden, in which he mocked the audience of the lectures given by Frans van Schooten, the Younger, as "equipped with sticks and leather aprons" and looking "very farcical."[64] Whether this anonymous account is an accurate description of the actual teaching happening in the lecture theatre might be difficult to ascertain. Yet, even taken at face value it can hardly be taken as evidence of a widespread derision of *Duytsche Mathematique* within the university. While the curators governing the university gave the subject its full support, the academic senate accepted it and its lectures within the university.

More than sixty years ago, Edgar Zilsel provided the image of the rise of science as the result of the handshake between the scholar and the craftsmen.[65] In allusion to this we might want to think of science as a result of the marriage between the scholar and the craftsmen – or at least between academic and artisanal families. Unlike the handshake, marriage created manifest links that would not be broken and nourished future development. This image has frequently been taken up by seventeenth century philosophers as new philosophy being the 'marriage between reason and experience,' a metaphor that probably has been taken somewhat more literally in the seventeenth century than is commonly done today.[66]

---

63 As Grämiger, *Verortungen von Wissen*, pp. 201–219, has pointed out, it is important to note that the existing representative functions of the fencing school remained with the inclusion of the teaching functions in practical mathematics.

64 Jean Nicolas de Parival, *De Vermaecklijckheden van Hollandt: een werck deses lands lof uytbasuynend* (Amsterdam: Samuel Imbrechts, 1661), 189: "In dit Bagijne-Hofff worden alle dagen (Behalven 's Woensdaeghs, en Saterdaeghs) van elf tot twaelf uuren, openbare Lessen gedaen in de Neerlandsche Tael, in de Mathematische Konsten, tot gerieve van alle ongeletterden, als Metselaers, Timmer-luyden, en diergelijcke meer; die haer dan met hoopen in die tijdt hier vinden; sonder-mantels, maer met hare stocken, en schoots-vellen, &c. versien; dat dan seer kluchtigh om sien is. Den Professor, die duytsche lessen voor haer doet, evenwel in sijnen gewoonlijcken aensienlijcken Professors-Tabbaert, ofte Rock, (soo wel als alle de andere Latijnsche Professoren de hare doen) is den Hoogh-geleerden en Wijdtvermaerden D. Franciscus van Schooten."

65 Edgar Zilsel, "The social roots of science," *American Journal of Sociology* 47 (1942): 544–562. For recent adaptation of Zilsel's thesis, see: Smith, *Body of the Artisan*; Pamela O. Long, *Artisan/Practitioners and the Rise of the New Sciences: 1400–1600* (Corvallis, OR: Oregon State University Press, 2011).

66 Cf. Henri Krop, "Het moeizame einde van een huwelijk (1687–1781): filosofen in de rol van echtscheidingsadvocaat," *Gewina* 30 (2007): 230–246.

Appendix: Signatories of Letters Relating to *Duytsche Mathematique* (c. 1607–1616)

Five undated letters concerning the *Duytsche Mathematique* can be found in the archives of Leiden University (Archief van Curatoren, Ms. AC 42, Bijlagen tot de Resolutiën, 1607–1618). Three of them relate to the vacancy in *Duytsche Mathematique*, so they must have been written between 1611 and 1615. Variations in name spelling and occupation between the letters are given in square brackets […]. English translations of self-identifications are put between quotation marks. Information from the Leiden *Album Studiosorum*; from Muller & Zandvliet's *Admissies* or from the 'Personen' on the website 'Erfgoed Leiden en omstreken' has been added, as well as information from other referenced sources. Information on city officials was taken from: E. Pelinck, "De functionarissen belast met de zorg voor de stadsbouwwerken te Leiden (1575–1818)," in: *Jaarboekje voor geschiedenis en oudheidkunde van Leiden en omstreken*, 1967, pp. 59–76. The first signatory of each letter is marked with an asterisk (*).[67]

1. Letter written by "auditors and practitioners of mathematical arts" (Auditores en Practisijnen der konsten van Mathema). They had heard the lessons of Frans van Schooten that he had given "upon order and request" of Van Ceulen and Van der Merwen both on the field ("ten platten lande") and in the private school in the house of Van Ceulen. Given his expertise they requested him to be appointed as professor. The list of signatories is published in Molhuysen, *Bronnen tot de geschiedenis der Leidsche Universiteit* (Den Haag, 1913), I, p. 391*. The letter must have been written some time after Van Ceulen's death on 31 December 1610.
2. Letter stating that the lessons of Van Ceulen and Van der Merwen had been "continued" after their death by Van Schooten. Given the "diligence, good instruction and demonstration as in the auditorium so in the field," the letter requests that Van Schooten should be allowed to continue in this. Given that it relates Van Schooten's teaching activities after Van Ceulen's death, it was probably written some time after the first letter. A possible occasion would be Maurits' first request to appoint Marolois in 1612.
3. Letter that is almost identical in wording to no. 2. The list of signatures in this letter differs more from the first letter than the list in no. 2, which suggests that it may have been written after the second letter.
4. Letter discussing the freedom from taxes. The letter states, "the curators having had promised the supplicants that they should enjoy such freedom from excises and imports as other *studiosi* of other faculties in the same university." This letter

67 I owe thanks to Huib Zuidervaart, who contributed additional information to this appendix.

also contains Van Schooten's signature which indicates that it would have been written when he was following the lectures in *Duytsche Mathematique*, not when he was giving them. The surveyors with known admission date who had signed the letter were admitted as surveyors between 1600 and 1604. Samuel Cloot's signature suggests that he signed the letter before 1607 when he started working as a military engineer in the Dutch army.

5. Letter requesting the 'auditors of the Dutch Mathematical auditorium' (Auditores vant Nederduytsche Mathematische auditorium) should be granted the 'freedom of the school' under certain conditions (discussed above). The letter was signed by university students as well as by craftsmen. The letter must have been written between 1612 (when Jacob Golius first arrived in Leiden) and 1616 (when Georg David Leysser had left Leiden), possibly in connection with Van Schooten's appointment, as this also clarified status issues relating to *Duytsche Mathematique*.

No.	Name	Letters signed					Remarks
		1	2	3	4	5	
1	Caesar van Achthoven	X	X				vaandrig (ensign) of the city of Leiden
2	Petrus Adolphus					X	student in mathematics from Norway, matriculated 4 August 1614, age 32
3	Govert Claesz. [van der Aer]		X	X		X	carpenter from Leiden, married in 1615
4	Jan Jacobsz. van Banchem	X	X			X	'sworn surveyor,' had enrolled as student in 1596, aged 13
5	Cornelis Jacobsz. van Banchem	X	X		X	X	'sworn surveyor,' no admission found
6	Jacobus Benthem [?]				X		
7	Pauls Boudewyns van Berlicom [Berlicum]	X	X	X			perhaps son of Boudewijn van Berlicum, councillor at the Raad van Brabant
8	Cornelis Beuysen	X					
9	Henderick Cornelisz. [van Bilderbeeck]	X	X	X			'bricklayer,' 'sworn surveyor,' admission Holland, 29 January 1602
10	Steffen van Blitterswyck	X*	X				student in geometry from Leiden, matriculated 25 July 1609, age 22
11	Wolfgang Theodoricus Borck		X	X			student in philosophy from Heidelberg, matriculated 21 November 1612, age 20; again in law on 12 October 1618
12	Adraen Dircxz. Bout	X	X				

(*cont.*)

No.	Name	Letters signed					Remarks
		1	2	3	4	5	
13	Gerardus Bouwens		X				
14	A[driaan de] Bruyn [Andriesz.]	X			X*		'sworn surveyor,' admission Holland, 20 September 1600
15	Albert Rodingius van Burch			X			
16	Daniel Cesarius	X	X			X	brick mason
17	Govert Claessen	X					'carpenter,' surveyor from Leiden, admission Holland, 24 December 1620
18	Harman Claesz.	X	X				'brick mason,' main stonemason of the city of Leiden
19	Claes Claesz.		X				
20	Samuel Cloot				X		curator of Maurits' collection of maps and mathematical instruments, son-in-law of Ludolf van Ceulen, worked as a military engineer from 1607–1609[a]
21	Pieter Cornelisz.				X		
22	Weijer Cornelisz.			X			'carpenter'
23	Henricus Custerus					X*	student in mathematics from Rostock, matriculated 30 July 1613, age 25
24	Jacob Davidt[s]	X	X	X			
25	Antonis Dircxsz.				X		probably a carpenter from Delfshaven, married in Leiden 14 March 1603
26	Gisbertus van Duivenvoorde			X			
27	Cornelis Frans	X	X				'carpenter'
28	Matthaeus Fritz					X	probably Matthaeus Pritse, student in law from Pommeren, matriculated 18 January 1614, age 18
29	Johan Faust von Fürstenberg					X	probably a student, but not inscribed in the *Album Studiosorum*
30	Adriaen Lourens Gael			X			brewer
31	Georgius [Joris] Geestecoorn			X		X	surveyor, admission Holland, 11 September 1615
32	Jacobus Golius					X	student in law from The Hague, matriculated 13 March 1610, age 14; again in law, 26 March 1618, age 21

No.	Name	Letters signed 1	2	3	4	5	Remarks
33	Claes Cornelisz. de Haes		X				
34	Nicolaus Hasius			X			student, worked as philosophy lecturer at Leiden University from 1613–1615
35	Jacob Heyns				X		
36	Pieter Hendericks					X	probably Pieter Henricxs van Bilderbeeck, surveyor from Leiden, admission 6 December 1616, son of no. 9, godfather of no. 58's daughter
37	Christian van Heule					X	surveyor from Leiden, admission 5 March 1621
38	Steffen van Heussen	X	X	X			from Leiden
39	Jacob Holtzapfel				X		'from Germany,' mostly likely a student
40	Huybertus [Cornelisz.] Hoogenboom	X					married in Leiden in 1612
41	Jan Jans[s]en	X	X	X		X	'carpenter'
42	Jan Jansen	X					'bricklayer'
43	Clemens Jansz.	X					'brick mason'
44	Franciscus Junius		X				student in philosophy from Heidelberg, matriculated 23 April 1608, age 17
45	Jan Joosen Koninck		X				'bricklayer'
46	Burchardus Koudt			X			
47	Güntherus a Landspergi [Günter von Landtsberg]		X	X			
48	Wolffgangus [Georgius] a Landtsbergi [Lansperch]					X	student in law, matriculated 14 February 1614, age 19
49	Jaques Lauwijck	X	X	X*			'schoolmaster'
50	Georgius David Leysser [Leyzer]					X	student in law, (Stirus Austriacus) matriculated 19 November 1613, age 20, had left Leiden for Paris before 29 September 1616[b]
51	Nicolas Libert		X				'brick mason'
52	Jochen Lodewijcx	X					'carpenter'
53	Laurens Louwmans					X	
54	Claes Meessen	X	X	X			'carpenter'
55	Claes Metersz.	X	X				'sworn surveyor,' no admission found
56	Jan de Milander				X		

(*cont.*)

No.	Name	Letters signed 1	2	3	4	5	Remarks
57	Noe[l] Jansoon du Mortier		X	X			surveyor from Leiden, admission 24 September 1621, before that working as a baker
58	Pieter Pietersze de Neyn		X				'brick mason,' son-in-law of no. 9, became city mason in 1632, worked also as an artistic painter, took drawing lessons with Esaias van de Velde in the period the letters were written
59	Christian van Peuls [?]		X	X			
60	Jan Janszoon le Pla		X	X			clothmaker in Leiden
61	Louris Jaspers du Pre			X			
62	Dirck Anthonisz. van Schalckwijck	X	X				'brick mason'
63	Frans van Schooten				X		see article above
64	Georgius Albertus a Seckendorff					X	student in law, matriculated 23 April 1613, age 20
65	Nicolaes [Claes] Petersz. Slabbinck			X	X		'sworn surveyor' from Leiden, admission Holland 21 June 1607
66	Johans Spehrfeltus [?]			X			
67	Cornelius Sprong	X					student in philosophy from Delft, matriculated 14 March 1609, age 29, in 1617 was engaged with the medical professors Reinier Bontius and Everardus Vorstius in a enterprise relating to draining the Harlemmermeer[c]
68	Johannes Fridericus a Stinck		X				
69	Andries van Swyeten	X	X*	X			married in Leiden 14 July 1618, probably identical with Andreas à Swieten, student in law from Leiden, matriculated 14 April 1615, age 27
70	Samuel Philips Taerling	X					
71	Winand Velbruygens [Winandus Velbruggen]		X				student in law from Bergen, matriculated 23 May 1611, age 17
72	Andries Verlyrerdius		X				

No.	Name	Letters signed					Remarks
		1	2	3	4	5	
73	Joannes van Venebergius			X			bricklayer
74	Joannes Chemnitz Vising			X			
75	Casper van Visser		X				
76	Johannes [Jan] de Vos	X					surveyor from Leiden, admission Holland, 14 June 1617
77	Pieter Willems [van Leyden]		X	X		X	carpenter
78	Jacob Willems	X					'carpenter'
79	Martinus Wimmers	X					'student' in 'Litterarum' from Nijmegen, matriculated 14 September 1607, age 13
80	Tielmannus Wyntgis	X	X				'student' in philosophy from Delft, matriculated 25 November 1609, age 18
81	Nicolaus de Wymysse [?]					X	
82	Jan Thielmansz. van Zeyl		X				'bricklayer'
83	Jan Otten van Zeyst	X	X	X	X		main carpenter of the city of Leiden, surveyor, admission Holland, 3 February 1604

a Cf. Zandvliet, *Maurits*, 78.

b *Die Autographensammlung des Stuttgarter Konistorialdirektors Friedrich Wilhelm Frommann (1707–1787)*, ed. Ingeborg Krekler (Wiesbaden: Harrasowitz, 1992), 733.

c Theodor H. Lunsingh Scheurleer et al., *Het Rapenburg: geschiedenis van een Leidse gracht*, vol. 5 (Leiden: Afdeling Geschiedenis van de Kunstnijverheid, R.U., 1990), 331.

CHAPTER 4

The Middelburg *Theatrum Anatomicum*: A Location of Knowledge and Culture in an Early Urban Context

Huib J. Zuidervaart

1 'Sites of Knowledge and Culture'

1.1 *Introduction*

On 17 May 1940 a German bombardment destroyed the centre of Middelburg, the capital of the Dutch province of Zeeland.[1] As through a miracle, a part of the city survived this inferno. However, by then the medieval hospital that once stood in this area – the *Gasthuis* – was already long gone, having been demolished in 1867. In its time the Middelburg *Gasthuis* had been one of the most prominent hospitals of the Netherlands.[2] Today only the former chapel of this hospital remains, now used as a church for Protestant worship. Furthermore, in the facade of the adjacent Nieuwstraat, a gate has been preserved that once gave access to the hospital (*figure 4.1*). Through this 'blind alley' one could also enter the Theatrum Anatomicum, a site founded in the years 1655–1658 for the dissection of anatomical bodies. This place was the first spot in Zeeland where an institutionalized form of scientific research was carried out, linked to a form of 'modern' empirical education.

1.2 *The* Theatrum Anatomicum *as a 'Site of Knowledge and Culture'*

In recent years the history of the seventeenth-century anatomy enjoys a renewed interest among historians of science and medicine. This not only

---

1 This paper is a shortened adaptation of my Dutch article "Het in 1658 opgerichte theatrum anatomicum te Middelburg. Een medisch-wetenschappelijk en cultureel convergentiepunt in een vroege stedelijke context," *Archief. Mededelingen van het Koninklijk Zeeuwsch Genootschap der Wetenschappen* (2009 [published 2010]): 73–140.

2 For the history of the Middelburg 'Gasthuis' as hospital, see: A.A. Fokker and J.C. de Man, *De gestichten en openbare gebouwen van Middelburg* (Middelburg: A.J. de Wilde, 1857), 17–20; D. Schoute, "De levensloop van een ziekenhuis. Geschiedenis van het Gasthuis te Middelburg," *Archief. Vroegere en latere mededeelingen voornamelijk in betrekking tot Zeeland* (1916): 137–286; J.L. Kool-Blokland, *De zorg gewogen. Zeven eeuwen Godshuizen in Middelburg* (Middelburg: Zeeuws Genootschap der Wetenschappen, 1990). For the chapel as church, see J.T. van Burg and H.J. Drayer, *500 jaar te gast. De geschiedenis van de Gasthuiskerk te Middelburg* (Middelburg: Christelijke Gereformeerde Kerk, 1994).

 | DOI:10.1163/9789004264885_005

FIGURE 4.1 Former entrance to the Middelburg *Gasthuis* hospital and the Middelburg *Theatrum Anatomicum*
PHOTO BY THE AUTHOR, 2009

because the development of anatomy itself is fascinating, but particularly because recent historical research has made it clear that modern science owes much to the early anatomists. The experience gained in anatomical practices, such as careful observation, the handling of scientific instruments, and the theoretical attempt to explain what one had observed, are competencies to which modern science is indebted a great deal.

Harold Cook, for instance, in his *Matters of Exchange: Commerce, Medicine and Science in the Dutch Golden Age* (2007) has made it very clear that the Dutch Republic has played a rather crucial role in the development of modern science. Dániel Margócsy, too, demonstrated that the culture of collecting, which flourished in this country with its worldwide commercial network, contributed in no small way to the rapid blossoming of science.[3] The early anatomical theatres in the Netherlands were a crucial part of the scientific infrastructure in the Dutch cities, and these locations are therefore important

3 Harold Cook, *Matters of Exchange. Commerce, Medicine, and Science in the Dutch Golden Age* (New Haven, CT: Yale University Press, 2007); Dániel Margócsy, *Commercial Visions. Science, Trade, and Visual Culture in the Dutch Golden Age* (Chicago: University of Chicago Press, 2014).

for our understanding of these developments. But, as historians have become aware, these sites were not used only for dissections, but encompassed also an important cultural component. The attention for anatomical theatres is a good example of the new interest in the spatial context of science. We use the Middelburg anatomical theatre as an example of such a combined scientific and cultural site, where activities took place and persons gathered in a way precursory to the manner in which people exchanged knowledge in the Dutch scholarly societies of the eighteenth century. The case of Middelburg has not been studied before, and is also instructive with regard to the interaction between well-known centres such as Leiden University, and more peripheral places.

1.3 *Anatomical Theatres*

In mid-sixteenth century Europe anatomical dissections became an influential form of practical medical training and gradually also of research. This interest in human anatomy served both the medical doctors as well as the surgeons. This led to two distinctive branches of practical anatomy, one set up in the context of university training and another performed within the surgeon guilds. Well known is the impact of the famous *De humani corporis fabrica* (Basel 1543), by the Padua professor Andreas Vesalius. This folio book was the first account of real dissections illustrated with numerous illustrations. This made the Italian city of Padua the centre of medical training and research for a long time. It was also here that in about 1570 a fixed anatomical site, an academic *Theatrum Anatomicum*, was created.

The first known anatomical dissection of a human body in the Netherlands was executed in 1550 by the brand new Amsterdam surgeon guild. The body of a criminal, called 'Suster Luyt' was dissected, and the skin of the cadaver was prepared and subsequently preserved in the surgeons' guild hall for many decades thereafter.[4] As early as 1555 an official approval for such dissections was granted to the Amsterdam surgeons' guild, in the name of the Spanish King Philip II, then as count of Holland, the sovereign of the country. In the following decades, when the focal point of anatomy shifted towards Northern Europe, the practise of dissecting was imitated in numerous places. By their nature, the location of these dissections, the anatomical theatres, became convergence points of new knowledge, supplementing the early royal cabinets of curiosities. Paula Findlen coined these places 'Sites of Knowledge,' locations

4 J.N. Keeman, "Dr Nicolaes Tulp and the surgeon's guild," in: *Nicolaes Tulp. The life and work of an Amsterdam physician and magistrate in the 17th century*, ed. S.A.C. Dudok van Heel et al. (Amsterdam: Six Art Promotion, 1991), 189–216, esp. 195–199.

that played an important role, both in the creation of new scientific ideas and in the acceptance thereof.[5] We will extend this phrase below to the seventeenth-century Dutch *Anatomical Theatre* as a 'Site of Knowledge *and* Culture.'

In the Netherlands the earliest academic *Theatrum Anatomicum* was established in 1590 at Leiden University, an institution founded only fifteen years before. In 1596 this temporary theatre was replaced by a more permanent arrangement, situated adjacent to the university library. In this way the new empirical source of knowledge – the anatomical theatre – was placed under the same roof as the traditional and long trusted classical source of knowledge. Because dissections only could be performed in the winter months, human and animal skeletons carrying vanitas attributes, referring to mankind's mortality, were placed in this theatre during summer. Other curiosities were also collected at this spot. Here, one could admire natural history objects, anatomical preparations, antiquities, Egyptian mummies, old textiles, rare stones, all kinds of prints and other paraphernalia, partly intended for instructional purposes, partly gathered with the idea to support mottoes like 'know thyself' and 'remember you will die.'[6] Soon this anatomical theatre became one of the crown jewels of Dutch scholarship: a place where knowledge, as well as devout contemplation played an important role. This Leiden *Theatrum Anatomicum* set an example that was imitated in most cities of the highly urbanized Dutch Republic. Eventually in the seventeenth-century no less than eighteen anatomical theatres were established in sixteen Dutch cities (*See Table 4.1*).[7]

Because the Dutch Republic lacked a stimulating royal court, expressions of science and art had to find other means of support than elsewhere in Europe, where kings and princes backed scientific initiatives through state patronage. In countries such as Italy, France or England, scholarly societies were founded already in the seventeenth century, privileged by the state sovereigns. Familiar examples are the *Accademia del Cimento*, founded in 1657, the *Royal Society of London*, founded in 1660, and the *Académie Royale des Sciences* in Paris, established in 1666. In the Netherlands such an official 'learned' society was not created until the mid-eighteenth century, and even then it was a privately funded

5 Paula Findlen, *Possessing Nature: Museums, Collecting, and Scientific Culture in Early Modern Italy* (Berkeley: University of California Press, 1996), chapter 3.

6 Tim Huisman, *'The Finger of God.' Anatomical Practice in 17th-Century Leiden* (Leiden: Primavera Press, 2008). For the contents of the Leiden collection of curiosities, see: H.J. Witkam, *Catalogues of all the chiefest rarities in the publick anatomie hall of the University of Leyden* (Leiden, 1980).

7 J.A.M. Slenders, *Het theatrum anatomicum in de Noordelijke Nederlanden, 1555–1800* (Nijmegen: Instituut voor Geschiedenis der Geneeskunde, 1989).

TABLE 4.1 Dutch cities with an Anatomical Theatre

No.	City	Founded	Founder	Highest offered type of education
1	Leiden-1	1590	University	University
2	Delft	1614	Surgeons' guild	Latin School
3	Amsterdam	1619	Surgeons' guild	Athenaeum
4	Utrecht	ca. 1620	Surgeons' guild	University
5	The Hague	1628	Surgeons' guild	Latin School
6	Franeker	Before 1629	University	University
7	Dordrecht	1634	Surgeons' guild	Athenaeum
8	Leiden-2	1638	Surgeons' guild	University
9	Rotterdam-1	1642	Surgeons' guild	Latin School
10	Breda	1644–1650	Illustrious School	Athenaeum
11	Harderwijk	1648	University	University
12	Groningen	1654	University	University
13	Middelburg	1655–1658	Surgeons' guild	Athenaeum
14	Rotterdam-2	1659–1662	Private theatrum of De Bils	Latin School
15	's-Hertogenbosch	1662	Surgeons' guild	Athenaeum
16	Haarlem	Before 1700	Surgeons' guild	Latin School
17	Arnhem	Before 1700	Surgeons' guild	Latin School
18	Kampen	Before 1700	Surgeons' guild	Latin School

initiative, although supported – or at least tolerated – by the various provincial or municipal governments.

So, in the seventeenth-century Netherlands, there was a sort of niche in the intellectual market. The five Dutch universities (Leiden, Franeker, Groningen, Utrecht and Harderwijk) were keen to fill this niche. According to the historian Klaas van Berkel, the relatively strong position of these universities explains why the formation of scholarly societies in the Netherlands occurred at relative such a late moment.[8] However, unofficial societies, where groups of like-minded people could meet, were already present in the early seventeenth

8 Klaas van Berkel, *In het voetspoor van Stevin. Geschiedenis van de natuurwetenschap in Nederland 1580–1940* (Meppel/Amsterdam: Boom, 1985), 84–85.

century. A familiar example is the *Collegium Mechanicum* founded in 1626 in Rotterdam at the instigation of the scholar-practitioner Isaac Beeckman.[9]

A closer look into the role of the anatomical theatres in the Dutch cities reveals that these sites were in fact locations of knowledge as well as culture. It was the Dutch sociologist and historian Jan Rupp who first noticed that the numerous Dutch anatomical theatres were an essential part of the Dutch seventeenth-century intellectual infrastructure. He investigated the anatomical theatres of Leiden, Delft and Amsterdam, and concluded that these theatres functioned not only as a place for dissection, but even more as an intellectual and cultural focal point.[10] At these locations not only surgeons and physicians, but also other intellectuals, artists and laymen could meet and discuss in a fruitful way books and scholarly artefacts, such as skeletons, anatomical preparations and other curiosity-provoking objects. All this with differences due to a *couleur locale*. Rupp investigated two aspects of the anatomical theatres: (1) their performance as a centre of knowledge and (2) their broader cultural impact. The latter was reflected by a range of qualities, for instance the presence of a library containing a variety of books, a collection of curiosities, containing naturalia as well as artificialia, and contacts with artists and poets. In short, in line with Rupp, we can label an anatomical theatre as a contemporary 'Site of Knowledge and Culture' if *in its own time* the following criterions were met:

1. First such an anatomical theatre should have facilitated meetings, where the practice of 'modern' (medical) science was stimulated. Such meetings should have been open for access to interested layman, perhaps after paying only a small entrance fee.
2. The site should have been a place, not only for confirmation of already known knowledge, but also for new research, the results of which should be discussed and published.
3. The presence of a library, containing not only medical books, but also more general works should underpin the site's role as a cultural convergence point.

---

9 Cornelis de Waard, *Journal tenu par Isaac Beeckman de 1604 à 1634*, vol. 2 (Den Haag: Martinus Nijhoff, 1942) 429, 445. See on other informal societies in the 18th-century, also my *Van Konstgenoten en Hemelse fenomenen. Nederlandse sterrenkunde in de achttiende eeuw* (Rotterdam: Erasmus Publishing, 1999).

10 J.C. Rupp, "Matters of life and death. The social and cultural conditions of the rise of anatomical theatres, with special reference to seventeenth century Holland," *History of Science* 28 (1990): 263–287; idem, "Het theatrum anatomicum: publiekscommunicatief fossiel of 'archetype,'" *Gewina. Tijdschrift voor de geschiedenis der geneeskunde, natuurwetenschappen, wiskunde en techniek* 25 (2002): 191–209.

4. The same would apply for a collection of curiosities, containing naturalia as well as artificialia.
5. Contacts with artists and poets were also a desideratum.

In his essays Rupp restricted himself solely to the anatomical theatres in Leiden, Delft and Amsterdam. Cook, too, discusses only the situation in these cities. Although this limitation may be understandable, because sources on the role of the anatomical theatres in other cities are scarce, this picture is still incomplete. Why would the role of an anatomical theatre have been different in cities such as Rotterdam, Dordrecht or Middelburg? In the seventeenth century these places had commercial and political connections hardly inferior to those of Amsterdam, Leiden or Delft. And if things in these cities were different, what then determined their *couleur locale*?

For example, the anatomical theatre of Rotterdam – founded in 1642 – complies also to the criterions mentioned above. This location was also explicitly designed 'after the exemplum of our neighbouring cities.'[11] Daniël de Moulin has shown that in the Rotterdam anatomical theatre public lectures were given from the very start. Since 1648 this site also featured a library and at least from 1664 there was also a collection of 'jewelry, furniture, books, instruments, and rarities.'[12] The Rotterdam anatomical theatre was clearly designed as a location with a cultural feature that went beyond pure anatomy. A similar story can be told for Dordrecht. Such an anatomical theatre was also in Middelburg, the capital of the province of Zeeland, the history of which location we will measure against the criterions mentioned above. Firstly, however, we will sketch briefly the history of this anatomical site and of the two *Collegia* attached to this place.

2 The Anatomical Theatre of Middelburg

2.1 *The City of Middelburg*

Mid-seventeenth-century Middelburg was one of the prominent cities of the Dutch Republic. With a population of around 20,000 inhabitants, of which a large contingent had roots in the Southern Netherlands, the city was a major seafaring and trading city. Many ships were built here, and Middelburg

---

11 D. de Moulin, "De natuurhistorische verzameling in het voormalige Theatrum Anatomicum te Rotterdam," *Rotterdams Jaarboekje* (1972): 129–139. See also: *Catalogus ofte naamlijst der voornaamste zaken, dewelke op de anatomiekamer der stadt Rotterdam te zien zijn* (Rotterdam 1740) and M.J. van Lieburg, "Het eerste theatrum anatomicum van Rotterdam (1642–1759)," *Rotterdams jaarboekje* (1976): 210–227 ; idem, "Het anatomisch kabinet der Clinische School te Rotterdam," *Rotterdams Jaarboekje* (1974): 251–270.

12 Ibidem.

accommodated not only the government of the province, but it was also the seat of the Chamber of Zeeland of both the East-India and the West-India trading companies. Late sixteenth and early seventeenth-century Middelburg was prosperous, as is demonstrated by a rich cultural life, especially in the realm of botany, literature and poetry. A chamber of rhetorics, 'Het Bloemken Jesse,' was active since the early sixteenth century. And throughout the century a large group of still-life painters was active, depicting flowers, fruit and other crops, starting early in the century with Ambrosius Bosschaert and his brother-in-law Balthasar van der Ast, going via painters such as Johannes Goedaert and François Rijckhals in the middle of the century, to Adriaen Coorte, selling both in Middelburg and Flushing around 1700.[13] Remarkably all these painters incorporated in their still-lifes small animals, mostly little insects, such as butterflies, bees, caterpillars, dragonflies, grasshoppers, beetles and so on. In itself these paintings give testimony of an attitude in which the careful observation of nature was self-evident. It is obvious that the study of these small, almost microscopic animals must have taken a considerable amount of time and energy. In the course of these observations the painter Goedaert even made the startling discovery that a larva first becomes a chrysalis, from which subsequently a butterfly emerges. Goedaert was one of the first in the world to illustrate this transformation. His three volume book *Metamorphosis Naturalis*, describing these stages in the development of insects, became a real bestseller, both in the Dutch Republic and abroad (*figure 4.2*).[14] His book in the Dutch

---

13 P.J. Meertens, *Letterkundig leven in Zeeland in de zestiende en de eerste helft der zeventiende eeuw* (Amsterdam: Noord-Hollandsche Uitgevers Maatschappij, 1943); Noortje Bakker et al., *Masters of Middelburg. Exhibition in the honour of Laurens J. Bol* (Amsterdam: K. & V. Waterman, 1984); Arjan van Dixhoorn, *Lustige geesten. Rederijkers in de Noordelijke Nederlanden (1480–1650)* (Amsterdam: Amsterdam University Press, 2009); H.J. Plankeel and T. de Jong, *Adriaen Coorte uit IJzendijke. Een eenzame stillevenschilder* (Hilvarenbeek, 2015); H.J. Zuidervaart, "'Scientia' in Middelburg rond 1600. De wisselwerking tussen academici, kooplieden en handwerkers in het vroegmoderne Middelburg: essentieel voor de creatie van 's werelds eerste verrekijker?," *Archief. Mededelingen van het Koninklijk Zeeuwsch Genootschap der Wetenschappen* (2017): 43–100 [issued in 2018].

14 Johannes Goedaert, *Metamorphosis Naturalis, ofte historische beschrijvinge van den oirspronck, aerd, eygenschappen ende vreemde veranderinghen der wormen, rupsen, maeden, vliegen, witjens, byen, motten ende dierghelijcke dierkens meer,* 3 vols, (Middelburg: Jacob Fierens, undated – vol. 1: 1660; vol. 2: 1667; vol. 3: 1669). Latin edition: vol. 1 [De Mey transl.] 1665; vol. 2 [Veezaert transl.] 1668; vol. 3 [De Mey transl.] c. 1670. English edition [Martin Lister transl.] 1682. Cf. G. Kruseman, "The editions of Goedaert's 'Metamorphosis naturalis,'" *Entomologische berichten* 16 (1956): 46–48; H.J. Zuidervaart, "Het Natuurbeeld van Johannes de Mey (1617–1678), hoogleraar filosofie aan de Illustere School te Middelburg," *Archief. Mededelingen van het Koninklijk Zeeuwsch Genootschap der Wetenschappen* (2001): 1–40, esp. 17–18.

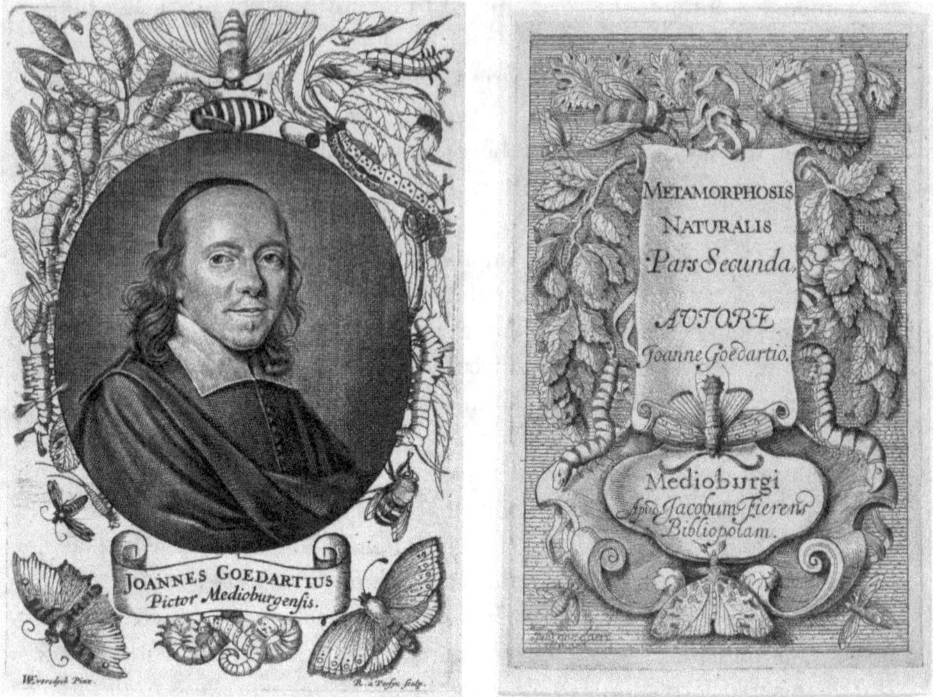

FIGURE 4.2 The painter-entomologist Johannes Goedaert (1617–1668), depicted on the frontispiece of his *Metamorphosis Naturalis* (1660–1663). Engraving by Reinier van Persijn after a painting of Willem Eversdijck
RIJKSMUSEUM, AMSTERDAM

language was translated in French, English and Latin. It is in this context that we must place the activities of the Middelburg *Theatrum Anatomicum*, or the 'Anatomy Chamber,' as contemporaries called it.[15]

2.2 *The Founding of an Anatomy Chamber (1655–1658)*

The earliest recorded anatomical dissection in Middelburg dates from 1622. But that event concerned mainly the manufacture of a skeleton for the education

15 Goedaert also used a magnifying glass in his study of insects. In that respect it is interesting to note that his aunt Sara Goedaert-Jansen was also the aunt of the well-known Middelburg optician Johannes Sachariassen, who falsely claimed that his father Sacharias Jansen had invented the telescope in 1590. Sara was Sacharias's only sister. For details, see H.J. Zuidervaart, "'Uit vaderlandsliefde': Pierre Borels *De vero telescopii inventore* (1656) en het negentiende-eeuwse streven naar een gedenkteken voor de 'ware uitvinder van de verrekijker,'" *Archief. Mededelingen van het Koninklijk Zeeuwsch Genootschap der Wetenschappen* (2007): 5–60, esp. note 21.

of new surgeons.[16] Ten years later, in 1632 real anatomical dissections started to take place in one of the upper rooms of the hospital. This 'chapel-room,' however, was not only rather difficult to reach, but also too near to the patients, who complained about the stench of the dead bodies. This was the main reason why the surgeons desired another place for these dissections.[17]

In an attempt to tackle these problems the dean of the guild, surgeon Jacob Evertsen, in 1655 successfully obtained from the city a small warehouse behind the hospital.[18] At first sight the date of this gift seems rather strange. Middelburg was barely recovered from a recent revolt which had expelled two of the most prominent mayors in 1651. Both men had just founded an Illustrious School (in 1650), which came into financial trouble after having lost their support. For this reason the Athenaeum could continue only as an extension of the local Latin School.[19] So why would the new city government bother to erect a new training facility in 1655, now for the surgeons' guild, paying all costs for the equipment of this location?

There were two reasons for this step: firstly, in 1654 Middelburg had suffered a syphilis epidemic, in the cure of which the surgeons' guild had played a crucial role.[20] Secondly, that same year a controversial, but famous anatomist moved to Middelburg; a man whose actions were of great, although not undisputed, importance for the development of anatomy in the Dutch Republic. His name was Louis de Bils, a nobleman of wealthy Flemish descent. De Bils claimed to have invented a new method of preparation of human corpses. He claimed being able to preserve these bodies for many years without loss of their natural color and shape. In 1648 he had settled in Sluis, in Zeeland.[21] There, De Bils produced a large collection of anatomical preparations, most

---

16 A.A. Fokker, "Losse bladen uit de geschiedenis van het chirurgijnsgilde te Middelburg," *Nederlands Tijdschrift voor Geneeskunde* 21 (1877): 333–372, esp. 353; Dirk Schoute, *Schets van het Middelburgsche Chirurgijnsgilde* (n.pl.; n.d. [=c. 1911]), 21.

17 Slenders, *Theatrum anatomicum*; Kool-Blokland, *De zorg gewogen*.

18 Kool-Blokland, *De zorg gewogen*, 97.

19 Willem Frijhoff, "Zeelands universiteit: hoe vaak het mislukte en waarom," *Archief. Mededelingen van het Koninklijk Zeeuws Genootschap der Wetenschappen* (1986): 7–41. See also: H.J. Zuidervaart, "Van Calvinistische orateurstoel tot Katholieke tabernakel. De wonderlijke lotgevallen van een katheder van de Middelburgse 'Illustre School' uit 1650," *Studium. Tijdschrift voor wetenschaps- en universiteitsgeschiedenis* 1 (2008): 62–69.

20 A.A. Fokker, *Onderzoek naar den aard van de epidemische en contagieuse ziekten die vroeger in Zeeland geheerscht hebben* (Middelburg: Altorffer, 1860); A.A. Fokker and J.C. de Man, *Levensberichten van Zeeuwsche Medici* (Middelburg: Altorffer, 1901), 69.

21 A.A. Fokker, "Louis de Bils en zijn tijd," *Nederlandsch tijdschrift voor geneeskunde* I–II (1865): 167–220; J.R. Jansma, *Louis de Bils en de anatomie van zijn tijd* (Hoogeveen: C. Pet, 1919).

of which he donated to the anatomical theatre of Leiden University, obviously in an attempt to obtain support, if not a position. For this precious gift – De Bils estimated his expenses at some 40,000 guilders – he was thanked by the newly appointed Leiden professor of anatomy Johannes van Horne, who issued a comprehensive testimonial in praise of De Bils's achievements. De Bils used this flattering testimonial in his first publication, published in 1655: the *Vertooch van verscheyde eyghene anatomische stucken* ['an account of some of my anatomical preparations']. In this pamphlet De Bils not only promoted his method of embalmment, but he also pretended to be an expert in medical theory. This pretention would spark a long controversy with numerous physicians in the Dutch Republic. Because De Bils did not reveal his methods, this secrecy unintendedly gave an enormous impetus to anatomical research in the Northern Netherlands.[22] In short, De Bils's presence in Middelburg seemed an excellent reason to create an anatomical site, this with the implicit wish to bind this famous anatomist to the city. To mark De Bils's presence in Middelburg, one of his local admirers published an illustrated account of De Bils's earlier decomposition of two Siamese twins (*figure 4.3*).[23] Unfortunately the refurbishment of the former warehouse into an anatomy chamber was delayed until 1658. By then De Bils already had left Middelburg, being appointed bailiff in Aardenburg, in the Flemish part of Zeeland.

2.3 *The Interior of the Middelburg Anatomy Chamber (1658)*

From a nineteenth-century description by Jan Cornelis de Man, then lector anatomy at the Middelburg Medical School, we know that the old anatomy chamber was a spacious, but cold and rather dark room with a stone floor. The room was nearly square: some seven meters long and wide, with a ceiling at a height of over four meters. From its interior only few details are known. A map from the early nineteenth century reveals a semicircular theatre, with a drain in the corner of the room.[24] At the very start in 1658 the amphitheater had not yet been completed. But a new instructive skeleton, as well as the beginning

---

22 Cook, *Matters*, 271–277; M.J. van Lieburg, "Nicolaas Zas (1610–1662). Een biografische bijdrage tot de geschiedenis van de chylus- en lymfetheorie," in: *De Novis Inventis. Essays in the history of medicine in honour of Daniel de Moulin on the occasion of his 65th birthday*, eds. A.H.M. Kerkhoff et al. (Amsterdam / Maarssen: APA-Holland University Press, 1984), 231–257.

23 Franciscus de Raet, *Anatomische beschrijvinge van een wanschepsel geboren op de Elderschans buyten Aerdenburg, ontleet, door jonck-hr. Louis de Bils* [...] *briefs-gewyse gesonden door d'heer doctor F. de Raet aen A. Parent* (Middelburg: Henrick Smidt, 1659).

24 Although this map concerns the shape of the theater after its refurbishment in 1755, it seems plausible that the 17th-century shape was maintained.

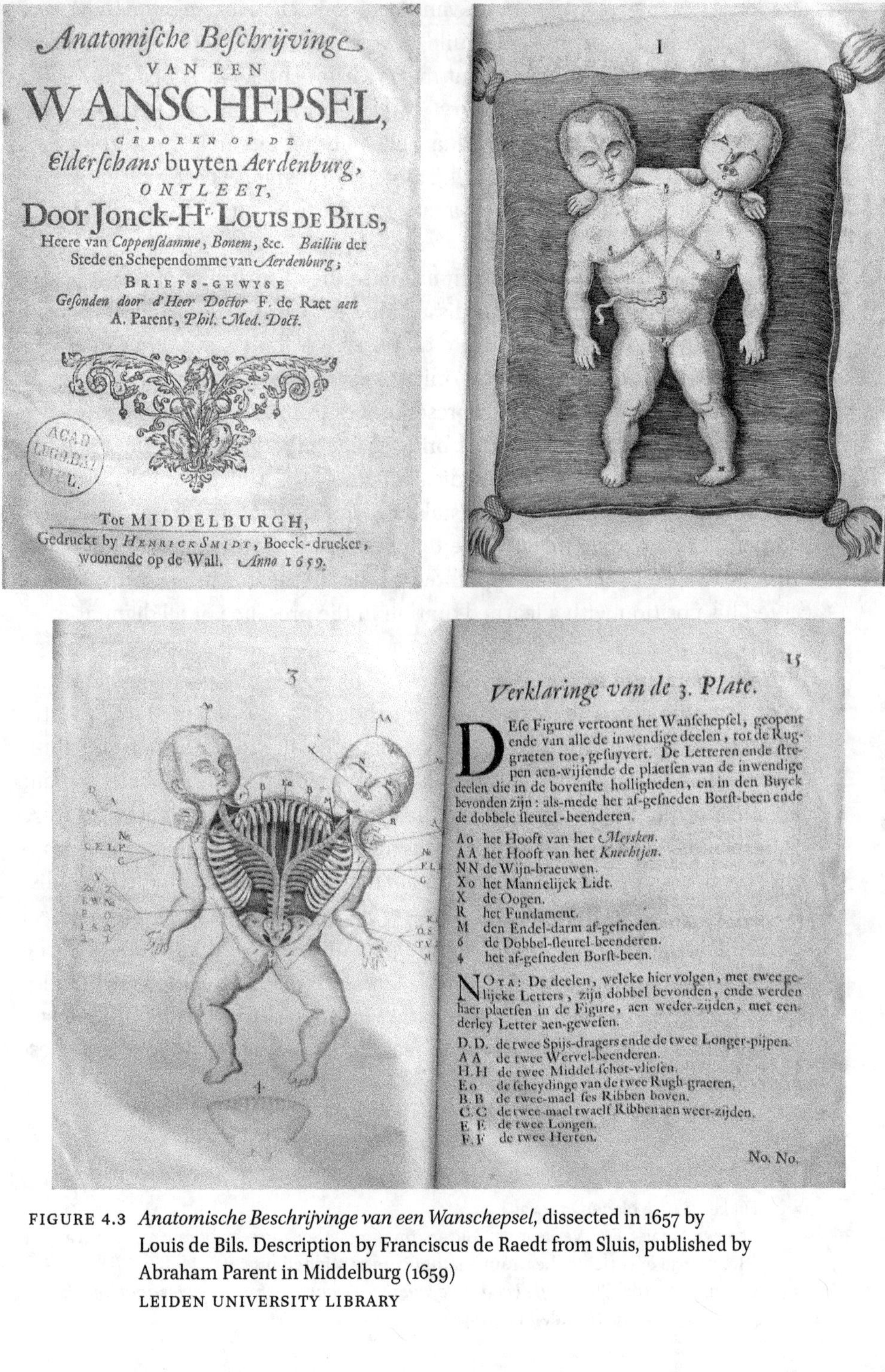

Anatomische Beschrijvinge
VAN EEN
WANSCHEPSEL,
GEBOREN OP DE
Elderschans buyten *Aerdenburg*,
ONTLEET,
Door Jonck-H^r. LOUIS DE BILS,
Heere van *Coppensdamme*, *Bonem*, &c. *Bailliu* der
Stede en Schependomme van *Aerdenburg*;
BRIEFS-GEWYSE
Gesonden door d'Heer Doctor F. de Raet *aen*
A. Parent, *Phil. Med. Doct.*

Tot MIDDELBURGH,
Gedruckt by HENRICK SMIDT, Boeck-drucker,
woonende op de Wall. *Anno* 1659.

15

Verklaringe van de 3. Plate.

DEse Figure vertoont het Wanschepsel, geopent ende van alle de inwendige deelen, tot de Rug-graeten toe, gesuyvert. De Letteren ende stre-pen aen-wijsende de plaetsen van de inwendige deelen die in de bovenste holligheden, en in den Buyck bevonden zijn: als-mede het af-gesneden Borst-been ende de dobbele sleutel-beenderen.

Ao het Hooft van het *Meysken*.
AA het Hooft van het *Knechtjen*.
NN de Wijn-braeuwen.
Xo het Mannelijck Lidt.
X de Oogen.
R het Fundament.
M den Endel-darm af-gesneden.
6 de Dobbel-sleutel-beenderen.
4 het af-gesneden Borst-been.

NOTA: De deelen, welcke hier volgen, met twee ge-lijcke Letters, zijn dobbel bevonden, ende werden haer plaetsen in de Figure, aen weder-zijden, met een-derley Letter aen-gewesen.

D.D. de twee Spijs-dragers ende de twee Longer-pijpen.
A A de twee Wervel-beenderen.
H.H de twee Middel-schot-vliesen.
E.o de scheydinge van de twee Rugh-graeren.
B.B de twee-mael ses Ribben boven.
C.C de twee-mael twaelf Ribben aen weer-zijden.
E E de twee Longen.
F.F de twee Herten.

No. No.

FIGURE 4.3 *Anatomische Beschrijvinge van een Wanschepsel*, dissected in 1657 by Louis de Bils. Description by Franciscus de Raedt from Sluis, published by Abraham Parent in Middelburg (1659)
LEIDEN UNIVERSITY LIBRARY

of a library, was present.[25] Simon van Beaumont, the city secretary who was appointed supreme dean of the guild, gave an expensive set of surgical tools made by the Rotterdam instrument maker Gisbrecht Servaes. Beaumont also donated the first books to the library: works of Adriaan van den Spiegel (1578–1625), professor of anatomy in Padua. Several other city magistrates followed his example. Former mayor Hendrik Thibaut, for instance, gave a real pièce de résistance: the famous anatomical atlas of Andreas Vesalius.

2.4 *The Formation of a* Collegium Anatomico Chirurgicum (*1658*)

As soon as the new anatomical dissection room was completed, a request was made to the city magistrates to authorize the formation of a Collegium Anatomico Chirurgicum, a joint initiative of surgeons and the *Doctores Medicinae*, led by a board with representatives from both professions.[26] Such an initiative was rather unusual. It bridged not only the status difference between the university trained medical doctors and the practically minded surgeons, but it was also an unmistakable sign that the theoretically oriented doctors assigned a greater value to empiricism. In July 1658 the physician Cornelis van de Voorde was appointed as the first lector in anatomy. He accepted his position with a learned oration in the presence of all dignitaries of the city (*figure 4.4*).

In this inaugural address Van de Voorde discussed the prerequisite of the art of anatomy.[27] As the main motive for the creation of an anatomy theatre in Middelburg, he mentioned the economic necessity to train good surgeons. After all, Middelburg ships sailed to all quarters of the world for the benefit and advancement of commerce. On those numerous ships the surgeons were solely responsible for all medical treatment. So, it was vital to give them a proper medical training. Moreover, the understanding yielded by anatomical dissections was also useful for anyone interested in the advancement of medical science. For that reason, Van de Voorde's lessons were also admissible to the general public, although Van de Voorde stipulated that his lessons would not equal those given at the Dutch universities. Van de Voorde would limit himself to knowledge necessary for practical healing. Therefore he presented his lessons in the Dutch language, and not in the scholarly Latin.

---

25 Fokker, "Losse bladen," 342, 354.

26 Zeeuws Archief, F.A. Verheye van Citters, inv. nr. 54b. See more elaborate my Dutch article "Het in 1658 opgerichte theatrum anatomicum te Middelburg."

27 C. van de Voorde, *Oratie, ofte Eerste inleydinghe, ghedaen inde chyrurgijns anatomie-kamer* [...] *den 17 july 1658* (Middelburg, 1659).

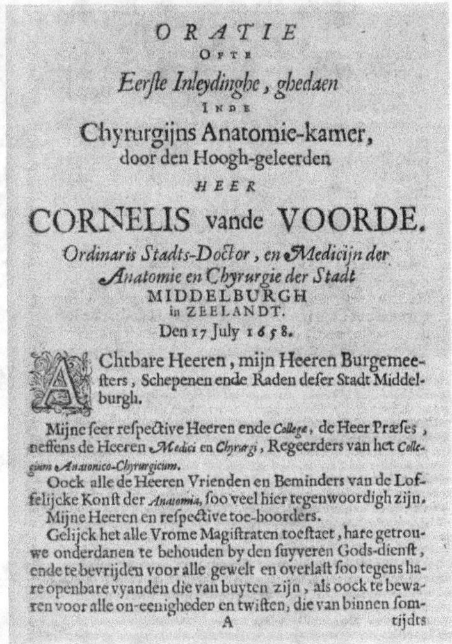

ORATIE
OFTE
Eerste Inleydinghe, ghedaen
INDE
Chyrurgijns Anatomie-kamer,
door den Hoogh-geleerden
HEER
CORNELIS vande VOORDE.
Ordinaris Stadts-Doctor, en Medicijn der Anatomie en Chyrurgie der Stadt
MIDDELBURGH
in ZEELANDT.
Den 17 July 1658.

Achtbare Heeren, mijn Heeren Burgemeesters, Schepenen ende Raden deser Stadt Middelburgh.

Mijne seer respective Heeren ende *Collega*, de Heer Præses, neffens de Heeren *Medici* en *Chyrurgi*, Regeerders van het *Collegium Anatomico-Chyrurgicum*.

Oock alle de Heeren Vrienden en Beminders van de Loffelijcke Konst der *Anatomia*, soo veel hier tegenwoordigh zijn.

Mijne Heeren en respective toe-hoorders.

Gelijck het alle Vrome Magistraten toestaet, hare getrouwe onderdanen te behouden by den suyveren Gods-dienst, ende te bevrijden voor alle gewelt en overlast soo tegens hare openbare vyanden die van buyten zijn, als oock te bewaren voor alle on-eenigheden en twisten, die van binnen somtijdts

A

FIGURE 4.4 *Left*: Cornelis van de Voorde (1628–1678), Medical Doctor educated in Leiden, Middelburg's first Lector in Anatomy. Engraving by the sculptor Pieter Philippe (1635–1702) from The Hague, in Van de Voorde's *Lichtende fakkel der cheirurgia ontstoken* (Middelburg 1664). The poem signed 'G.T.' is most likely written by Galenus Trezel, Medical Doctor in Middelburg since 1661
SOURCE: COLLECTION KONINKLIJK ZEEUWSCH GENOOTSCHAP DER WETENSCHAPPEN – ZELANDIA ILLUSTRATA, ZEEUWS ARCHIEF, MIDDELBURG
Right: Title page of Van de Voorde's *Oratie* (1658)
SOURCE: ZEEUWSE BIBLIOTHEEK, MIDDELBURG

Because only the cold of the winter could delay the decay of a dead body, it would take until the winter of 1658–1659 before Van de Voorde was able to carry out his first public dissection.[28] But the medical lessons given every fortnight were a great success. Already in the fall of 1659 the *Collegium Anatomico Chirurgicum* requested the city council permission to construct a larger set of benches. A skilled carpenter made a design with "subsellia," or handrails to lean against, comparable to those in the Leiden *Theatrum Anatomicum*.[29] Van de Voorde had studied in Leiden, so it is plausible that this site formed the inspiration for the Middelburg variant. Earlier that year the Leiden professor Johannes van Horne had visited Middelburg, and that fact probably stimulated

28 Schoute, *Schets*.
29 Zeeuws Archief, Archief Verheye van Citters, inv. nr. 54b.

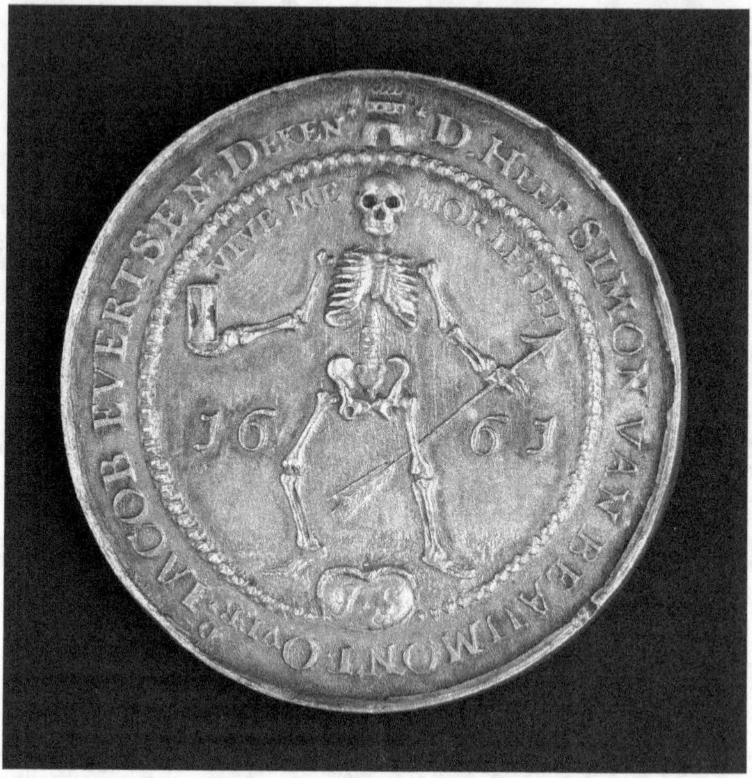

FIGURE 4.5 Medal of the Middelburg surgeons' guild designed in 1661 by Matthijs Hooft, a sculptor working at the Mint of Zeeland. A skeleton with an hourglass and arrow with the text VIVE ME | AMOR LETHI ('Live mindful of death'). For this anatomically correct skeleton displayed with vanitas symbols, the skeleton composed in 1657 for the Middelburg anatomy chamber probably was the model.
SOURCE: COLLECTION KONINKLIJK ZEEUWSCH GENOOTSCHAP DER WETENSCHAPPEN, ZEEUWS MUSEUM, MIDDELBURG

this application.[30] Comparable to Leiden, the best places in the Middelburg Theatre were reserved for the city magistrates, directly followed by the board and the members of the *Collegium Anatomico Chirurgicum*. The back rows were retained for the apprentices and the general public. Admission to a meeting was only granted after showing an entrance medal (*figure 4.5*). Members

30 The trip of Johannes van Horne (1621–1670) to Middelburg is mentioned in F.M.G. de Feyfer, "Die Historische Ausstellung der Natur und Heilkunde in Leiden," *Janus* (1908): 88–102, esp. 93. Van de Voorde's *Lichtende fakkel der cheirurgia ontstoken* (Middelburg: François Krook, 1664) contains a poem signed 'J.V.H.' It is tempting to identify this with his Johannes van Horne.

of the surgeons' guild and the collegium received such a medal after paying a yearly fee. Special guests, or 'persons of quality,' were granted free admission, after showing a smaller medal made from lead.[31] Others could attend a dissection by paying four shilling ("stuivers"), or one fifth guilder.

2.5 *A New – Richly Furnished – Guild Room for the Surgeons (1668)*

The lectures and dissections became such a success that around 1666 a second property could be acquired at the end of the hospital's garden, diagonally adjacent to the anatomy chamber. Here, a new guild room could be built at the end of a blind alley. This structure was completed in 1668 (*figure 4.6*).[32] In this new building a nice representative space was created, equipped with various rich ornaments, such as an ornate fireplace, stained glass windows and gold leather draperies. After the demolition of the building, in 1869, the former lecturer of the Middelburg Medical School received "a not unkind copper fountain" from this guild room.[33] The rich decoration of the building demonstrates the optimal self-confidence of the surgeons' guild. The mere fact that the guild could finance everything from its own resources proves that the revenues of the lessons and public dissections must have been considerable. Where in the years 1655–1659 the installment of the anatomy chamber needed to be fully funded by the city, the funding of a luxurious guild chamber in 1668 was no problem at all.

2.6 *The Formation of the* Collegium Medicum Medioburgense *(1668)*

The obvious success of the surgeons' guild triggered a new development. Partly related to the completion of the new guild chamber as a genteel meeting place, a second society was established, this time exclusively restricted to the local medical doctors. This new *Collegium Medicum Medioburgense*, under the Latin motto *Majestate et Fide* ("Majesty and Faith") was founded by Anthony Everaerts, physician of the Walloon hospital and the oldest serving medical

31 The medal of 1661 sums up the names of the board of the guild. The legend reads: D. HEER SIMON VAN BEAUMONT OVER[D][EKEN]: IACOB EVERTSEN DEKEN. On the reverse side are the names: IOOS SEYS ˙ IASPER BRAEMS BELE[s] (*Beleders*) ˙ CORNELIS HENDRICX BOEC[R]. (*Boekhouder*). Cf. Jacob Dirks, *De Noord-Nederlandsche Gildepenningen* (Haarlem: Erven F. Bohn, 1878), 328–329 and Marie G.A. de Man, "Een onbeschreven penning van het chirurgijns- of Cosmus- en Damianus-gilde te Middelburg," *Tijdschrift van het Kon. Ned. Genootschap voor Munt- en Penningkunde* 20:2 (1912): 124–136.

32 Cf. W.O. Swaving, *Catalogus van het Stedelijk Museum van Oudheden (Oudheidkamer) te Middelburg. Vierde aangevulde druk* (Middelburg: Altorffer, 1910), nr. 61.

33 J.C. de Man, *De Geneeskundige School te Middelburg, hare lectoren en leerlingen. 1825–1866*, 2 vols. (Middelburg: D.G. Kröber, 1902).

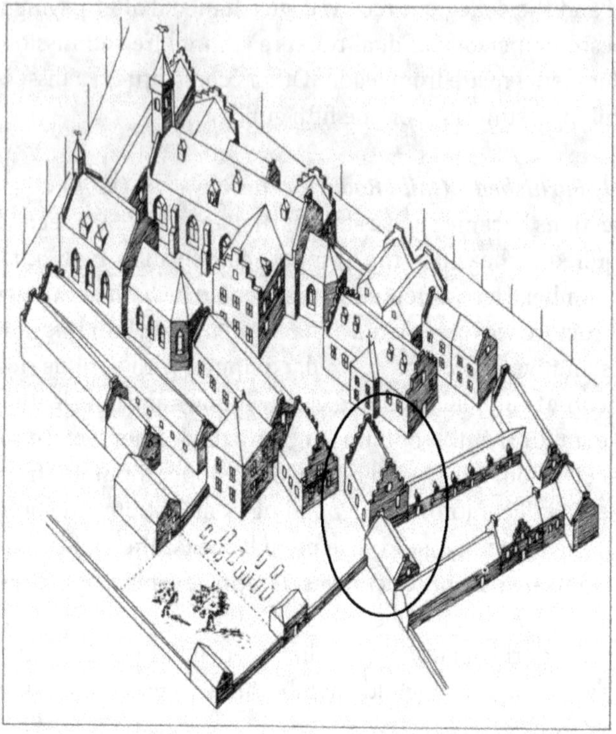

FIGURE 4.6 Reconstruction of the former Middelburg Hospital Complex at the corner of Lange Delft and the Nieuwstraat, based on the Map of Middelburg by Cornelis Goliath (1657).
In the circled area the two buildings of the surgeons' guild were located. At the top: The 'Potters' room (acquired in 1655), which housed the anatomical theatre. Below: the shed demolished in 1668 for the construction of the new guild room. After its completion the library and the natural history collections were accommodated here.
SOURCE: KOOL-BLOKLAND, *DE ZORG GEWOGEN* (1990)

doctor (*figure 4.7*). The board of the new collegium consisted of Everaerts and the two other city doctors, completed by the lecturer in anatomy, Cornelis van de Voorde. Everaerts presented this collegium as a re-establishment, because in the beginning of the seventeenth century a similar *Collegium Medicum Medioburgense* had been in existence for a very short time.[34] With this iniative

34 In 1613, the city physician Karel Frank is mentioned as the oldest member of this earlier *Collegium*, of which only the four city doctors had been member. Further almost nothing

FIGURE 4.7 The medical doctor Anthony Everaerts (1627–1679), founder of the *Collegium Medicum Medioburgense* (1668), holding a book marked 'Helmont,' indicating his attachment to the anti-Galenic work of Jan Baptista van Helmont (1579–1644). Portrait, made in 1705 by Everaerts' grandson Anthony Rethaan (1685–1708), son of Johan Pieter Rethaan (1648–1708) and Margaretha Everaerts (1657–1693), after a now lost original.

SOURCE: RIJKSDIENST VOOR HET CULTUREEL ERFGOED, NR C-1577

Everaerts aimed to unite all medical doctors practicing in Middelburg. This new Middelburg Collegium *Medicum Medioburgense* soon developed into a thriving society with a membership of in total twenty-three medical doctors.

2.7 *Purpose and Statutes of the Society of Medical Doctors*

With this initiative, Everaerts, evidently sought to restore the primacy of the medical doctors in affairs concerning illness and health. The success of the earlier *Collegium Anatomico Chirurgicum*, with a mixed composition of physicians and surgeons, had become something of a threat to the position of the profession of Medical Doctor. Originally, the physicians were involved with the illnesses of the internal organs of the human body, while external wounds and fractures were the domain of the surgeons. But in anatomical dissections these two domains were intermingled. Moreover, as a consequence of the numerous anatomical and related physiological discoveries in the 1650s and 1660s, classical Galenic-Aristotelian medicine staggered on its pedestal. Everaerts himself, for instance, had become an adept of the iatro-chemical school, represented by Jan Baptista van Helmont, Antonius Deusing and Franciscus de le Boë Sylvius. These new insights called for a redefinition – and a new demarcation line – between the domain of the theoretically trained physician and the practical working surgeon. This demarcation line was felt as a problem, especially in Middelburg, where many ship's surgeons were trained. At sea a ship's surgeon was the only authority in all issues concerning sickness and health. At home, in the city, the additional value of an academically trained medical doctor in relation to a surgeon needed a clear enhancement.

The preserved 'secret' statutes of the *Collegium Medicum Medioburgense*, formulated in scholarly Latin, clearly demonstrate this intention.[35] These

---

of its activities is known. Cf. Fokker and De Man, *Levensberichten*, 53–54; P.J. Meertens, *Letterkundig leven*, 441 and C. de Waard, in: *Nieuw Nederlands Biografisch Woordenboek* II, column 1049.

35 The original statutes and list of members of the *Collegium Medicum Medioburgense* surfaced in 1798 in the auction of the library of Egbert Philip van Visvliet, son of Meinard van Visvliet, lector of anatomy in Middelburg from 1741–1750. This manuscript was bought by Leonardus Louijssen Boom, a medical doctor in Middelburg, who bequeathed it to the city after his death in 1808. Unfortunately this richly illustrated manuscript was lost in 1940 during the bombardment of Middelburg. A copy of the text and a Dutch translation, made in the 19th century by Jan Cornelis de Man, survived in the archive of the 'Vereeniging van Genees- en Heelkundigen in Zeeland,' no. 62. A second copy is in the University Library Amsterdam, in the collection of Adriaan Abraham Fokker, "Inspecteur Geneeskundig Staatstoezicht in Zeeland" from 1854–1878. The Dutch translation has been published in Zuidervaart, "Het theatrum anatomicum te Middelburg," 114–116. For the augmented list of members, see there 117–123.

regulations reveal that the main objective of the physicians' collegium was to construct and maintain the identity of the profession of medical doctor, according to pre-arranged social and legal standards. Firstly the *Collegium Medicum Medioburgense* instructed itself to robustly maintain the city regulation from 1624 that every physician in Middelburg should hold a university degree in medicine.[36] Further it was forbidden to steal away each other's patients. Mutual tariff agreements were agreed upon, including regulations concerning the supply of medicines through pharmacies or private channels. But above all it was required to uphold the status of the medical profession itself. It was, for instance, strictly prohibited that members of the physicians' collegium were engaged in beard shaving or any other form of 'degrading' work. With this clause a clear demarcation line was drawn with the surgeons, for whom beard shaving was an often practised sideline of income. This implied however no personal aversion to the surgeons, for at least three of the medical doctors in the *Collegium Medicum Medioburgense* had started their careers as surgeons. No, it was a package of measures implying a form of early discipline formation, intended to stipulate a clear distinction between the two professions.

However, the physicians' collegium had another interesting ambition, as is revealed by one of the most interesting passages of its regulations. At every meeting the 'Assessor' was obliged to enter on the agenda either one of Hippocrates's aphorisms, or a practical *casus*, or any other 'Medico-Physical' issue. Certainly included in this last phrase were the results of anatomical dissections and other experiments performed within the Middelburg anatomy chamber, or elsewhere. The members should discuss these topics 'freely and with all security.' In short, one of the objectives of the *Collegium Medicum Medioburgense* was the exchange of expertise and connoisseurship on the basis of equality. In this the Middelburg doctors followed an example set by the *Collegium privatum Amstelodamense*, a research-oriented medical society, founded in Amsterdam only three years before.[37]

With its multiple objectives – identity building, early discipline formation, representing interests, sharing of knowledge, discussion and research – the

---

36 For this regulation of 1624, see Collection Fokker (UBA). Cf. J.C. de Man, "De wettelijke bepalingen op het vroegere leveren van medicynen door doctoren te Middelburg," *Archief uitgegeven door het Zeeuwsch Genootschap der Wetenschappen* 6 (1888): 1–42.

37 In 1665 this Amsterdam group had just published a first collection of anatomical studies. G.A. Lindeboom, "Het Collegium Privatum Amstelodamense (1664–1673)," *Nederlands Tijdschrift voor Geneeskunde* 119 (1975): 1248–1254 and H.L. Houtzager, "Het Collegium Privatum Amstelodamense en de wetenschappelijke twist tussen Reinier de Graaf en Jan Swammerdam," *Maandblad Amstelodamum* 78 (1991): 73–83.

Collegium Medicum Medioburgense represents a nice example of an intermediate form between a professional regulatory body (such as a guild) and a learned society (where scientific findings are presented and discussed).[38] The early form of a scholarly society is also reflected in the provisions that regulated the internal organization of the physicians' *Collegium*, such as the establishment of a board, a determination of the hierarchy of the members, the frequency of the meetings, the method of (democratic!) decision making, and also the demand for absolute secrecy, complete with serious provisions for an adequate penalty for violations of the *Collegium*'s rules. On the one hand these provisions are similar to agreements made in the guilds, but on the other hand they were supplemented by objectives and regulations which would emerge only much later (in the second half of the eighteenth century), in the so-called *dilettantengenootschappen* ('societies of dilettantes'), a type of learned society that by that time was populated by Enlightened scientific enthusiasts. On the strict regulations in this type of societies Wijnand Mijnhardt has remarked:

> The origin and meaning of the tight regulations that often were accompanied by a complicated system of fines, are difficult to interpret. It is plausible that such regulations were a necessary condition in groups whose members were confronted with this kind of proto-democratic decision making procedures for the first time. Seen in that light, the 'dilletantengenootschappen' served as schools for modern decision making.[39]

As it now has become clear that such regulations existed already in seventeenth-century Middelburg, the origin of the strict rules of the eighteenth-century Dutch learned societies has become more evident.

---

38 On the topic of the formation of learned societies in the Netherlands, see the debate between Vermij and Mijnhardt: Rienk Vermij, "Genootschappen en de Verlichting. Enkele overwegingen," *De achttiende eeuw* 25 (1993): 3–24; Wijnand W. Mijnhardt, "Genootschappen en de Verlichting: een repliek," *De achttiende eeuw* 26 (1994): 101–114, followed by Rienk Vermij, "Nieuwe wijn in oude zakken? Iets over plaats en functie van genootschappen in de maatschappij van het ancien régime," *Tijdschrift voor geschiedenis* 24 (1999): 24–46.

39 "Herkomst en betekenis van de strakke reglementering die vaak met gecompliceerde boetesysteem gepaard ging, zijn nog moeilijk te duiden. Het is aannemelijk dat een dergelijke reglementering in gezelschappen, waarvan de leden vaak voor het eerst met dit soort proto-democratische besluitvormingsprocedures werden geconfronteerd een noodzakelijke voorwaarde was. In dat licht bezien zouden de dilletantengenootschappen teven als leerschool voor moderne besluitvorming gediend kunnen hebben." Wijnand W. Mijnhardt, *Tot Heil van 't Menschdom* (Amsterdam: Rodopi, 1985), 57.

2.8 *The Demise of the Zeeland* Collegia *around 1680*

The last entry of a member in the physicians' collegium occurred in 1679. When three years later the a new physician was appointed as doctor of the Middelburg *Gasthuis* hospital, the *Collegium Medicum Medioburgense* apparently had ceased to exist. The name of the new doctor is not inscribed on the collegium's memberships list. This fits with the date on a (now lost) painting of a Middelburg anatomy lesson, where the caption states that the anatomy chamber was closed in 1679. Apparently this closure ran parallel to the demise of the physicians' collegium.

The demise of the anatomy chamber and the connected collegia can partly be explained by the death of a number of key players in Middelburg society. Not only president Everaerts died in 1679, but around this time several other members and other intellectuals also passed away, one of them being the lecturer Van de Voorde. However, other developments also contributed to the closure of the anatomy chamber. Around this time the intellectual climate in Middelburg changed rapidly. How this happened is explained by the Middelburg publisher and bookseller Willem Goeree, son of the former city doctor Hugo Willems Goeree († c. 1643), after his move to Amsterdam. In the first book Goeree published in his new home town, he included a preface dedicated to the citizens of Middelburg. In it Goeree described firstly how the prosperity of Middelburg had increased – "thanks to the advice and assistance of wise and learned men" – but then the leaders of the orthodox branch of the Calvinist Church had seized power in the city government. Among others, this had led to the closure, in 1678, of the just newly re-opened Middelburg Illustrious School.[40] The local *coup d'état* to which Goeree refers was the result of a conflict started in 1676, after the appointment of the liberal theologian Wilhelmus Momma as one of Middelburg's ministers and as professor at the revived Illustrious School. But this appointment was done without the approval of the conservative Classis of Walcheren, the supreme governing body of Zeeland in theological affairs. This conflict of interests soon evolved into a political game, in which the Middelburg intellectuals eventually suffered defeat.[41]

---

40 [Willem Goeree], *Voorbereidselen tot de bybelsche wysheid en gebruik der heilige en kerkelijke historien uit de alderoudste gedenkkenissen der Hebreen, Chaldeen, Egiptenaars, Romeinen* (Amsterdam: Goeree, 1690), "Voor-reden." See on this period also Roelof van Gelder, *Naar het Aards Paradijs. Het rusteloze leven van Jacob Roggeveen, ontdekker van Paaseiland (1659–1729)* (Amsterdam: Balans, 2012).

41 Cf. H. Boonen, "De strijd tegen W. Momma: een bladzijde uit de kerkgeschiedenis van Middelburg," *Kerkhistorisch archief* 4 (1866): 287–338. See also the Dutch version of this article "Een medisch-wetenschappelijk en cultureel convergentiepunt."

3 Significance

3.1 *The Zeeland Anatomy Chamber: an Intellectual and Cultural Convergence Point?*

What was the significance of the Middelburg anatomy chamber and its two related *Collegia*? Can this medical site indeed be regarded in a cultural manner as described by Rupp for Leiden, Delft and Amsterdam? Was the Middelburg anatomical theatre a similar 'Site of Knowledge and Culture'? Rupp investigated two aspects of the anatomical theaters: their performance as a centre of knowledge and their broader cultural context and importance. Therefore let us check these criterions, starting with the knowledge aspect.

3.1.1 A Public Meeting Place for Scientific Meetings

From the history of the Middelburg anatomical theatre it is obvious that the Middelburg dissections were accessible for anyone, in exchange for a small fee. Moreover, the financial success of the anatomy chamber also proves that in the 1660s there must have been plenty of paying spectators. Such public access was certainly *not* granted for the formal meetings of the *Collegium Medicum Medioburgense* (1668), where strict secrecy was required for all its members. So with regard to the point of public admittance, at least part of Rupp's criterion has been met.

3.1.2 A Location for New Research and Scholarly Publications

Another important question is whether or not the Middelburg anatomy chamber and its two related *Collegia* generated stimuli for new medical research, the results of which were published to be known to colleagues elsewhere. This criterion is also fulfilled. Let me sum up the research efforts of a number of key players in the Middelburg anatomical scene. What was the purport of their scientific publications?

3.1.2.1 *Ferdinand Gruiwardt (1628–1701)*

The Middelburg medical scholars focussed particularly on their core business: anatomy and the training of surgeons. The physician Ferdinand Gruiwardt, for instance, published in 1659 not only the *Oratie* of lector Cornelis van de Voorde, but a year later he published also an up-to-date edition of a popular surgeons' textbook, entitled *Examen der Chyrurgie* ('the surgeons examination'). Gruiwardt dedicated this book to his father-in-law, the former city doctor Johannes Wilhelmi (1598–1661), then the 'oldest practitioner' of medicine in Zeeland.[42] Originally this textbook was composed by the late Middelburg surgeon Cornelis Herls in 1625. It was a practical textbook, especially written for surgeons working at the ships of the East and West India Companies and the Dutch admiralties.[43] Gruiwardt revised the book completely according to state-of-the-art insights. For instance his edition was based on Harvey's 'circulation of the blood.' Gruiwardt's book of 1,000 copies was sold out quickly, and a reprint with a print run of 1,500 copies was made after three years. An illegal Amsterdam reprint soon followed, reportedly with a print run of up to 3,000 copies. Later editions appeared in 1672, 1680, 1700, 1703 and 1723.[44] So Gruiwardt's book was a real bestseller. However, because of Gruiwardt's move to Goes in 1666, his involvement with the Middelburg medical scene stopped after that date. Others however, followed his example, some of whom obtained a similar commercial success.

3.1.2.2 *Cornelis van de Voorde (1628–1678)*

The newly appointed Middelburg lector Cornelis van de Voorde published his own surgical text book, the *Lichtende fakkel der cheirurgia* ('The shining torch of surgery'), in 1664. In this book Van de Voorde presented the content of his lessons in the Middelburg anatomy chamber.[45] This quarto-edition was twice as voluminous as Gruiwardt's octavo-edition of the *Examen der Chyrurgie*. Van de Voorde's book, too, was issued four times between 1664 and 1680, being expanded with every reprint. In his work Van de Voorde revealed himself as a keen observer. He was, for example, the first person able to distinguish between

---

42 J. Banga, *Geschiedenis van de vroegere geneeskunde en van hare beoefenaren in Nederland* (Leeuwarden: W. Eekhoff, 1868), 504. F. Gruiwardt, *Ampliatie van de documenten en stucken, raeckende de deductie van Ferdinandus Gruiwardt* [...] *over 't recht van sijn huys-vrouw en kinderen* ([n.p.] 1668), 21–22. Cf. A.A. Fokker, "Het leven van Ferdinand Gruiwardt (1628–1701), den auteur van Zeeuwsch treurtoneel. Uit de nagelaten papieren van Adr. A. Fokker," *Nieuwe bijdragen voor rechtsgeleerdheid en wetgeving, nieuwe reeks* 6 (1880): 48.

43 Cf. A.E. Leuftink, *De geneeskunde bij 's lands oorlogsvloot in de 17e eeuw* (Assen: Van Gorcum, 1953); Idem, *Harde heelmeesters. Zeelieden en hun dokters in de 18e eeuw* (Zutphen: Walburg Pers, 1991; 2008[2]).

44 Cornelis Herls and Ferdinand Gruiwardt, *Examen der chyrurgie. Nu doorgaens verbetert, en vermeerdert* [...] *beneffens een aenhanghsel van den omloop des bloedts, en kort Examen van de doses der purgatien, vomitorien, en opiaten. Alle jonge chyrurgijns seer nut, en dienstigh, insonderheyt die haer naer Oost- ofte West-Indien begeven. Aen-hanghsel des reviseurs van den om-loop des bloedts* (Middelburg: Françoys Kroock, 1660).

45 Van de Voorde, *Lichtende fakkel*, dedication, iii. This book was issued four times: Middelburg: F. Krook, 1664; Amsterdam: J. Vinckel and J. van Velsen, 1668 and 1676; With additions by A. de Heide and the title *Nieuw lichtende fakkel* (Middelburg and Amsterdam: W. Goeree, 1680).

a malignant tumor and "water cancer," a gangrene caused by malnutrition, for which disease Van de Voorde coined the name "noma."[46] New developments were also included. For instance, in 1668 (interestingly the founding year of the *Collegium Medicum Medioburgense*) Van de Voorde intensified his involvement in chemistry and the art of distillation to such a degree that this activity earned him the nickname of *Chymische dokter* ("chemical doctor").[47] Some results of this new path of medicine were also included in Van de Voorde's *Chirurgijns Zee-compas verdeeld in XVI streken*, finished in 1679.[48]

3.1.2.3 *Antony Everaerts (1628–1679)*

Everaerts (also known as Everaers), the founder of the physicians' collegium, had been a proper researcher for a long time. Already in 1654, shortly before the establishment of the Middelburg anatomy chamber, he conducted a study to the cause of the syphillis epidemic that had struck the city that year. In this treatise Everaerts proved that the disease was transmitted through the suckling by women of different children.[49] And later, in the 1660s, Everaerts conducted a research programme into the procreation of animals.[50] Together with his colleague Everardus van Hoorn, he dissected a wide range of animals: rabbits, but also dogs and other – often still living – animals. In his *Novus et genuinus*

---

46 Van de Voorde, *Lichtende fakkel* (ed. 1676), 547; C. van de Voorde and A. de Heide, *Nieuw lichtende fakkel*, 538. Cf. K.W. Marck, "Noma: de Nederlandse geschiedenis van een inmiddels bijna vergeten aandoening," *Nederlands Tijdschrift voor Geneeskunde* 145 (2001): 2482–2487.

47 Banga, *Geschiedenis*, who refers to Van de Voorde and De Heyde, *Nieuw lichtende fakkel*, 713, 744.

48 C. van de Voorde, *Chirurgijns Zee-compas verdeeld in XVI streken* (Middelburg: W. Goeree, 1679), issued as an appendix of the 1680 edition of the *Lichtende fakkel*. Reissued in an adaptation by the physician P. Boon (Amsterdam: Andries van Damme, 1719).

49 Anthonius Everardi, *Antiqui morbi recrudescentis per suctricem inducti, cum Gallico vel Indico collatio* (Middelburg: H. Smetius, n.d. [ca. 1656]). This booklet is dedicated to the Middelburg physician Everardus van Hoorn († ca. 1665). According to Van der Aa this text was translated into Dutch and published as *Herwekte sooghster-siekte* (Middelburg, 1661), 12°. A copy has not been found. However, a Dutch translation is included in Stephan Blankaart's *Venus belegert en ontset; oft, Verhandelinge van de pokken, druipers, chankers, klap-ooren, en des selfs toevallen. Met een grondige en zekere genesinge* (Amsterdam: T. Van Hoorn, 1685; 1688²; 1696³). See also the German version: "Die neuerweckte Säuger-Seuche in ihrer Vergleichung mit der Frantzosen- oder indianischen Kranckheit [...] vorgestellt von [...] Antonio Everaers" in: *Die belägert- und entsezte Venus. Das ist chirurgische Abhandlung der so genannten Frantzosen auch Spanischen Pocken Drüppert, Sjankert, Klap-Ohren, &c.* (Leipzig: Fritsch, 1699), 476–544.

50 Peter Eldering, "Middelburgs biologisch onderzoek in de 17e eeuw," *Archief. Mededelingen van het Koninklijk Zeeuws Genootschap der Wetenschappen* (1986): 87–102, esp. 95–96. See about Everaerts: Banga, *Geschiedenis*, 506–507.

hominis brutique animalis exortus of 1661 Everaerts was one of the world's first anatomists who noticed a time difference between the moment of conception and the visible formation of an embryo. He dedicated this book to the Middelburg city government. The book also contains a Latin poem by a certain 'J.W.,' which signature probably stands for the above mentioned Johannes Wilhelmi.[51]

Even more sensational, however, was Everaerts's anatomical study of a "monstrosity" born in Middelburg. In this Latin booklet, published in 1662 and dedicated to the "famous" medical Middelburg *Collegium Anatomico Chirurgicum*, Everaerts presented an illustrated description of the decomposition of a premature miscarriage.[52] He regarded his observations as support for a recent theory of De Bils on the role of so-called "milk" containers. In this respect it is interesting that to illustrate his findings, Everaerts sought the assistance of the well-known Middelburg painter and entomologist Johannes Goedaert (*figure 4.8*).[53] That Goedaert and Everaerts were acquainted makes sense, as Everaerts was well-known for his love of still-life paintings. Precisely in this genre Goedaert had acquired some fame.[54] Moreover, just in this period Goedaert was engaged in the publication of his *Metamorphosis Naturalis*, his study into the development of insects. In this publication Goedaert was assisted by the Middelburg minister (and physician) Johannes de Mey, as well as by the publisher Jacques Fierens, the same man who had issued Everaerts's booklet.

---

51 Anthonio Everardi, *Novus et genuinus hominis brutique animalis exortus* (Middelburg: François Kroock, 1661).

52 Anthonius Everardi, *Lux è tenebris affulsa ex viscerum monstrosi partus enucleatione* (Medioburgi: J. Fierens, 1662; 1663[2]). This publication is discussed by Thomas Bartholin in a letter to Steno, dated Leiden, 26 August 1662, printed in Thomas Bartholin, *Epistolarum medicinalium, Centuria IV, variis observationibus curiosis & utilibus referta* (Hagae Com.: P. Gosse, 1740), epistola xxvi, 105–106. See also: Thomas Bartholin, *De Hepatis Exautorati Desperata Causa* (Hafnia: D. Paulli, 1666), 87–93. Everaerts's researches were collected in 1686 and reprinted as *Cosmopolitae Historia Naturalis, comprehendens Humani Corporis Anatomiam, & Anatomicam delineationem, ab ipsis primis Foetus rudimentis in Utero* (Lugduni Bat.: J. Kellenaar, 1686). Goedaert's drawing of the "monstrum" is also reproduced in the appendix of Gerardus Blasius's adaptation of Fortunius Licetus's *De monstris* (Padua: Frambotti, 1668), 300–306.

53 Cf. L.J. Bol, "Een Middelburgse Brueghelgroep, IX, Johannes Goedaert, schilder-entomoloog," *Oud Holland* 74 (1959): 1–19; idem, "Johannes Goedaert, schilder-entomoloog II," *Tableau* 7 (1985): 48–54. See also: Eric Jorink, "De logica van de luis. Johannes Goedaert (1617–1668) en Johannes Swammerdam (1637–1680)," *Kunstschrift* 52:1 (2008): 10–17.

54 Everaerts's death occurred at an auction of still life paintings in Antwerp. See for the Middelburg School of still life painters, Bakker et al., *Masters of Middelburg*. For Goedaert, see 79–80; 186–191; 258–259.

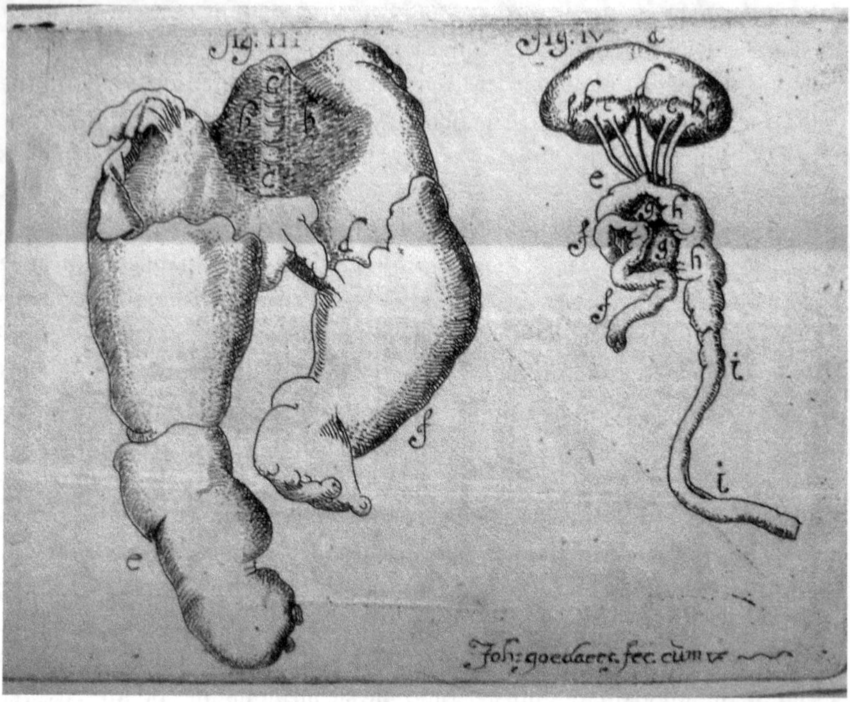

FIGURE 4.8 Drawing by Johannes Goedaert of an anatomical disection by Anthony Everaerts of a human 'Monstrosity.' From: Anthonio Everardi, *Lux è tenebris affulsa ex viscerum monstrosi partus enucleatione* (Medioburgi, 1662)
SOURCE: LEIDEN UNIVERSITY LIBRARY

All these gentlemen arguably had an interest in the new natural sciences. De Mey translated Goedaert's work into Latin, assisted by a fellow minister Paulus Veesaerdt, probably the brother of the Middelburg surgeon Jacob Veesaart.[55] The latter joined the surgeons' guild in 1660, after being trained by Van de Voorde.[56] And Van de Voorde too, was connected to Goedaert. Van de Voorde not only trained Goedaert's son as a ship's surgeon,[57] but in his *Chirurgijns Zee-compas* he claimed to have instructed Goedaert how to study insects.[58] In

55 Cf. Zuidervaart, "De Mey," note 82.

56 See the list of Middelburg surgeons, printed in Fokker, "Losse bladen."

57 In the late 1660s Goedaert's son, "a beautiful young man and celebrated surgeon," had the bad luck of being captured with his ship by "the Turks." Johannes de Mey's solidarity with the Goedaert family is emphasized by his efforts to redeem this son from slavery. Cf. Zuidervaart, "De Mey," 18.

58 Van de Voorde, *Chirurgijns Zee-compas*, 589. Van de Voorde claims to have stimulated and instructed "his good friend" Goedaert in his insect studies ("daartoe ik hem niet weinig aanleiding en bequaamheid heb gegeven").

return Goedaert had sold to Van de Voorde a recipe for the preparation of absinthe, which he had brought home from a travel to France. Van de Voorde believed that this liquor could cure numerous ailments, and according to later rumours Van de Voorde earned a lot of money via the prescription of this "medicine."[59]

3.1.2.4 *Johannes Verbrugge* (*c. 1633–1681*)

In 1676, in the third edition of his *Lichtende fakkel*, Van de Voorde added a tract composed by Johannes Verbrugge, who as surgeon of the *Gasthuis* hospital was one of Van de Voorde's close colleagues. In doing so he followed the example of his former colleague Gruiwardt, who already in 1672 added Verbrugge's *Chirurgijns Scheepskist* (a handy list of supplies for ship's surgeons) to his adaptation of Herls's *Examen der Chyrurgie*. As author in his own right, Verbrugge also did well. Around 1670 he published his own *Chirurgijns of heelmeesters reys-boeck*, of which book a fourth printing was issued in 1677. Shortly afterwards followed his *Vestibulum chirurgiae, ofte den inganck der heelkonst* (1674) and his *Heel-konstige examen ofte Instructie der chirurgie* (1677). The last book was by far the most successful. This manual for the preparation to a surgeons' exam was reprinted numerous times, under various titles, until 1766. The Amsterdam publisher Jan Claesz. ten Hoorn was so excited about Verbrugge's success that he invited him to extend an ophthalmic book by the French physician Jacques Guillemeau from 1585. This adaptation appeared in 1678.[60]

3.1.2.5 *Jacob du Buisson* (*c. 1630–after 1672*)

In 1671 another Middelburg surgeon found his way to the printing press. This time it concerned the Middelburg plague master Jacob du Buisson. Originally, in 1648 Buisson had embarked on a study in medicine at Utrecht University, but he never finished this education. In his student years and for his own use, he had translated a Latin book on the plague, written by his Utrecht professor IJsbrand van Diemerbroeck into Dutch. His Middelburg friends persuaded him to have it printed and so the *Traktaat van de Peste* was published in 1671 by the Middelburg printer Pieter van Goethem. This book, too, was reprinted, even as late as 1711.

---

59 Pieter de la Rue, *Geletterd Zeeland*, (Middelburg: M. en A. Callenfels, 1741), 63; Banga, *Geschiedenis*, 507–509.

60 *Hondert en dertien Gebreken en genesinge der oogen, eertijdts beschreeven door Mr. Jaques Guillemeau. En nu vermeerdert door Mr. Johannes Verbrigge, Chirurgijn en Gasthuys-Meester binnen Middelburgh in Zeelandt* (Amsterdam: Jan Claesz. ten Hoorn, n.d. [1678]).

3.1.2.6 *Antonius de Heyde (1646–1705)*

Perhaps the most fundamental Middelburg researcher was the physician Antonius de Heyde (also written as 'De Heide'). His case is a nice example of a mutual exchange beyond the borders between a physician and a surgeon. De Heyde came from a family of surgeons. Originally he was also trained in this profession by Van de Voorde and Cornelis Hendricx, one of the deans of the Middelburg surgeons' guild.[61] After a one-year medical study at the University of Leiden, under the supervision of Franciscus de le Boë Sylvius, De Heyde returned to Zeeland, where in 1668 he immediately joined Everaerts's Collegium of medical doctors. In doing so, De Heyde did not renounce his background as a surgeon. The surgeon Johannes Verbrugge, for instance, remained always his "esteemed friend."[62]

In the years 1675–1677, at the request of the Middelburg book printer Willem Goeree (who later became his brother-in-law), De Heyde translated two influential books on pharmacy written by the Oxford professor Thomas Willis.[63] This English physician had studied the effect of drugs in the human body, and in that context Willis had also performed anatomical dissections.[64] Two years later, after Van de Voorde's death, Goeree requested De Heyde also to finish a new edition of the *Lichtende fakkel*. This posthumous adaptation became far more than a mere reprint. De Heyde introduced footnotes in Van de Voorde's work, in which he commented from his own experience on the cases his former teacher had discussed. In line with these numerous additions the word 'New' was added to the title. The book now was issued as *Nieuw lichtende fakkel der chirurgie*.

---

61 G.A. Lindeboom, "Anton de Heide als proefondervindelijk onderzoeker," *Tijdschrift voor de Geschiedenis der Geneeskunde, Wiskunde, Natuurwetenschappen en Techniek* 6 (1983): 121–134; H.J. Zuidervaart, "De Middelburgse arts, chirurgijn en chemicus Antonius de Heyde/Heide (1646–1705) meer in context geplaatst," *Zeeland. Tijdschrift van het Koninklijk Zeeuwsch Genootschap der Wetenschappen* 28 (2019) 86–99.

62 Van de Voorde and De Heyde, *Nieuw lichtende fakkel*, 571. Van de Voorde, too, refers to a surgeon (Jacob Everts[en], as his "most distinguished friend." Cf: Van de Voorde, *Chirurgijns Zee-compas*, 748.

63 Thomas Willis, *Nieuwe en geneeskundige verhandeling vande fermentatie ofte rysing,* [...] *Beneffens een tractaat van des scheurbuiks oorsprong volgens de nieuwe chymische gronden* [...] *uyt het Latijn vertaalt* (Middelburg: W. Goeree, 1676); idem, *D'Algemeene en bysondere wercking der genees-middelen* [...] *uyt het Latijn vertaalt, en met noodighe aanteyckeningen verrijckt door A. d' Heide, M. Docter* (Middelburg: W. Goeree, 1677). See for the identification of De Heyde as the translator of the first book, Saskia Klerk, *Galen Reconsidered. Studying drug properties and the foundations of medicine in the Dutch Republic Ca. 1550–1700* (Ph.D. dissertation, Utrecht University, 2016), 178.

64 Cf. H. Isler, *Thomas Willis 1621–1675. Doctor and Scientist* (New York: Hafner, 1968).

This scholarly work stimulated De Heyde to investigate the 'basics of medicine' more thoroughly "than usually is done by its practitioners."[65] The result of his research was a pharmaceutical textbook, called the *Nieu Ligt der Apothekers* (the new light of the pharmacists), printed in 1682. Based on his recent understanding of pharmaceuticals and related chemicals De Heyde promoted a severe simplification of the use of medical drugs that were en vogue at the time. In this chemical approach to medicine, De Heyde clearly followed the footsteps of his Middelburg teacher Van de Voorde and his Leiden professor De le Boë Sylvius.[66] But De Heyde also referred to the work of local colleagues. He mentioned, for instance, chemical experiments done by "a certain accurate friend," who was experienced in *stof-scheiding* ("separation of substances").[67] Older work, noted down in his *gedenk-kladden* ("memory notebook") was also remembered.[68] In the meanwhile anatomical dissections were not neglected. Just like Everaerts had done before, De Heyde frequently dissected animals: dogs, cats, sheep, chickens and frogs. In his view it was allowed to experiment 'unashamedly' on animals. In this way one could try out a way of handling, later to be used to the benefit of humans.

In his researches De Heyde assigned an indispensable role to the microscope.[69] During his study in Leiden in the late 1660s the newly invented single-lens microscope was introduced in anatomical research and fellow students, as for instance Jan Swammerdam, made good use of this instrument.

---

65 Antonius de Heide, "Opdracht, aan al degene die de vordering der genees-konst ter herten gaat," in: *Nieu ligt der apotekers, aanwijsende de onkennis ontrent de kragt der Genees-middelen, en verbeterende grove mis-slagen in 't voorschrijven en bereiden der Genees-middelen gemeenlijk begaan* (Amsterdam: Heirs of Joannes Janssonius van Waasberge, 1682). De Heide derived his title from the frontespiece of a book already published in 1657 (*Pharmacia Galenica & Chymica. Dat is apotheker ende alchymiste ofte distilleer-konste. Begrijpende de beginselen ende fondamenten derselver* (Amsteldam: Van Ravesteyn), that, however, has a completely different content. In 1683 this earlier book was combined with that by De Heide and published by the Amsterdam publishers Jan ten Hoorn and Jan Bouman as a pirate edition. See further Klerk, *Galen reconsidered*, 192, 197–198.

66 J.C. de Man, *Antonius de Heide, med. doctor te Middelburg: ontdekker der later zoo beroemd geworden trilhaarbeweging* (Middelburg: D.G. Kröber, 1905). Gruiwardt too characterised De Heyde in 1688 as a "good chemist." Cf: F. Gruiwardt, *Medicinale en chirurgicale observatiën* (Amsterdam: T. Ten Hoorn, 1688), 78.

67 De Heide, *Nieu Ligt*, "Opdracht" and 115.

68 Van de Voorde and De Heyde, *Nieuw lichtende fakkel*, 476: experiment performed by De Heyde on 25 July 1678.

69 Ibidem, "voorreden": "... 't welk sonder hulpe van vergrootglasen niet ontdekt kan worden." ("... which can not be detected without the help of magnifying glasses"). See for his use of the microscope also: De Heide, *Nieu Ligt*, 109–110, 164.

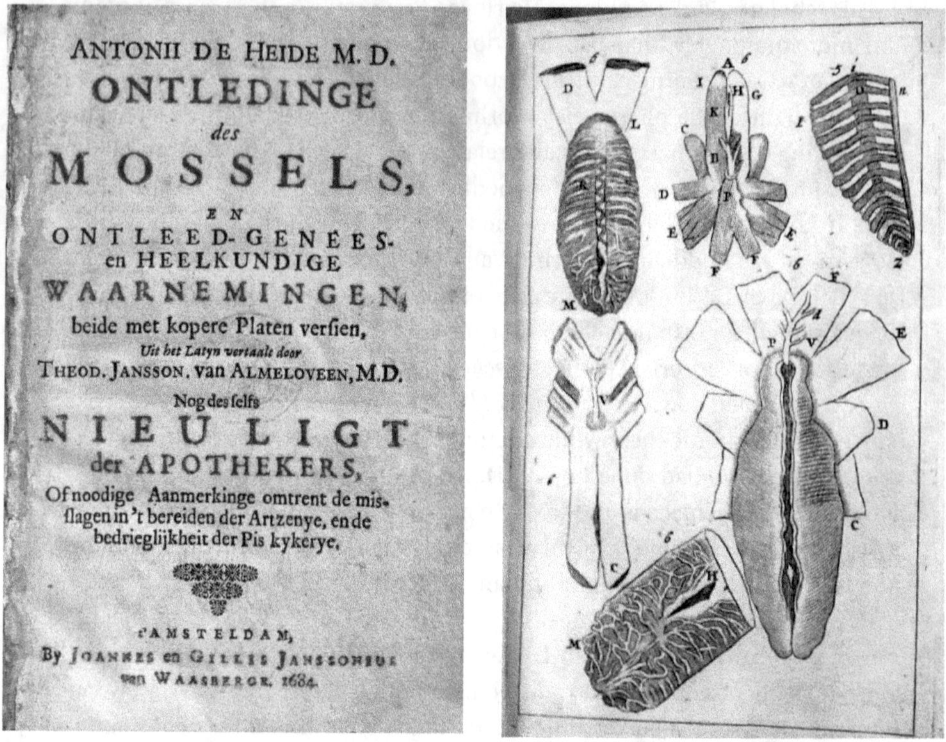

FIGURE 4.9 Antonius de Heyde's microscopic research into mussels. From: *Ontledinge des mossels* (1684), the Dutch translation of De Heyde's *Anatome mytuli* (1683)
SOURCE: AMSTERDAM UNIVERSITY LIBRARY

De Heyde, too, used this novel optical technique with great success. With the microscope he unearthed a large number of microscopic novelties, of which his discovery of the ciliary movement in mussels is his best known achievement (*figure 4.9*).[70] De Heyde can also be considered a pioneer in the field of

70 A. de Heide, *Anatome mytuli, Belgicè mossel, structuram elegantem eiusque motum mirandum exponens, subjecta est centuria observationum medicarum* (Amsterdam: Janssonius van Waesberge, 1683; 1684^2). In 1684 this booklet was translated into Dutch by the Utrecht physician Theodoor Janssonius van Almeloveen (who was related to De Heyde's wife) as: *Ontledinge des mossels en ontleed-, genees- en heelkundige waarnemingen beide met kopere platen versien* [...] *nog desselfs Nieu ligt der apothekers* (Amsterdam: Joannes & Gillis Janssonius van Waesberge, 1684). An enlarged reprint was published in 1686: A. de Heide, *Experimenta circa sanguinis missionem, fibras motrices, urticam marinam &c. Accedunt ... observationes medicae, nec non anatome mytuli. Editio nova* (Amsterdam: Janssonius van Waesberge, 1686). Cf. De Man, *Antonius de Heide*; Lindeboom, "De Heide"; idem, "De ontdekking van het trilhaarepitheel door Antonius de Heide (1683)," *Nederlands Tijdschrift voor Geneeskunde* 124 (1980): 839–841.

bone formation. His work on the regeneration of bone in frogs, published in 1682, is regarded as the starting point of the study of osteogenesis.[71] With these publications De Heyde was the third Dutchman (after Jan Swammerdam and Antoni van Leeuwenhoek) who revealed the results of original microscopic research to the public.[72] De Heyde even developed his own microscopic techniques. This enabled him to examine the flow of blood in a living specimen.[73]

3.1.3 The Library of the Surgeons' Guild Room

A major source that reveals the cultural function of the Middelburg surgeons' guild room is the printed catalogue of its library, issued by the Middelburg publisher Hendrik van Hoekke in 1746.[74] This list of 146 books is accompanied by the names of (almost) all the donors: a total of 105 persons and two institutions. An analysis of this list shows that the library largely (i.e. 86%) consisted of books published prior to 1680, that is to say issued before the (first) closure of the Middelburg anatomy chamber (*see figure 4.10*).[75] So it can be concluded that the largest part of library was constituted in this first period. The catalogue also provides interesting data on the composition of the donors (*see table 4.2*).

It turns out that not even half of the books were donated by people with a medical-pharmaceutical background. Members of the municipal magistrate, pastors, *et cetera* had donated one third of the books. Even Johannes Rapper, sexton of the New Church (the man acted also as beadle of the Illustrious School) gave two books, including a copy of Joost van den Vondel's poetry. Having said this, it is remarkable that none of the 53 known members of the contemporary Middelburg chamber of rhetoric in the period 1650–1680 can be found among the donors.[76] Apparently, the more literary inclined citizens of Middelburg were not involved in the scientifically oriented activities in the anatomy chamber.

---

71 De Heyde, "Ontleed- Genees en Heelkundige Observatien of Waarnemingen," inserted in his *Nieu Ligt*, 83 ff. Cf. Leonard F. Peltier, *Fractures. A history and iconography of their treatment* (San Francisco: Norman Publishing, 1990), 211.

72 Edward G. Ruestow, *The microscope in the Dutch Republic. The shaping of discovery* (Cambridge: Cambridge University Press, 1996), 291, note 61.

73 Lindeboom, "De Heide."

74 *Naam-register der Boeken vereert aan de Chirurgyns Ontleed-Kamer der Stad Middelburg* (Middelburg: H. van Hoekke, 1746). Zeeuws Archief, Archive Verheye van Citters, inv. nr. 54b. A copy is also in the Amsterdam University Library.

75 The Middelburg Anatomy Chamber was re-opened in 1706. Cf. Zuidervaart, "Het theatrum anatomicum te Middelburg."

76 List of members provided by Arjan van Dixhoorn's *Repertorium van rederijkerskamers in de Noordelijke Nederlanden 1400–1650*, online at http://www.lustigegeesten.nl.

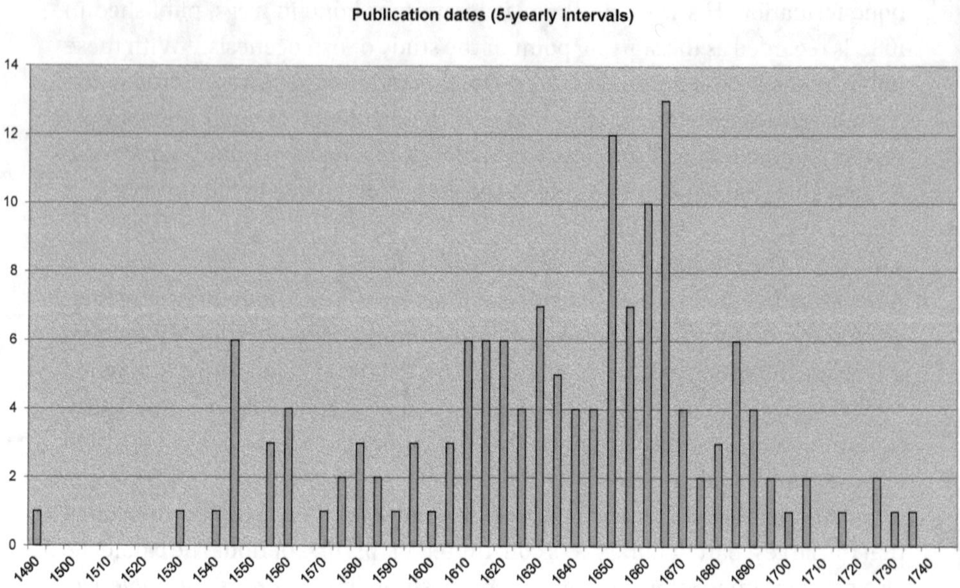

FIGURE 4.10 Overview of the publication dates listed in the catalogue of 1746 (bundled in five-year intervals)

The composition of the library is also revealing. It appears that only about half of the books had a medical-pharmaceutical character (*see table 4.3*). About an eighth of the books was related to theological subjects and a similar proportion was devoted to mathematics or natural philosophy. The rest were travelogues, atlases, history tracts and even pure literary books. The library clearly possessed a 'learned' character: nearly two-third of the books was written in Latin (*see table 4.4*).[77]

Evidently the catalogue starts with a Bible, purchased by the surgeons' guild itself. Middelburg's magistrate donated John Ray's *Historia universalis plantarum*, a standard work in the field of taxonomy. Further highlights were, an incunable edition of Galen's works of 1490, a cartographic aid (the *Zeekaartboek begrijpende de hele wereld*) by Govert Willemsz. from 1587,[78] and a large *Atlas*

77 The library of the surgeons' guild of Leeuwarden (composed in the 18th-century) had a similar 'learned' character. This surgeon's library is the only other one of which a catalog (1820) has been preserved (containing 232 books). Cf. M.J. van Lieburg, "Het Collegium Medicum te Leeuwarden," in: PBF. *De Provinsjale Biblioteek fan Fryslân. 150 jaar geschiedenis in collecties*, ed. J. van Sluis (Leeuwarden: Tresoar, 2002), 81–86.

78 Cf. C. Burger Jr., "Oude Hollandsche zeevaart-uitgaven: het groote zeekaartboek van Govert Willemsz.," *Tijdschrift voor boek- en bibliotheekwezen* 9 (1911): 69–79.

TABLE 4.2 Composition of the private donors to the surgeon's library of the Middelburg Chamber of Anatomy

Category	Number of persons	Percentage	Cumulative %
Pharmacist	2	2%	
Medical Doctor	20	19%	
Surgeon	24	23%	
Related persons to one of these categories	2	2%	
			46%
Publisher or Book trader	4	4%	
Kunstschilder	1	1%	
Painter – Artisan			
Sexton	1	1%	
Clergyman	4	4%	
Magistrate	26	25%	
			34%
Unidentified	10	10%	
Unknown	11	10%	
			20%
	105	**100%**	**100%**

by Mercator and Hondius, published in 1634. Among the anatomical works were several standard works. Cornelis van de Voorde, for instance, gave the *Opera Omnia* by the Italian anatomist Gabriel Falloppius, a man who from 1551 until 1562 had taught at Padua, in the archetype of all anatomical theaters. But the collected works of a more 'modern' natural philosopher such as René Descartes were also present.

3.1.4 The Collection of Naturalia and Artificialia

The guild room was not only furnished with a library, it was also a space where a collection of curiosities was gathered, mostly natural history items. An inventory has not survived, so unfortunately it is not known what exactly was collected here. Dirk Schoute, who studied the now destroyed archives before

TABLE 4.3 Number of books in the surgeon's library, sorted by genre

Genre	Number	Percentage	Total
Medical & Pharmaceutical			*53%*
Medicine	44	30.1%	
Anatomy	16	11.0%	
Surgery	11	7.5%	
Botany	6	4.1%	
Pharmacy	1	0.7%	
Scientific & Applied			*12%*
Natural Philosophy	14	9.6%	
Navigation	3	2.1%	
Non-scientific			*35%*
Theology	19	13.0%	
Church History	3	2.1%	
General History	12	8.2%	
Classical Antiquity	7	4.8%	
Travelogues	5	3.4%	
Literature	2	1.4%	
Law	1	0.7%	
Bibliography	2	1.4%	
	146		**100%**

TABLE 4.4 Number of books in the surgeon's library, sorted by language

Language	Number of books	Percentage
Latin	93	64%
Dutch	45	31%
French	6	4%
English	2	1%
	146	**100%**

World War II, mentions a wondrous and monstrous jawbone, donated in 1675.[79] Furthermore, in one of his books, De Heyde reports about a deformed femur, prepared by the surgeon Johannes Verbrugge, stored at the surgeons' guild room, "where many beautiful items were kept."[80] De Man, who had seen both sites with his own eyes, wrote in 1857 that all the natural history items were disposed of, after the abolition of the surgeons' guild in 1795.[81] In his time, the Middelburg Medical School still preserved a few specimens from the former anatomy chamber, among others some "snakes in liquor," a few pathological specimens and some skeletons.[82] De Man also remembered four paintings, depicting Hippocrates, Galen, the apostle Paul and an unknown celebrity. These paintings – according to De Man with no "artistic value" – had been sold off for three guilders. Finally, until 1940 one large object from this collection was preserved in the Middelburg Town Hall. This was a large shoulder blade of a whale, decorated with a painted image of a lesson in anatomy. A photograph published by Schoute is all that is left. Unfortunately the picture is not very distinct, but the unmistakable seventeenth-century dress of the persons depicted, suggest that an anatomy lesson was shown led by Cornelis van de Voorde (*figure 4.11*).[83]

3.1.5 Contacts with Artists and Poets

The last criterion formulated by Rupp concerned the involvement of painters and poets with the anatomists. Here, too, the fact that the archives are lost poses a problem. It is known, however, that in the sixteenth century the Middelburg surgeons' guild had a tradition to be involved in a play, performed in a procession. This event took place eleven days after *Corpus Christi*, when a local relic, a piece of the Holy Cross, made a circuit through the city. Although

---

79 Schoute, *Schets*, 45.

80 Van de Voorde and De Heyde, *Nieuw lichtende fakkel*, 571.

81 De Man, *Geneeskundige School*, 37–38.

82 Ibidem. See also the *Catalogus der boek- en plaatwerken en van de verzameling praeparaten, toebehoorende aan de voormalige geneeskundige school te Middelburg* (n.p.; 1873), 28.

83 Swaving, *Catalogus*, nr. 129. The painter would be a certain O.J. Steenberg, about whom nothing further is known. The legend stated: "Anatomie-camer begonnen door Cornelis van de Voorde, a° 1658 en gesloten 1679. – Opnieuw ontsloten door P.M. [=Philippus Maximiliaan] Helvetius in 1693 en gesloten in 1702. – Op nieuw geopend door A. Helvetius 1716." Schoute, *Schets*. The identification of the depicted anatomist as the lecturer Arnoud Helvetius (appointed in 1716) can be rejected on the following grounds: first the seventeenth-century clothing, and secondly the normal posture of the anatomist depicted. The said Helvetius was a very small man. Possibly this was due to a hereditary condition, as his grandfather Johannes Fredericus Helvetius was described as being a dwarf.

FIGURE 4.11 Photograph made in 1911 of a large (more than one meter high) shoulder blade of a whale, depicting a lesson in anatomy, probably painted by O.J. Steenberg. Around the anatomist (presumably Cornelis van de Voorde) some gentlemen can be seen. At the top the image of a skull above two crossed bones can be discerned. The caption reads: '*Anatomie camer begonnen door CORN. V. D. VOORDE anno 1658 en gesloten 1679 – op nieuw ontslooten door P.M. HELVETIUS 1693 – geslooten 1702, Opnieuw geopend door A. HELVETIUS 1716.*' ('Anatomy camer started by Cornelis van de Voorde, anno 1658 and closed in 1679 – Opened again by P.M. Helvetius 1693 – closed 1702, opened again by A. Helvetius 1716'). The object was lost in May 1940 due to the bombardment of Middelburg.
FROM: SCHOUTE, *SCHETS* (1911)

the habit of making processions stopped in 1572 with Middelburg's transition to the side of the Reformation, this demonstrates a former openness towards the arts.[84] It is plausible to suppose that such attitude remained in later decades. From the publications related to the Middelburg anatomical theatre mentioned above, we do know that dedicatory poems were made, and probably were recited on the spot.[85]

Concerning the involvement of artists, there is the same problem of a lack of sources. The actual participation of Johannes Goedaert (1617–1668) is already discussed. It is further known that the painter and art dealer Laurens Bernards († 1676) donated a book to the surgeons' library, but further we have to guess. For instance, was the painter Henry Berckman (1629–1679) involved, whose family member Johannes Berckman († 1693) was a surgeon of the Middelburg *Gasthuis*? And what about Zacharias Blijhooft, an equally important painter, who worked in Middelburg from 1658 until his death in 1681?[86] In his hand several detailed studies of flowers are known, as well as a number of portraits, among which a depiction of the Calvinist minister and natural philosopher Johannes de Mey, in his professorial gown of the Middelburg Illustrious School.[87] Fascinating also is Blijhooft's portrait of a yet unidentified scholar, now preserved in the Hermitage in Russia. The sitter is probably a Middelburg physician or apothecary, pointing to a book with a Latin inscription on which a branch of a bush rests. In the background figures a skull and a person in a study.[88] These works suggest that Blijhooft was close to our group of Middelburg intellectuals. Still, the number of artists involved with the Middelburg anatomy chamber seems to be not that large. When in 1680 Antonius de Heyde wanted to illustrate a deformed femur from the Middelburg surgeons' guild room, he was not able to find a draughtsman skilled enough to do the job. Nevertheless, three years later, he succeeded in finding one; however, in line

84 Lo van Driel, "Jezus en de overspelige vrouw. Een 16e-eeuws processiespel van de Middelburgse chirurgijns," *Voortgang. Stichting Neerlandistiek VU*, 19 (2000): 43–85.

85 Poems are known made by "J.V." (=Johannes Verbrugge?), "J.H." and a certain Johan Raaf, printed in Gruiwaert, *Examen* (1660); by "J.W." in: Everardi, *Novus et genuinus hominis* (1661); by the physician Galenus Trezel and "J.V.H.," printed in: Van de Voorde, *Lichtende Fakkel* (1664); by "J.H." (in Dutch) in Buisson, *Traktaat van de Peste* (Middelburg: Pieter van Goethem, 1671).

86 Caroline Stroeks, *Zacharias Blijhooft fecit* (Doctoraalscriptie kunstgeschiedenis, Universiteit Leiden, 2008).

87 Cf. Zuidervaart, "De Mey."

88 See for this portrait of a man aged 54 (?), dated 1661: https://rkd.nl/explore/images/149878. A candidate for the sitter is Petrus Burs (1602–1667), city physician of Middelburg 1636–1665, and in 1658 the first president of the Middelburg *Collegium Anatomico Chirurgicum*. See my Dutch article "Het in 1658 opgerichte theatrum anatomicum te Middelburg."

with Leeuwenhoek's habit of not mentioning the names of his collaborators, De Heyde too, remained silent about the name of this artist.[89]

3.2 *Conclusion: The Middelburg Anatomical Theatre in Nationwide Perspective*

What significance can be attributed to the Middelburg anatomy chamber? Can it indeed be called a 'location of knowledge and culture,' in the sense described in the beginning of this article? We have demonstrated that its function during the years before its demise in 1679 was indeed more than purely anatomical. The public dissections, the discourse between physicians, surgeons and other intellectuals, the sociability that came with it are all resembling the conventions of the later cultural societies of the eighteenth century. The Middelburg library, the collections of natural history specimens, as well as the paintings do fit into the model of cultural convergence outlined by Rupp for the anatomical theaters in Leiden, Amsterdam and Delft.

With this finding, we may conclude that the so-called 'absence' of official scholarly societies in the Northern Netherlands is not so much due to a lack of interest for the developing 'new' science. The Dutch infrastructure of scientific discourse simply found different way of organizing itself. In fact, the founding of the *Collegium Medicum Medioburgense* in 1668 occurred in the same decade as the founding of the *Royal Society of London* (1660) and the *Académie Royale des Sciences* in Paris (1666). In the Dutch Republic, society was structured through local lines, where the city, not the state, provided the organizing principles.[90] So, in scientific discourse, the various anatomical theatres fulfilled an important role, overlooked for a long time. The different structure of the Dutch Republic caused a different crystallisation of the intellectual infrastructure in the seventeenth-century. It is no coincidence that the first official scientific society in the Netherlands was founded in 1751, only a few years after the instalment (in 1747) of the Princes of Orange-Nassau as hereditary Stadtholders of the Dutch Republic. The new court culture associated with this instalment promoted a new development, with other organizing principles.

Having said this, what nationwide impact can be attributed to the Middelburg anatomists? From the viewpoint of anatomical research, it must be said that their impact was rather limited, at least more limited than those

89 Van de Voorde and De Heyde, *Nieuw lichtende fakkel*, 571; De Heide, *Nieu Ligt der Apothekers* (1682), 87 and fig. VI.

90 Only in affairs concerning the highest form of education, the founding of a university, did the provinces have the sole right of founding. See for instance C.A. Davids, "Universiteiten, illustre scholen, en de verspreiding van technische kennis in Nederland, eind 16e–begin 19e eeuw," *Batavia Academica* 8:1 (1990): 3–34.

of the anatomical theaters in Leiden, Amsterdam or Delft. Today the research conducted by Everaerts, Van de Voorde and De Heyde is only scarcely remembered. Where at the present time many people still know the name of microscopists such as Van Leeuwenhoek and Swammerdam, the reputation of De Heyde as an early Dutch microscopist is known only to a select group of historians.[91] One explanation for this rather modest position with regard to anatomical research may be that the Middelburg anatomy chamber primarily served born Zeelanders, who returned to this province after their academic studies. Further, the Calvinist modesty, so characteristic to Zeeland, did not help to promote their research, or to point out to visiting foreigners the richly decorated guild room with its adjacent anatomy chamber as a location worth seeing. No record exists of any visitor admiring the anatomical preparations and other curiosities exhibited in the Middelburg anatomy chamber. The English natural historian John Ray, for example, who visited Middelburg in March 1663, was not brought to the local anatomical preparations, while shortly afterwards, in Brussels, Ray admired five anatomical bodies, all prepared by Louis de Bils. Ray was also eager to visit the anatomical theatres in Delft and Leiden.[92] The French physician and *conseiller du Roy* Balthasar de Monconys, who in 1663 also visited Zeeland's capital, paid a visit to the painter-entomologist Johannes Goedaert, as well as to his publisher Jacques Fierens; but the anatomical site remained hidden to him.[93] The same can be said for the English physician Edward Browne during his visit in 1668. In his travelogue Browne summed up several anatomical curiosities he saw in Leiden and Amsterdam, among which were some prepared by De Bils. But in Middelburg Browne only admired two living eagles, kept in a cage at the town hall (reflecting the presence of eagles in the city's coat-of-arms).[94] Similar superficial things were seen by the Tuscan

91 Ruestow, *The microscope in the Dutch Republic*, 291, note 61.

92 John Ray, *Travels through the Low Countries* [1663], second edition (London 1738), 9, 19, 25 and 32. See also the travel notes from Ray's companion Phillip Skippon, who mentions the visit in Middelburg to the cabinet of natural rarities of "in one's Cliver's house" (a member of the well known Cliever family): See: *Voyages and Travels, some now first printed from original manuscripts*, vol. 6 (London, 1732), 384–385. For their visit to the anatomical theatre in Delft, see 391; the same in Leiden, 400.

93 Balthasar de Monconys, *Journal des voyages* [...] *où les sçavants trouveront un nombre infini des nouveautez* (Lyon: Boissat et Remeus, 1665), 109–113; (Paris, 1695^2), vol. 2, 216–222.

94 Edward Brown, *Naauwkeurige en gedenkwaardige reysen* [...] *door Nederland, Duytsland,* [*etc.*] (Amsterdam: J. ten Hoorn, 1696), 37. For Leiden and Amsterdam see 12 and 23–24. Cf: C.D. van Strien, *British travellers in Holland during the Stuart Period. Edward Browne and John Locke as tourists in the United Provinces* (Leiden: Brill, 1993) and the same, *Touring the Low Countries. Accounts of British travellers 1660–1720* (Amsterdam: Amsterdam University Press, 1998).

prince Cosimo de Medici, who visited Middelburg in March 1668.[95] It therefore seems that the Middelburg anatomy chamber was primarily intended for local use, not only then but also in later years. In 1857 even De Man remarked that 'our anatomical rooms were no landmark that could impress strangers.'[96]

Still for contemporary surgeons the nationwide impact of the Middelburg anatomy chamber was very important. Well into the eighteenth century Middelburg remained the most important center for the production of textbooks for surgeons.[97] Therefore, from the educational viewpoint it may be said that for some time Middelburg was the Dutch epicentre of medical-nautical knowledge.

All things considered, our investigation into the history of the Middelburg anatomical theatre leads to the conclusion that the people connected with this location not only did generate scientific work of substance, but also made important educational contributions. With this site even some painters and poets were connected. Therefore, in line with Rupp's earlier conclusions for the anatomical sites of Amsterdam, Delft and The Hague, the Middelburg anatomy chamber can certainly be coined a true 'location of knowledge and culture.'

---

95 *De twee reizen van Cosimo de Medici, prins van Toscane, door de Nederlanden (1667–1669). Journalen en documenten*, ed. G.J. Hoogewerff (Amsterdam: J. Müller, 1919), 128–130, 187–188, 215–216; Lodewijk Wagenaar and Bertie Eringa, *Een Toscaanse prins bezoekt Nederland. De twee reizen van Cosimo de Medici, 1667–1669* (Amsterdam: Bas Lubberhuizen, 2014).

96 "Onze lokalen waren geen merkwaardigheid die men vreemdelingen vertonen kon." De Man, *Geneeskundige School*, 38.

97 Zuidervaart, "Het theatrum anatomicum te Middelburg."

PART 2

Connections

CHAPTER 5

Breyne's Botany: (Re-)locating Nature and Knowledge in Danzig (circa 1660–1730)

Alette Fleischer

> If my small garden plot can be next sumer [sic] be enlarged and made happy but with some branch of other of yr tulip bearing tree 1.2.3. and of liquid amber and of others that yr Indy's produce; I shall acknowledge myself Sir, to be extremely obliged by yr favour sir if you shall please transmit to me any of the dryed plants in the next spring, you may send them to [skipper] Wilhelm Hen. Cornellison and may be further directed to my sonne Jacob Breynius junior in Dantzick.[1]
>
> JACOB BREYNE (Danzig) to JAMES PETIVER (London), circa 1692

The Danzig-based tradesman and amateur botanist Jacob Breyne (1637–1697) wrote this undated letter to his English friend, the apothecary James Petiver (circa 1663–1718). Breyne requested several specimens for his garden that originated from America. The Liquid amber tree was introduced in England in 1683 and the white magnolia – the 'tulip bearing tree' - in 1687.[2] The merchant desired several dried plants from the Chelsea Physic Garden of which Petiver was in charge. Breyne would keep the dried and pressed specimens in special portfolios, known as herbaria, in his library. Petiver shipped everything to Danzig (now Gdansk, Poland) using the services of a Dutch skipper who frequently sailed between England, Holland and the Baltic region.

Jacob Breyne and his son the physician Johann Philipp (1680–1764) were avid botanists and gardeners. The Breynes lived in Danzig, a city where they could

1 British Library (BL), MS Sloane 4066 f. 271–2, Jacob Breyne to J. Petiver, n.d.

2 It is unclear to which type of tulip tree Breyne is referring. The *Liriodendrum tulipifera*, with its yellow flowers, was introduced in England in 1629. Breyne possibly refers to the newly introduced Magnolia. P.J. Jarvis, "North American Plants and Horticultural Innovation in England, 1550–1700," *Geographical Review* 63 (1973): 477–499, notably 497–498; Penelope Hobhouse, *Plants in Garden History* (London: Pavilion Books, 1994) 126–7.

 | DOI:10.1163/9789004264885_006

oversee the family's trade in drugs and dyes, specializing in the export of Polish cochineal from Danzig to cousins based in Amsterdam and Leiden (Holland). Polish cochineal – like its better-known Latin American counterpart – was used to make red dyes and paint for the textile industry. Via the global entrepôt of Amsterdam, these dyes ultimately arrived in Leiden, the principal center of cloth production in the Republic.[3] The Breynes were in control of a section of the family business, living in the Baltic city for commercial purposes, sailing regularly to the Republic for their trade. Meanwhile, the family settled down in their new home town, Danzig.

In early modern history, trading in natural goods, such as dyes and drugs, went hand in hand with an interest in natural history as one had to obtain knowledge of natural materials and its uses. Father and son's passion for plants fit in this picture. They experimented with plants in gardens and went on botanizing excursions around Danzig. At the same time, they accumulated and exchanged plants and *naturalia* from every corner of the earth. The Breynes tapped into the trading network of the Dutch Republic and took advantage of the regular shipping service between Danzig and Amsterdam and that of the Dutch East India Company to Asia. Their skills as negotiators of natural goods and natural knowledge, making use of the existing trading network, enabled the Breynes to communicate materials and knowledge around Europe and beyond. Thus, over time, the family used their commercial contacts to accumulate and disperse botanical objects and knowledge. Their trade allowed them to move up the intellectual ladder, in Danzig and in the rest of Europe.

From the mid-sixteenth century onwards, Danzig, situated along the Baltic Sea, was a bustling port. Ships regularly sailed between Holland, Denmark, Sweden, Estonia, and Russia, carrying grain (wheat and rye), copper, wood, fur, and flax via Danzig to Amsterdam.[4] In return, Dutch ships brought cheap cargo, for instance salt, brick, floor flags, roof tiles, luxury goods such as German and

---

3 Judith H. Hofenk de Graaff, "Veranderend kleurstofgebruik in de Leidse textielververij in de zestiende en zeventiende eeuw," *Textielhistorische Bijdragen* 36 (1996) 5–19.

4 For more on the Baltic trade: eds. Remmelt Daalder, Els van Eyck van Heslinga, J.Th. Lindbald, Peter Rogaar, and Peter Schonewille, *Goud uit Graan; Nederland en het Oostzeegebied 1600–1850* (Zwolle: Waanders, 1998); J.Th. Lindblad, "Evidence of Dutch-Swedish Trade in the 17th Century," in: *Baltic Affairs, Relations between the Netherlands and North-eastern Europe 1500–1800*, eds. J.Ph.S. Lemmink and J.S.A.M. van Koningsbrugge (Nijmegen: Institute for Northern and Eastern European Studies, 1990) 205–228; Milja van Tielhof, *De Hollandse graanhandel, 1470–1570, Koren op de Amsterdamse Molen* (Den Haag: Stichting Hollandse Historische Reeks, 1995). Milja van Tielhof, *The 'mother of all trades': the Baltic grain trade in Amsterdam from the late 16th to the early 19th century* (Leiden: Brill, 2002); the contributions in: eds. W.G. Heeres, L.M.J.B. Hesp, L. Noordegraaf, and R.C.W. van der Voort, *From Dunkirk to Danzig, Shipping and trade in the North Sea and the Baltic, 1350–1850* (Hilversum: Verloren, 1988).

French wines, Asian spices, and textiles from England and India. Through its geographical location, Danzig became an important node in a mercantile network that connected Western Europe and Asia with the cities along the Baltic Sea. This important harbour allowed for the circulation of (processed) materials and knowledge to nearby and distant parts of the world via its extensive trading network.

Trade brought not only intellectual and commercial goods to and from the city. In its wake followed artists, artisans and engineers, remaining in the Baltic area for a designated period, which their family trade dictated. Already in the second half of the sixteenth century, artisans from the Low Countries left the turmoil between Catholics and Protestants to find employment elsewhere. In the Baltic area, their skills were in demand and much appreciated. In 1591, engineer Simon Stevin (1548–1620) came to Danzig to advise on the renovation of the harbour, and between 1592 and 1595, artist and architect Herman Vredeman de Vries (1527–circa 1607) and his son Paul constructed several buildings in Danzig.[5] Due to the imported materials, stonemasons and architects created an appearance similar to that of a town in Holland, with the tall, slender houses and gabled rooftops. (*figure 5.1*)

The mercantile network linked Danzig not only to Amsterdam but also to various other sites along the vast trading routes as is argued here. This paper does so by sketching the mercantile and botanical activities of the family Breyne at a seemingly peripheral location of knowledge. However, Danzig was an important harbour for the Dutch Republic. Until recently, historians of science tend to focus more on more distant sites (Batavia, Ceylon, Japan) than on less exotic areas closer to home. However, for more than a century, Danzig was a lively location for the production and consumption of knowledge and materials. This was furthered thanks to the commercial and botanical activities of the family Breyne. Marketing commercial and botanical wares went along similar paths and methods. Making use of interconnecting networks of traders, naturalists, correspondents, artisans and family members, the Breyne family created their own location of knowledge, as will be showcased here.

5 F. Skibinski, "Early-modern Netherlandish sculptors in Danzig and East-Central Europe; A study in dissemination through interrelation and workshop practice," *Nederlands Kunsthistorisch Jaarboek* 63 (2013): 111–135; Charles van den Heuvel, *'De Huysbou': A reconstruction of an unfinished treatise on architecture, town planning and civil engineering by Simon Stevin* (Amsterdam: KNAW, 2005) 73; B. Woelderink, "Het bezoek van Simon Stevin aan Dantzig in 1591," *Tijdschrift voor de geschiedenis der Geneeskunde, Natuurwetenschappen, Wiskunde en Techniek* 3 (1980): 178–186; ed. Peter Furhing, *De wereld is een tuin; Hans Vredeman de Vries en de tuinkunst van de Renaissance* (Gent/Amsterdam: Ludion, 2002).

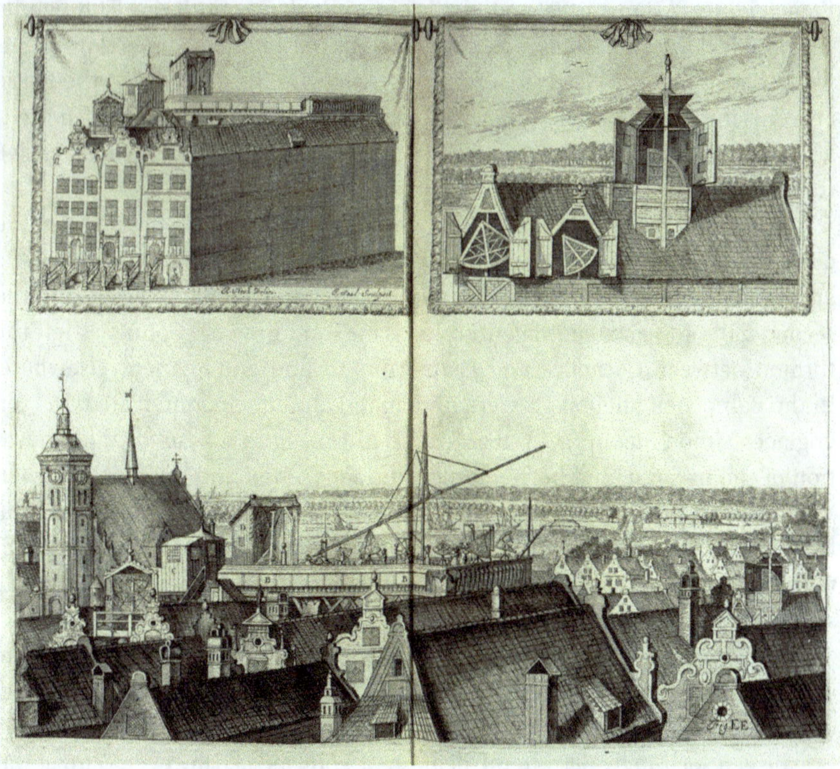

FIGURE 5.1 View of Danzig and its Netherlandish architecture. With Johannes Hevelius's house and roof-top observatory by Andreas Stech (draughtsman) and Isaak Saal (engraver). From: J. Hevelius, *Machina Coelestis Pars Prior* (Danzig, 1673)
SOURCE: SPECIAL COLLECTIONS, LEIDEN UNIVERSITY LIBRARY

In the last decades, historians have begun to map the geographies of science, tracing the cycles of accumulation and exposing the various ways of how and where knowledge and materials were produced and consumed. For instance, David Livingstone, Lissa Roberts and James Secord come to understand that knowledge, practices and materials circulate between different sites and are adapted to meet local requirements.[6] From these locations, the reworked

6 A number of historians and sociologists of science have commented on this process where human and environmental spaces are standardized in order to accommodate the production of scientific knowledge. See David N. Livingstone, *Putting Science in its Place. Geographies of Scientific Knowledge* (Chicago/London: University of Chicago Press, 2003); L.L. Roberts, "Centres and Cycles of Accumulation," in: *Centres and cycles of accumulation in and around the Netherlands during the early modern period*, ed. L.L. Roberts (Münster: LIT Verlag, 2011) 3–27; James A. Secord, "Knowledge in Transit," *Isis* 95 (2004): 654–672; Sven Dupré and

knowledge, practices and materials could again be transported elsewhere. 'Circulation' is intricately linked to 'location,' since without a location the circulated knowledge, practices and materials cannot emerge or take root.

This volume invites historians to take a closer look at different locations of knowledge and to further understand where knowledge and materials found fertile soil and from where it could re-enter the network of circulation. The term 'location' is neutral, since it can incorporate a centre of accumulation, a place at the so-called periphery, and "colonial territories" as historians David Wade Chambers and Richard Gillespie have suggested.[7] For a location of knowledge to become successful, it needs to be part of a wider web of other locations in which materials, knowledge, practices, and people can circulate.

A reliable shipping network was an essential part in the circulation process during the seventeenth century, so as to connect the various locations together, near or far. Trade and natural knowledge went hand-in-hand as historians Harold Cook, Claudia Swan, Dániel Margócsy, and others have extensively argued.[8] Contained in objects or on paper, exotic materials and knowledge began to stack up in private libraries, cabinets of curiosity and (botanical) gardens. Looking at Dutch seventeenth century history, historians tend to concentrate on the Dutch East India Company (VOC) that transported amateurs and physicians and their botanical and medical observations between the Republic and the Far East (i.e. Japan, Indonesia).[9]

Christoph Lüthy (eds.), *Silent Messengers: The Circulation of Material Objects of Knowledge in the Early Modern Low Countries* (Münster: LIT Verlag, 2011).

7 Historians Chambers and Gillespie have promoted the use of the term 'location' in their article on technoscience in the non-Western world i.e. colonial territories. David Wade Chambers and Richard Gillespie, "Locality in the History of Science: Colonial Science, Technoscience, and Indigenous Knowledge," *Osiris* 15 (2000): 221–240.

8 For further reference, see: Alette Fleischer, "Trading Places: (ex)changing nature and knowledge at Cape of Good Hope, circa 1652–1700," in: *Centres and cycles*, ed. L.L. Roberts, 110–127; Harold J. Cook, *Matters of Exchange, Commerce, Medicine, and Science in the Dutch Golden Age* (New Haven, CT/London: Yale University Press, 2007); Claudia Swan, "Collecting Naturalia in the Shadow of Early Modern Dutch Trade," in: *Colonial Botany, Science, Commerce, and Politics in the Early Modern World*, eds. Londa Schiebinger and Claudia Swan (Philadelphia: University of Pennsylvania Press, 2005) 223–236; Daniel Margocsy, "'Refer to folio and number': Encyclopedias, the Exchange of Curiosities, and Practices of Identification before Linnaeus," *Journal of the History of Ideas* 71 (2010) 63–89; See also various contributions in: eds. Pamela Smith and Paula Findlen, *Merchants and Marvels: Commerce, Science and Art in Early Modern Europe* (New York/London: Routledge, 2002).

9 There is a wealth of publications on collecting knowledge and nature from the East. To name a few: Klaas van Berkel, *Citaten uit het boek der natuur opstellen over Nederlandse wetenschapsgeschiedenis* (Amsterdam: Bert Bakker, 1998); eds. Ellinoor Bergvelt and Renée Kistemaker, *De wereld binnen handbereik. Nederlandse kunst- en rariteitenverzamelingen, 1585–1735* (Zwolle/Amsterdam: Waanders, 1992); Johannes Heniger, *Hendrik Adriaan van Reede tot*

This essay wants to contribute to this volume by arguing that the notion of locations of knowledge and material production can be situated anywhere as long as it is connected to a network that allows a regular stream of communication, people and goods. Historian James Secord points out that knowledge making is not bound by time or space, it can happen at any site at any given moment.[10] The idea 'location of knowledge' connects nearby and distant places through different networks, and in the case of this argument the focus is on the production, consumption and circulation of natural knowledge of the family Breyne in Danzig.

1 Danzig and the Dutch

Around 1650 Danzig was, as an autonomous city, part of the Kingdom of Poland. It was situated in an area called Kashubia that, in turn, was part of the Pomerania or Borussian/Prussian region. Danzig consisted of mostly tradesmen, artisans, engineers, and scholars of Polish, German and Dutch descent. The number of inhabitants significantly increased, from approximately 40,000 by the end of the sixteenth century to circa 100,000 in the mid 1650s. Several thousands of the newcomers came from the Dutch Republic. Most of the inhabitants were Lutherans. German was the official language, but most spoke Polish. The city's Gymnasium Academicum sive illustre (founded in 1558) was a pre-university, and it housed the faculties of Theology, Philosophy, Medicine, and Law.[11] It boasted students like the astronomer Johannes Hevelius (1611–1687), botanist Jacob Theodor Klein (1685–1759) and Jacob Breyne.

Drakenstein (1636–1691) and Hortus Malabaricus; a contribution to the history of Dutch colonial botany (Rotterdam/Boston: Balkema, 1986); The various contributions in: Schiebinger and Swan, *Colonial Botany*; Richard Grove, "Indigenous Knowledge and the Significance of South-West India for Portuguese and Dutch Constructions of Tropical Nature," *Modern Asian Studies* 30 (1996): 121–143. To a lesser extent the trading route to South America: Maria Angélica da Silva and Melissa Mota Alcides, "Collecting and Framing the Wilderness: the Garden of Johan Maurits (1604–79) in North-East Brazil," *Garden History, Journal of the Garden History Society* 30 (2002): 153–176; P.J.P. Whitehead and M. Boeseman, *A Portrait of Dutch 17th century Brazil; Animals, plants, and people by the artists of Johan Maurits of Nassau* (Amsterdam/Oxford/New York: North-Holland Publishing Company, 1989).

10 Secord, "Knowledge in Transit," 654–672.

11 Dariusz M. Brycko, "The Danzig Academic Gymnasium in Seventeenth Century Poland," in: *Church and School in Early Modern Protestantism. Studies in Honor of Richard A. Muller on the Maturation of a Theological Tradition*, eds. Jordan J. Ballor, David S. Sytsma, and Jason Zuidema (Leiden: Brill, 2013) 339–346.

Trading ties between Holland and Danzig date back to the fifteenth century. The trade consisted mostly of cereals and was very successful over a long period of time.[12] The Dutch named this profitable trade the *Moedernegotie*, which means 'the mother of all trades.'[13] Ships loaded with Baltic grains (wheat and rye) regularly sailed to Holland. Dutch merchants exported salt, red brick and tiles, all cheap and heavy cargo. Along with the more obvious commercial activities, this mercantile network allowed for an exchange of people, knowledge, goods, and paperwork, which entered and exited the important harbour of Danzig. Furthermore, it took Dutch traders, physicians, architects, and engineers to the Baltic area to oversee and expand their commerce or to build a career.[14]

Mainly economic and maritime historians, as the wealth of publications shows, discuss the history of Danzig.[15] Their research traces the cereal trade, which brought merchants, artists and engineers to Danzig. Polish/Pomeranian fortune-seekers and youngsters, in turn, came to the Low Countries to work for the VOC, hoping to be trained as a merchant or to enrol in courses at the various Dutch universities.[16] The Baltic grain trade, in junction with the East India trade, enabled the transplantation and transformation of people, matter and information from one location to the other whereby Dutch amateurs, botanists, physicians, and surveyors inquired into local nature, interacted with local people and exchanged materials, practices and knowledge.[17]

---

12 Tielhof, *Graanhandel*, 86–106.

13 Tielhof, *The 'mother of all trades.'* Tielhof translated the Dutch word 'Moedernegotie' to 'mother of all trades' in her well-received book. I use her translation.

14 Peter Schonewille, "Nederlanders en de Oostzeecultuur," in: *Goud uit Graan*, ed. Daalder et al., 114–129, 120.

15 Lemmink and Koningsbrugge, *Baltic Affairs*; Heeres and Hesp, *From Dunkirk to Danzig*; Jonathan Israel, *The Dutch Republic, Its Rise, Greatness, and Fall 1477–1806* (Oxford: Oxford University Press, 1995); Maarten Prak, *Gouden Eeuw. Het raadsel van de Republiek* (Nijmegen: SUN, 2002); A.Th. van Deursen, *De last van veel geluk. De geschiedenis van Nederland 1555–1702* (Amsterdam: Prometheus, 2004); and the following websites: http://www.soundtoll.nl/index.php/nl/onderzoek/literatuur; http://www.geheugenvannederland.nl/?/nl/collecties/de_nederlandse_oostzeehandel_in_de_periode_1600-1850/bibliografie.

16 Jan de Vries and Ad van der Woude, *Nederland 1500–1815, De eerste ronde van moderne economische groei* (Amsterdam: Balans, 2005), notably pages 482–485. Danzig was an important harbour for the grains trade to Holland, with profits of approximately one million guilders per annum; see also: M. Bogucka, "Dutch Merchants' Activities in Gdansk in the First Half of the 17th Century," in: *Baltic Affairs*, eds. Lemmink and Koningsbrugge, 19–32; Stefan Kiedron, "Voorwaarts naar het land van de orakels! Oosteuropeanen aan Noordnederlandse universiteiten na de Opstand," *De Zeventiende Eeuw* 10 (1994): 73–78.

17 This occurred, for instance, at the Cape of Good Hope. See: Alette Fleischer, "Trading Places: (ex)changing nature and knowledge at Cape of Good Hope, circa 1652–1700," in: *Centres and cycles*, ed. Roberts, 110–127.

To secure their on-going trade, Amsterdam merchants worked with local agents who acted as go-betweens to oversee their negotiations and control the shipments. Usually these middle men had an interest in the business as a way to establish mutual trust and ensure trade. Instead of agents, some trading families urged one of their members, usually a younger brother, to move to Danzig and take care of business on behalf of the entire family. With the volume of goods shipped to the Low Countries growing, the number of Dutch moving to Danzig over the course of the seventeenth century increased.[18] Holland's merchants brought their trade practices and methods of bookkeeping. They introduced regularly published commodity price lists in Dutch, just as was done in Amsterdam.[19] As goods were exported, traders, their knowledge and their practices moved into Danzig. Steadily, the newcomers helped to shape Danzig.

Equally, trade brought people, knowledge and materials to the Low Countries. Adventurous young men from Polish Prussia went to the Republic to study at Dutch universities or tried making a living by joining the VOC.[20] For instance engineer and architect Nicolaus Goldmann (1611–1665) left Wroclaw to study Law in Leiden. He remained there to teach engineering and fortification privately to Dutch and foreign scholars, amongst whom Tilman van Gameren (1632–1706) of Utrecht and the Polish nobleman Jerzy Sabastian Lubomirski (1616–1667) were students. In circa 1660, the architect and painter Gameren accepted an invitation from Lubomirski and traded Holland for Poland, where he remained for the rest of his life constructing churches, houses and palaces. Gameren was not the first artist to move to the Baltic area and Poland. Already a century earlier, various Dutch and Belgian sculptors and stonemasons preceded him.[21] A little later, around the turn of the century, the Danzig-born Daniel Gabriel Fahrenheit (1686–1736) moved to Amsterdam to train as a merchant. Since he was more interested in natural philosophy, he left the family

---

18 Tielhof, *Graanhandel*, 178.

19 Tielhof, *The mother of all trades*, 175.

20 For Polish men in the service of the Dutch East India Company: Roelof van Gelder, *Het Oost-indisch avontuur. Duitsers in dienst van de VOC* (Nijmegen: SUN, 1997) 54–55.

21 Skibinski, "Early-modern Netherlandish sculpturs"; Jeroen Goudeau, *Nicolaus Goldmann (1611–1665) en de wiskundige architectuurwetenschap* (Groningen: Philip Elchers, 2005), Eymert-Jan Goossens and Koen Ottenheym, *Tilman van Gameren (1632–1796). Een Nederlandse architect aan het hof in Polen* (Amsterdam: Stichting Koninklijk Paleis, 2002) 34–37. For more on Dutch art and Eastern Europe: Thomas DaCosta Kaufmann, *Toward a Geography of Art* (Chicago: University of Chicago Press, 2004) 114–153. For more on Danzig's art, architecture and scientific illustrations, the contributions in the exhibition catalogue T. Grzybkowska (ed.), *Aurea porta rzeczypospolitej: Danziger Kunst vom 15. bis zum 18. Jahrhundert* (Gdansk: Grzegorczyk, 1997).

business to become a glass blower and instrument maker in Amsterdam, eventually also teaching experimental philosophy.[22]

Within and outside an early modern trading city like Danzig, there are various communities in operation simultaneously, from small scale (family and neighbours) to world wide (overseas contacts). Such as family circles that connected the city to other parts of the world. These family circles were part of trading networks, where commerce, money and trust helped to shape an international community. They were also part of locally constructed circles of immediate family, of neighbourhoods, of acquaintances, of colleagues. A city-dweller could be part of multiple communities, and this helped to give shape to the societal identity of an early modern city like Danzig.[23] The interaction between Danzig and the Dutch Republic occurred on various commercial and scholarly levels and involved men and women from different backgrounds as it touched the fields of natural history, trade, architecture, masonry, instrument making, and art. Danzig functioned as a location where (botanical) materials, practices and knowledge were consumed and produced. It contributed to the rise of natural knowledge in Danzig, in Holland and even reached the farthest ends of the Dutch trading realm, such as South Africa, Indonesia and Japan. Students of nature in the Far East were well aware of the latest publications, while in turn they contributed to the expanding botanical knowledge in Europe, as we shall explore below. Significant quantities of exotic flora discovered along the mercantile network were collected, shipped, planted, as well as described and depicted in books printed in Danzig. The botanical activities and connections of the Breyne family, their network and their important publications on exotic and local plants demonstrated Danzig was intricately tied to this vast Dutch network and thus could become a location of knowledge.

---

22 Pieter van der Star, *Fahrenheit's Letters to Leibniz and Boerhaave* (Amsterdam: Rodopi, 1984). Huib J. Zuidervaart, *Van 'Konstgenoten' en Hemelse Fenomenen. Nederlandse Sterrenkunde in de Achttiende Eeuw* (Rotterdam: Erasmus, 1999) 62–63. Fokko Jan Dijksterhuis, "Low Country Opticks; The optical pursuits of Lambert ten Kate and Daniel Fahrenheit in early Dutch 'Newtonianism,'" in: *Newton and the Netherlands, How Isaac Newton was Fashioned in the Dutch Republic*, eds. Eric Jorink and Ad Maas (Amsterdam: Amsterdam University Press, 2012) 159–183.

23 Fabrizio Nevola and David Rosenthal, "Urban Communities in Early Modern Europe (1400–1700): A Research Review," http://www.ahrc.ac.uk/documents/project-reports-and-reviews/connected-communities/urban-communities-in-early-modern-europe-1400-1700-a-research-review/ (accessed January 2018).

2 Breynes's Botany

The family Breyne originated from the province of Brabant in the South of the Netherlands in the sixteenth century. Jacob Breyne's Protestant great-grandfather Matthias originally came from Antwerp, Flanders. After the Siege of Antwerp (1585), when the Spanish troops had re-conquered the city and restored the Catholic order, the family moved to the Dutch Republic (Leiden and Amsterdam). During this time of unrest, one of Matthias' grandsons from the Republic moved to Danzig via Hamburg in order to oversee the family's dyes and drugs trade. Jacob Breyne senior married a local girl of Dutch origin, Maria Anna Moorman. In 1637, his son Jacob was born.[24]

Jacob Breyne senior shipped Polish cochineal to Leiden via Amsterdam. This insect was – and still is – used to make red dye and Breyne supplied this dye to relatives involved in the textile industry in Holland.[25] Trading in dyes and drugs (i.e. dried medicinal plants) went hand-in-hand with an interest in botanical knowledge and practices. At the Danzig Academic Gymnasium, young Jacob received lessons in botany from physician Christian Mentzel (1622–1701), who taught at the pre-university educational institution from 1648 to 1650. During this time, Mentzel and Breyne set off on excursions in the fields around Danzig to find and collect local plants.[26] In the early 1650s, Jacob was sent to the Low Countries to be trained as a merchant by his uncle Pieter Breyne in Leiden. He combined this training with a study in botany under professor of medicine Adolf Vorstius (1624–1663) at the University of Leiden without registering as a student. Upon his father's demise in 1655, Jacob had to return to Danzig to take over the family business.

The Breyne family owned several houses in Danzig's city centre and an estate with a large botanical garden at Brabank, near the city and close to the

---

24 Jos Kuijlen, "De Danziger Botanicus en Koopman Jacob Breyne (1637–1697) en zijn betekenis voor de Hollandse Plantkunde," *Tijdschrift voor de Geschiedenis der Geneeskunde, Natuurwetenschappen, Wiskunde en Techniek* 5 (1982): 116–118. Kuijlen's short biography lacks any references and notes. See also Breyne's vita in: Jacob Breynius, *Prodromus Fasciculi Rariorum Plantarum Primus et Secundus* (Danzig: Thom. Joh. Schreiberi, 1739); Mary Gunn and Enid du Plessis, *The Flora Capensis of Jakob and Johann and Philipp Breyne* (Johannesburg: Brenthurst Press, 1978) 21–22; Helmut Roob and Cornelia Hoff, *Jacob und Johann Philipp Breyne, zwei Danziger Botaniker im 17. und 18. Jahrhundert* (Gotha: Forschungsbibliothek, 1988) 9–10; Alicja Kurkowa, *Jakub i Jan Filip Breynowie, Studium z dziejów kultury książki, XVII i XVIII wieku* (Wrocław: Ossolineum, 1989).

25 Dirk Jaap Noordam, "Textielondernemers en het Leidse patriciaat, 1574–1795," *Textielhistorische Bijdragen* 36 (1996): 20–36.

26 Mentzel had compiled four volumes of *Theatrum rerum naturalium Brasiliae* in 1660, based on drawings the Dutch artist Albert Eeckhout made in Brazil. In 1682 he published the botany book *Pinax botanonymos polyglottos*.

river Motlaw. This enabled the merchant to continue with his botanical pursuit. Jacob Breyne collected both local and exotic flora. He accumulated books on botany and published several books on this topic. Botany remained a lifelong passion, which Breyne passed on to his youngest son. While Jacob had to step into his father's footsteps, Jacob's son had the opportunity to study medicine and botany, something that Jacob would have liked for himself. Thanks to their commercial enterprise, the Breyne family progressively moved up the scholarly ladder, as will be shown here.

From 1655, Jacob Breyne regularly made journeys around Danzig and travelled into the Kashubian countryside to collect local flora. After drying and pressing the collected plants, Breyne compiled an herbarium, a type of portfolio, into which he glued all the botanical specimens. His first herbarium, compiled in 1659, was simply labelled *Herbarius vivus*.[27] It contained dried local plants, which the amateur botanist annotated with names, mostly in Dutch, and sometimes added information about the plant's habitus. This manuscript functioned as a type of botanical diary meant for personal use only.[28] (*Figure 5.2*)

In 1673, Breyne compiled a second herbarium. This four-part manuscript, named "Plantae rariores Borussicae et Cassubicae," consisted of exsiccates (dried plants) gathered in the Pomeranian region. Carefully compiled and bound, the plants were neatly glued to the pages, written with clear handwriting and the names in Latin. Breyne created several versions of this type of herbarium, which was intended as gifts for friends, patrons and fellow botanists.[29] One manuscript was sent to London, to his English contact, the botanist and physician James Petiver (c. 1658–1718), in 1697. The two men exchanged letters, seeds and books for a number of years. Breyne's mercantile network expanded also into England. One of his skippers, Mr. Cornelissen, was often called for to carry not only commercial wares but also botanical goods.[30]

Breyne informed Petiver in a letter accompanying the herbarium that:

> I am now fully resolved to publish my Viridarium of Prussia & Cassubia, as soon as ever I shall take leave of my weaknist to take a journey one summer more through Prussia & Cassubia to take a view and reckonize

---

27 Breyne's herbarium is in the collection of Naturalis Biodiversity Center, Leiden (botany section).

28 For more on types and uses of herbaria, see: Brian Ogilvie, *The Science of Describing, Natural History in Renaissance Europe* (Chicago/London: University of Chicago Press, 2008) notably 165–173.

29 Another version of this herbarium of Prussian and Kashubian plants is in the collection of Naturalis Biodiversity Center, Leiden (botany section).

30 BL MS Sloane 3321, f. 14 and 4067 f. 81. Mr. Cornelissen is mentioned by Breyne.

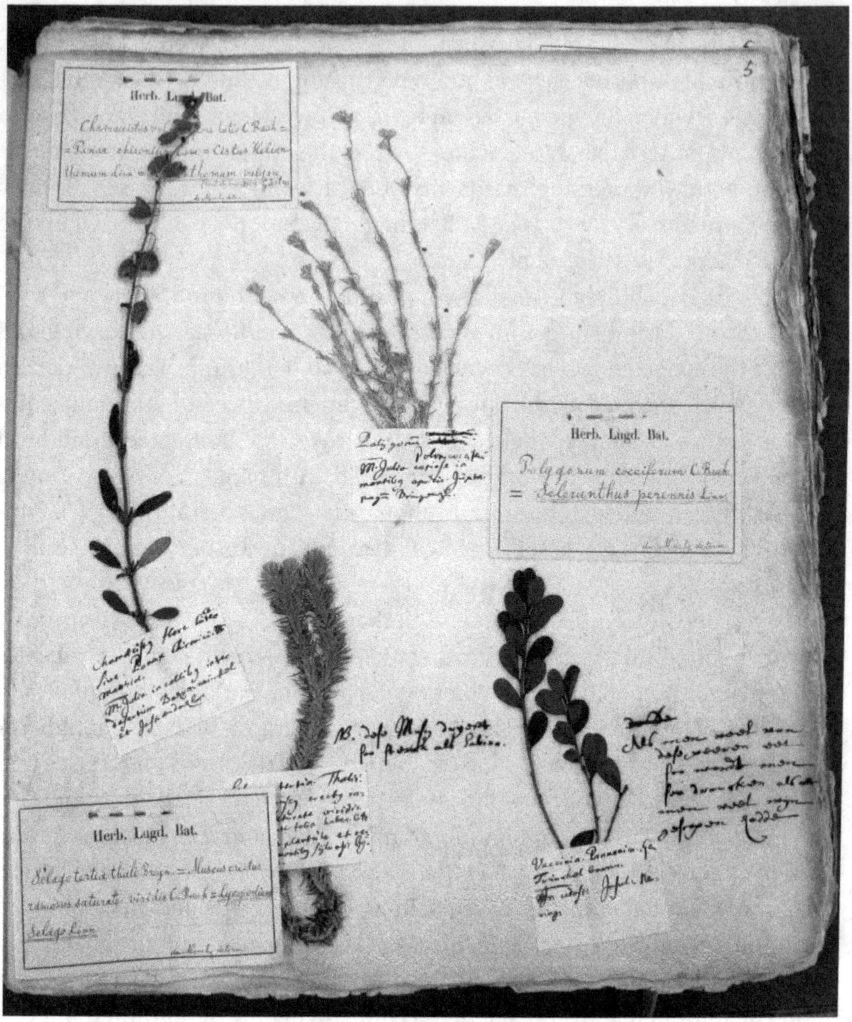

FIGURE 5.2 Page from Jacob Breyne's *Herbarius vivus* (Danzig, 1659)
COURTESY NATURALIS BIODIVERSITY CENTER, LEIDEN (BOTANY SECTION)

> those plants which I had determined to have this year and had made some beginning too with [...] wonderfull success as to the observations, tho' my health was not answerable to succeed for it suffer much improvement withby.[31]

He concluded his letter with the following: 'The Cassubian & Brussian plants not described I earnestly begg that you will not suffer them to be published,

31 BL MS Sloane 4066, f. 271, Jacob Breyne to J. Petiver, n.d.

because I do design and figure them myself for the publique.'[32] His death in 1697 prevented this publication. Petiver stored the received specimens and bound it together with other folios with exsiccates gathered by other botanists.[33] (*Figure 5.3*)

Jacob Breyne discovered an ardent student of nature in his son Johann Philipp Breyne (1680–1764). In 1698, Johann Philipp began his study in medicine under professor Herman Boerhaave (1668–1738) in Leiden and he received his doctorate in 1702. Johann Philipp published two dissertations in Leiden, one on the ginseng root *Botanico-medica de radice gin-sem, seu nisi, et chrysanthemo bidente Zeylanico acmella dictlo* (Leiden, 1700) and one on fungi called *Dissertatio medica inauguralis de fungis officinalibus et eorum usu in medicina* (Leiden, 1702). Between 1702 and 1704, Johann Philipp spent around a year touring in England and another travelling to Spain, Portugal, Italy, Austria, Saxony, France, and Germany. The young doctor then returned to Danzig in 1704, where he set up his practice as a physician. Breyne junior remained a passionate botanist, which manifested in his gardens and botanical publications. The ascent up the scholarly ladder that had begun with Jacob was continued by his youngest (and only surviving) son. The young J.P. Breyne benefitted from the commercial and botanical contacts that his father had constructed and maintained in Danzig, in the Republic and in the rest of Europe.

In 1702, Johann Philipp went to England and visited James Petiver and the already famous Hans Sloane (1660–1753). Jacob Breyne had introduced his son to Petiver and Sloane by letter. A year later, Johann Philipp was elected fellow of the Royal Society, nominated by Petiver. After London, the young man travelled to Oxford, Bath, Bristol, Badminton, and Cambridge. In each town, he met local botanists and tried to botanise plants along the way. Johann Philipp wrote to Petiver that

> I have observed at Badminton [which was the estate of the Duchess of Beaufort, a garden lover and a patroness of botanists] many fine plants that I never saw before, but could not make any collection, by reason of ye short time I did stay there. I hope ye vessels that last arrived from ye West Indies have brought you a great many of raretyes, which I have great mind to see.[34]

---

32 BL MS Sloane 4066, f. 272, Jacob Breyne to J. Petiver, n.d.

33 J.E. Dandy, *The Sloane Herbarium, an annotated list of the Horti Sicci composing it: with biographical accounts of the principal contributors* (London: British Museum, 1958) 58–59. Pages from Breyne's MS are kept in the Natural History Museum, H.S. 231.

34 BL MS Sloane 3321, fol. 101, Johann Philipp Breyne to J. Petiver, circa 1702.

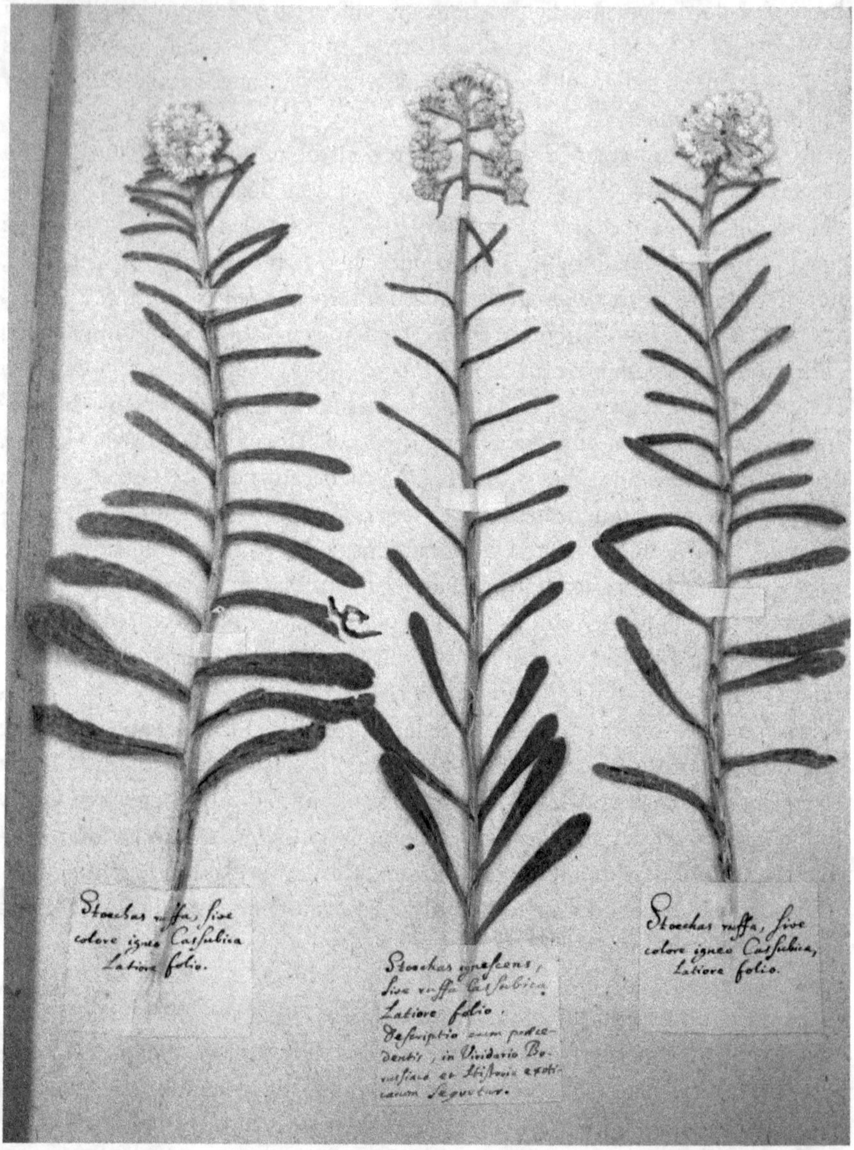

FIGURE 5.3 Page from Jacob Breyne's *Plantae rariores Borussicae et Cassubicae* (Danzig, 1673)
COURTESY NATURALIS BIODIVERSITY CENTER, LEIDEN (BOTANY SECTION)

For Breyne junior, Southern English flora was just as interesting for him as the rarities from – for instance – Jamaica. He was more familiar with the plants around Danzig and those that came from the East Indies.

Johann Philipp Breyne finished his trip in Portsmouth. From there, he had to wait for a ship to bring him back to Amsterdam. He wrote to Petiver on 31 May 1703 that this harbour was "not very agreeable to me; so forever I divert me selv with simpling, as ofter ye wether permits it, at sea and land, where I meet with many good plant ..."[35] Since it took several weeks before the wind allowed the Dutch squadron to arrive, he ordered his friend to send the following books:

> Lassels Treatis or Description of Italie, Dr Lister's journey to Paris with ye remarks on it, Mr Ray's of old English words, Dr. Arbuthnots Papers again Doctor Woodward, Drs Cole Tractatus de casu quodam epileptico and 2 copies of Mr Ray's methodus.[36]

As William Cole's book appeared in 1702 and the other books were published between 1697 and 1699, it can be said young Breyne was fully up-to-date on the publications of his fellow scholars.

For about 40 years, Hans Sloane, James Petiver and J.P. Breyne kept in touch through numerous letters, some of which were published in the *Philosophical Transactions*.[37] Petiver's reputation as a mediator of natural materials was questionable, as he had a mercantile attitude towards natural history and materials. He did not always reciprocate in delivering precious *naturalia* to his contacts nor was he very accurate with his documentation and ways of storing

35 BL MS Sloane 3321, fol. 120, Johann Philipp Breyne to J. Petiver, 1703.

36 Richard Lassels, *An Italian voyage, or, A compleat journey through Italy in two parts; The second edition with large additions, by a modern hand* (London: R. Wellington and B.B. Lintott, 1698); Martin Lister, *A Journey to Paris in the Year 1698* (London: Jacob Tonson, 1699); John Ray, *A Collection of English Words not generally used, with their significations and original, with catalogues of English birds and fishes; and an account of the preparing and refining such metals as are gotten in England* (London: Christopher Wilkinson, 1691); John Arbuthnot, *An Examination of Dr. Woodward's Account of the Deluge, &c. With a comparison between Steno's philosophy and the Doctor's, in the case of marine bodies dug out of the earth* (s.l., 1697); William Cole, *Consilium œtiologicum de casu quodam epileptico: quo respondetur epistolœ* (London: H. Rhodes, 1702); John Ray, *Methodus Plantarum Nova, Brivitatis Et Perspicuitatis Causa Synoptice in Tabulis Exhibita* (London: Faithorne and Kersey, 1682).

37 Dandy, *Sloane Herbarium*, 97. Between 1694 and 1713 Johann Philipp exchanged letters with James Petiver. BL MS Sloane 3321, ff. 9–13, 24, 53, 87, 91 (L), 101, 106, 116, 120, 122, 124, 138, 177, 266, 270. Roob and Hoff, *Jacob und Johann Philipp Breyne*, the Breynes received circa 38 letters from Sloane (70) and circa 12 letters from Petiver (67).

objects.[38] However, Petiver remained a steady contact for Johann Philipp, although young Breyne was often worried whether the London apothecary had received or shipped the desired books, parcels and letters.[39]

Besides Petiver, Johann Philipp received letters and parcels from other fellow physicians living on Europe's mainland. For instance, Johann Jakob Scheuchzer (1672–1733) from Zürich, who he considered "a very curious and laborious man in ye natural history of Switzerland," had shipped to him a "specimen Lithografia Helvetiae ... with some natural things."[40] J.P. Breyne extended his father's botanical network by corresponding with, for instance, the botanist Carl Linnaeus in Sweden, whom he may have met in the Low Countries. The correspondence with Petiver, Sloane, Linnaeus, and others showed that this shared interest in natural history was tied to the circulation of the most exotic materials and the latest publications.

As a fellow of the Royal Society, Johann Philipp's letters did not only address botany but also other forms of natural history. For instance, he mediated knowledge gathered by other natural inquirers based in Danzig. From his friend Daniel Gottlieb Messerschmidt (1685–1735) of Danzig, J.P. Breyne received samples of mammoth's teeth. Messerschmidt had collected various *naturalia* in Siberia in the late 1710s, a journey commissioned by Czar Peter the Great.[41] Young doctor Breyne presented these findings in 1728 to "some learned Gentlemen in Dantzig." Upon Messerschmidt's return to Danzig in 1730, he brought even more materials with him. In 1735, Johann Philipp shared this with the Royal Society, via a letter to Hans Sloane:

> That those Teeth and Bones of Elephants were brought thither by no other Means but those of a Deluge [...]. In such Manner, not only the

38 Dániel Margócsy, *Commercial Visions; Science, Trade and Visual Culture in the Dutch Golden Age* (Chicago: University of Chicago Press, 2014), chapter one; Eric Jorink, "Sloane and the Dutch Connection," in: *From Books to Bezoars, Sir Hans Sloane and his Collection*, eds. Alison Walker, Arthur MacGregor, and Michael Hunter (London: British Library Publishing, 2012) 57–70. Both state Petiver had his eye on commerce, for instance by returning ordinary materials after having received exotic naturalia, or that he ignored to ship requested objects to others.

39 BL MS Sloane 4039, f. 369, J.P. Breyne to Petiver. A letter from Amsterdam, 3 October 1704. "I received your letter of Juni last bij mij arrival in this place bij which I understood, you have not receiv'd my letter from Rome, which makes me a little uneasy, for I sent you one observation, which I thought verij worthy your curiosity."

40 BL MS Sloane 3321, f. 106–7, Johann Philipp Breyne to J. Petiver, circa 1702. Johann Jacob Scheuchzer, *Specimen Lithographiæ Helveticæ curiosæ, quo lapides ex figuratis Helveticis selectissimi aeri incisi sistuntur et describuntur* (Tiguri [=Zürich]: D. Gessner, 1702).

41 Anke te Heesen, "Accounting for the Natural World, Double-Entry Bookkeeping in the Field," in: Schiebinger and Swan, *Colonial Botany*, 237–251.

> holy Scripture may serve to prove natural History; but the Truth of the Scripture which says that Noah's Flood was universal, a thing which is doubted by many, may be proved again by natural History.[42]

With his ideas on the nature of this "elephant," the Danzig physician contributed to a topical debate on natural history and the Bible, one which took place in Sloane's circles.[43] J.P. Breyne, wealthy and well-connected, was in a position to collect, explain and transmit natural knowledge that was either found locally or retrieved from distant and nearly inaccessible areas to contacts residing elsewhere.

J.P. Breyne followed through on the trading networks that his father and the widespread family had constructed. Family ties and commerce connected Danzig to the Republic and botany connected the Danzig family with distant locations, in the Far East and in the rest of Europe. Jacob Breyne had initiated a botanical network with contacts in Holland and England. James Petiver was one of Jacob's correspondents with whom Johann Philipp could expand a botanical exchange. Part of their network formed the familiar commercial channels, with recurring ships and skippers and a fixed sailing route. A regular and reliable shipping service enabled a dependable distribution of parcels, persons and papers. The Breyne family formed a valued node along which plants and knowledge were accumulated and dispersed between East and West. Relying on their network, father and son Breyne circulated botany via letters, prints and books, as discussed in the following section.

3 The Breynes Botanical Publications

Jacob Breyne published several books about plants during his life. The first was the *Exoticarum aliarumque minus cognitarum Plantarum centuria prima* (in short: *Centuria*) of 1678, with 101 entries and illustrations. Breyne published this ca. 240-page folio himself and printed it at his own printing house in order to safeguard it from illegal versions. This 'coffee table book' was dedicated to his Dutch patron and friend, the statesman and curator of the Leiden Hortus Botanicus, Hieronymus van Beverningh (1614–1690) who owned a country estate with a vast garden near Leiden. Breyne's book was well received, both by

42 Johann Philipp Breyne, "Observations, and a Description of some Mammoth's bones dug op in Siberia, proving them to have belonged to Elephants," *Philosophical Transactions* 40 (1737–1738): 124–138.

43 Jill Cook, "The Elephants in the collection: Sloane and the History of the Earth," in: *From Books to Bezoars*, eds. Walker, MacGregor, and Hunter, 158–167.

amateurs and botanists including physician Andreas Cleyer, Jan Commelin (1626–1692), the director of the Hortus Botanicus of Amsterdam and later Carl Linnaeus, especially for the introduction of unknown Cape plants and the description of the tea shrub.[44] The *Centuria* represented Breyne's accumulated findings of his 1660s tour through the Low Countries, of his botanical excursions around Danzig and of his exotic materials collection.

Most of the plants discussed in the *Centuria* originated from the Cape of Good Hope, the Far East (Ceylon, Java and Japan), South America (Peru), and some from the Mediterranean region. It featured depictions and descriptions, together with information regarding the original habitus and in which Dutch garden Breyne saw it grow. When applicable, Breyne referred to other botanical publications in order to complete the reader's thirst for knowledge. The *Centuria* introduced around 20 Prussian and Kashubian plants that Breyne considered noteworthy for his readers. He included the *Caryophylata Montana* that grew in his own garden in Danzig and the *Pulsatilla Polyanthos*, a plant his brother Pieter Breyne had found growing near Warsaw.[45] In short, flora from near and far appeared in the *Centuria* (*Figure 5.4*).

This paper likes to suggest that from Breyne's point of view, Danzig botany was indigenous knowledge, since it was found in his region. Whereas seen from the perspective of his botanical partners in Holland and England, these plants were just as unfamiliar as those from overseas. The different ways of perceiving what is local and exotic was not clear cut, as it depended on the viewpoint of the collector/perceiver. By publishing *Centuria*, Breyne offered the opportunity to botanists elsewhere to compare samples of Kashubian vegetation with flora found along the VOC-routes.[46] Later in his life, the Danzig amateur contemplated publishing a book that dealt solely with Polish plants. Unfortunately, Breyne died in 1697, before he even began the production of this new book.

The *Centuria* epitomised the intricate connections between Dutch private gardens, such as the country estates of Leeuwenhorst owned by statesman Gaspar Fagel (1634–1688) and Oud-Teylingen, the residence of Hieronymus van Beverningh, and the botanical gardens of Amsterdam and Leiden. Furthermore, it was tied with the Dutch East and West India Companies in Asia, South Africa

---

44 Kuijlen, "De Danziger Botanicus"; Heniger, *Hendrik Adriaan van Reede*, 165.

45 Breyne, *Centuria*, 131, 135.

46 Alix Cooper, *Inventing the Indigenous, Local Knowledge and Natural History in Early Modern Europe* (Cambridge/New York: Cambridge University Press, 2007). Amongst natural historians and botanists at this time, there was a growing awareness regarding local flora that paralleled the inquiry into exotic botany.

FIGURE 5.4 Frontispiece by Andreas Stech (draughtsman) and Lambert Visscher (engraver) of Jacob Breyne, *Exoticarum aliarumque minus cognitarum Plantarum centuria prima* (Danzig, 1678)
COURTESY NATURALIS BIODIVERSITY CENTER, LEIDEN (BOTANY SECTION)

and America, since merchant ships brought exotic plants, seeds, bulbs, exsiccates (dried plants), images, and descriptions to and from collectors and botanists in the Republic and beyond. Letters and parcels were exchanged between Batavia (Andreas Cleyer), Japan (Willem ten Rhyne), London (Hans Sloane and James Petiver), Italy (Paolo Boccone), and Paris (Joseph Pitton de Tournefort). Via different trading posts, botanical knowledge and materials came into the hands of amateurs, physicians, botanists, merchants, and artists. These people, in turn, transformed nature into gardens, drawings, herbariums, or books. This allowed botanists and amateurs to examine and enjoy unfamiliar flora and botanical knowledge.

Since the Danzig-Dutch merchant regularly travelled to Holland for business, he combined these trips with garden visits and meetings with other botanical enthusiasts. Breyne made notes of the cultivated plants he saw growing, obtained plant samples and gathered bulbs and seeds.[47] When it was not possible to collect actual plant samples, he resorted to amassing exsiccates and paintings made by flower painters, such as Stephan Cousijns of The Hague.[48] Furthermore, Breyne received botanical writings and plant specimens from his good friend Willem ten Rhyne, a Dutch physician who lived on the island Deshima in Japan. Ten Rhyne regularly sent parcels and letters to his network of (amateur) botanists. Breyne received, amongst other items, a four-page Latin text on the growth and maintenance of the Japanese tea and a parcel containing seeds of various South African plants, as well as information on shape, colour and use.[49]

As an addition to the *Centuria*, Breyne incorporated Ten Rhyne's extensive contribution on the Japanese tea, called *Excerpta ex Observationibus Japonicis, Physicis, &c, Wilhelmi ten Rhyne De Fructice Thee.*[50] Breyne could have had various reasons to print this text. First of all, Ten Rhyne was, in Breyne's words "summo amico meo" (my best friend), so it may have been a special favour to Ten Rhyne.[51] Secondly, tea drinking became popular thanks to returning VOC

47 Roob and Hoff, *Jacob und Johann Philipp Breyne*, 77.

48 Heniger, *Hendrik Adriaan van Reede tot Drakenstein*, 71; Gunn and du Plessis, *Flora Capensis*, 16. Most plants described were taken from living or dried specimens. Sometimes it was based on a watercolour or painting.

49 Breyne, *Centuria*, 115; Roob and Hoff, *Jacob und Johann Philipp Breyne*, 78; and Naturalis Biodiversity Center, Leiden (botany section): MS. *Apparatus Botanicus, qui olim studiis inseaviit visi D. Hieron. Van Beverning, XVII Voll.* Beschrijvinge op wat wijs en manier den Theeboom geraijt en de Thee gemaakt wordt; Exerpta ex Itinerario Wilh. Ten Rhijne, de plantis sunta Promont. Bonae Spei ab ipso collectis.

50 Breyne, *Centuria*, appendix: De frutice thee [...] Jacobum Breynium transmissarum.

51 Kuijlen, "De Danziger Botanicus," these men were childhood friends.

employees. In Europe, there was a growing desire to understand more about the tea plant and its possible medicinal qualities. It so happened that physician Cornelis Bontekoe published his treatise on tea (*Tractaat van het excellenste Kruyd Thee*), propagating tea as a cure-all in the same year as the *Centuria*, 1678.[52] Breyne most likely knew that Bontekoe was to publish such a treatise. Living in Danzig, the so-called periphery, Jacob Breyne was certainly knowledgeable in the latest books and developments in the field of botany, in Japan and in Holland, which motivated the Danzig botanist to add an appendix with his friend's findings regarding the tea shrub.

In 1680, Breyne published a second book. The text-only folio *Prodromus fasciculi rariorum plantarum* was printed in Danzig by David Fridericus Rhetius under Breyne's supervision. The *Prodromus* was meant to serve as a prequel to another type of *Centuria*. The book listed names and locations of the exotic plants he collected during his 1670 trip to Holland. In the introduction, Breyne described a visit to the Hortus Botanicus in Leiden where he studied the newly arrived plants. The amateur botanist also mentioned the names of all the "illustrious and noble men" who helped him to publish this book. These were wealthy statesmen, diplomats and merchants who owned country estates, the places where they pursued botany. Breyne named several, amongst others Beverningh, nobleman Simon van Beaumont (The Hague), botanist-alderman Jan Commelin (Amsterdam), apothecary Frans van Sevenhuizen (The Hague), and silk merchant Philips de Flines (Haarlem).[53] These men could easily obtain, through their wealth and/or connections, plant samples from all corners of the world.[54]

The *Prodromus* presented exotic plants in alphabetical order, each with a short description. Per entry, Breyne listed in which garden he had seen it grow and whether the plant flowered, bore fruit or carried seeds at the time of his observation. Sometimes the amateur botanist named the country of origin and, when applicable, referred to other publications. Breyne's publication

52 Cook, *Matters of Exchange*, 293–300 on Bontekoe; Christoph Schweikardt, "More than just a Propagandist for Tea: Religious Argument and Advice on a Healthy Life in the Work of the Dutch Physician Cornelis Bontekoe (1647–1685)," *Medical History* 47 (2003): 357–368.

53 Jacob Breyne, *Prodromus Fasciculi Rariorum Plantarum* (Danzig: D.F. Rhetius, 1680), see preface. These were the same names amateur botanists such as Leonard Plukenet had listed in his *Speculum Herbarum* compiled after George London's visit to Dutch gardens. See Natural History Museum London, Sloane Herbarium HS 91. See also Mary Gunn and L.E. Codd, *Botanical Exploration of Southern Africa* (Cape Town: Balkema, 1981) 28–30.

54 Cook, *Matters of Exchange*, 304–317. Cook discusses at length how Dutch amateurs and merchants obtained prized plants from the Cape and Asia.

represented plants imported into Dutch gardens, as well as flora collected and described by VOC servants stationed in various trading posts. In his hometown, the Danzig merchant catalogued, assembled and examined every aspect of a plant (leaves, flowers, stems, roots, etc.) using live samples, drawings and dried specimens in combination with the notes he made in Holland and the information he received through his mercantile network.

In 1689, Breyne published a third book, *Prodromus fasciculi rariorum plantarum secundus*, aided by his son Johann Philipp who was nine years old at the time. This volume was based on the first *Prodromus*, but this extended version included 33 engravings by, amongst others, Isaak Saal and the Dutchman Jan Veenhuysen. The *Prodromus secundus* included Jacob Breyne's findings of his garden tour in the summer of 1688 to the Low Countries.[55] In August 1688, Breyne senior, still in Holland, received a parcel from a recently returned VOC servant Georg Meister. Meister was a German gardener who had spent several years in Asia. Meister handed Jacob Breyne a "herbarium of circa 300 paintings of Indian and Japanese vegetation," together with a collection of seeds and plants. In Batavia, the German gardener had received instructions from his superior, physician and botanist Andreas Cleyer (1634–ca 1697/1698) to deliver this parcel personally to Breyne.[56]

The German-born physician and botanist Cleyer was in the service of the VOC and stationed in Batavia since 1662 until his death in 1697 or 1698. For a few years Cleyer was the supervisor for the Dutch on Deshima, Japan, where he illegally accumulated plants and plant knowledge. Stationed in the Far East, the physician was nevertheless familiar with Breyne's botanical publications, brought to him via the VOC shipping network. Moreover, Cleyer and Breyne had the same botanical contacts, sharing patrons like Beverningh and Fagel, as well as scholars of the Hortus Botanicus in Leiden. By exchanging botanical knowledge and materials, they enforced their position in this web of botanists, physicians and wealthy amateurs.

Cleyer had provided the parcel with Japanese specimens, as he considered Breyne "the great amateur of herbs and all vegetation, who at his own expense had it [the *Centuria*] described and printed." It is unknown whether Cleyer

---

55 Jacob Breyne, *Prodromus fasciculi rariorum plantarum secundus* (Danzig: D.F. Rhetius, 1689). Breyne possibly came to Holland together with the ships carrying the July grain harvest.

56 George Meister, *Der Orientalisch-Indianische Kunst- und Lust-Gärtner* (Weimar: Kiepenheuer, 1973) [original Dresden, 1692] notably 13 and 141; Wybe Kuitert, "Georg Meister, a Seventeenth Century Gardener and his Reports Oriental Garden Art," *Japan Review* 2 (1991): 125–143.

owned a copy of the *Centuria*, but he was aware of its existence.[57] Jacob Breyne acknowledged the receipt of Cleyer's present in the *Prodromus secundus*, and he provided additional plant knowledge gathered by his contacts, such as the problems or successes of planting and growing imported seeds and bulbs.[58] The Danzig merchant-botanist thus gathered and received seeds, drawings and dried specimens, both directly from his botanical friends in the East and via his Dutch connections. At this Baltic location of knowledge, through a network of traders, physicians, gardeners, apothecaries, painters, and amateur botanists, Breyne accumulated local and exotic plants, which he turned into highly-regarded publications.

After his father's demise in 1697, Johann Philipp Breyne proceeded with acquiring botanical specimens and publishing books. In a letter dated 11 November 1705 to James Petiver in England, Johann Philipp wrote that he wanted "to publish following books 1) centuria with additions; 2) Flora Prussianum on the natural history of Prussia."[59] In 1731, J.P. Breyne published his treatise *Historia naturalis cocci radicum tinctorii* in Danzig.[60] It featured the Polish cochineal with descriptions of the male and female and its life cycle. This scale insect was used to make red dye for textiles and thus an important component for the Breyne family business in the sixteenth and seventeenth centuries. Breyne's treatise indicated trade and local natural history went hand-in-hand in Danzig's busy harbour. The merchants depended on the export of local products, and at the same time it fuelled the inquiry into local natural history.

Around 1739, Breyne junior reprinted – not the *Centuria* as he had proposed earlier – but the two *Prodromi* books. He combined it in one large volume with additional pictures, and he included his dissertation on ginseng. In honour of his father, it was aptly titled *Jacobi Breynii, Gedanensis, Prodromi fasciculi rariorum plantarum primus et secundus*, and it contained a vita of Jacob Breyne together with a drawing made by Johann Philipp's friend Daniel Gottlieb Messerschmidt. For his study of the ginseng root, during the time that Breyne was a student in Leiden, he used the available materials in the gardens of Gaspar Fagel and Hieronymus van Beverningh, both wealthy patrons and garden owners of his father. The botanical knowledge that Johann Philipp picked up as a student was transported to Danzig, where, years later, he printed his findings

---

57 Meister, *Der Orientalisch-Indianische Kunst- und Lust-Gärtner*, 141.

58 Breynius, *Prodromus secundus*, 101.

59 BL MS Sloane 3321, fol. 177–8, Johann Philipp Breyne to J. Petiver, 1705.

60 Johann Philipp Breyne, *Historia naturalis Cocci Radicum Tinctorii quod polonicum vulgo audit* (Danzig: C. a Beughem, 1731).

in the *Prodromus*. From Danzig's location of knowledge, Breyne relocated his findings on the ginseng root to collectors in the Republic and England.

Through the production of books, father and son Breyne made exotic and unfamiliar (Danzig) local flora visible and known to others. This tension of what is local and what is exotic is intricately linked to the discussion of what is centre and what is periphery. For their botanical correspondents, these plants were exotic and Danzig situated along the periphery of Western Europe. Using the VOC shipping routes and the *Moedernegotie*, this effort allowed for the transformation and dispersal of nature, enabling others to become acquainted with local and foreign flora. Breyne's 'noble and illustrious men,' and other amateurs, botanists and medical men accumulated botanical knowledge and materials in their houses and gardens. Visiting botanists and amateurs could examine, exchange and enjoy *in situ* nature's treasures. Simultaneously in Danzig, Jacob and Johann Philipp Breyne turned their botanical collection into transportable books. The Breyne family thus brought local and exotic flora to a wider audience, enabling those who used these books to further their knowledge of nature.[61] Regarding the science of botany in the eighteenth century, Danzig was an important location of knowledge.

The books were published in Danzig at the authors' own expense, using their own or local printing presses. The Breyne family wanted to control and oversee the printing process, partially to avoid piracy and plagiary. On the other hand, having control over the publishing process enabled the family to control the transformative process whereby written texts and plant materials were turned into typeface and prints. For this transformative process, Jacob and J.P. Breyne employed artists who reworked each plant into a drawing that was then transformed into a copperplate, as the next section illustrates. The Breynes employed German, Dutch and Polish artists and artisans such as the aforementioned Stephan Cousijns, Andreas Stech, Isaak Saal, as well as Jacob's three granddaughters to produce and duplicate their botanical findings. The transformative act of re-presenting nature into a reproducible image took place in Danzig from which these volumes circulated via various trading routes to Holland, England and beyond.

---

61 On the role of physicians as inquirers and dispersers of natural history, see: Harold J. Cook, "Physicians and natural history," in: *Cultures of Natural History*, eds. N. Jardine, J.A. Secord, and E.C. Spary (Cambridge, MA: Cambridge University Press, 2000) 91–105. On the constitution of natural history through exchange and commerce of "matters of fact:" Cook, *Matters of Exchange*, and Harold J. Cook, "The cutting edge of a revolution? Medicine and natural history near the shores of the North Sea," in: *Renaissance and Revolution; Humanists, scholars, craftsmen and natural philosophers in early modern Europe*, eds. J.V. Field and Frank James (Cambridge, MA: Cambridge University Press, 1993) 45–61.

4 Producing Knowledge and Nature in Danzig

For his publications, Jacob Breyne depended on the skills of artists to transform nature into pictures. Painter Andreas Stech did most of the plant drawings from flora found in Breyne's garden and herbaria. Eight illustrations were based on flower paintings, since Breyne did not possess these specimens from the Cape of Good Hope, South Africa. For the publication of the *Centuria*, Beverningh, Breyne's patron, commissioned the painter Cousijns to make watercolours of flora growing in Dutch gardens.[62] In Danzig, either Stech or Saal turned Cousijns' paintings into drawings.[63] Breyne employed local artisans to produce comprehensible and reproducible illustrations based on his materials of plants, exsiccates and drawings. Working with Danzig-based artisans allowed the botanist to oversee the transformation of nature into illustrations (*Figure 5.5*).

For the *Centuria*, Danzig engraver Isaak Saal turned nearly all drawings into copperplate engravings. The Amsterdam engraver Jan Veenhuysen made one and the Danzig artisan Johann Benssheimer another. The Haarlem-born Lambert Visscher created the title page using the 1670 design of Stech as the foundation. Both Veenhuysen and Visscher must have divided their time between Amsterdam and Danzig between circa 1675 and 1690 since they worked also as illustrators for the astronomer Johannes Hevelius.[64] Engravers like Saal would cover the back of a drawing with chalk and place it on a copperplate. The next step was to trace the lines of the drawing with a burin. On the plate appeared a sketch in chalk. This formed the basis of an engraving.[65] It took two steps to transform a plant into a picture.[66]

---

62 Jacob Breyne, *Exoticarum aliarumque minus cognitarum Plantarum centuria prima* (Danzig: D.F. Rhetius, 1678). See caps. XLIIX [sic], LX, LXI, LXII, LXXXIV, XCIII, XCIV, XCV.

63 D.O. Wijnands, E.J.A. Zevenhuizen, and J. Heniger, *Een Sieraad voor de Stad. De Amsterdamse Hortus Botanicus 1638–1993* (Amsterdam: Amsterdam University Press, 1994) 70–74.

64 Janet Vertesi, "Instrumental images: the visual rhetoric of self-presentation in Hevelius's Machina Coelestis," *British Society for the History of Science* 43 (2010): 209–243, notably 220–228. Inga Elmqvist Söderlund, *Taking Possession Of Astronomy. Frontispieces And Illustrated Title Pages In 17th-Century Books On Astronomy* (Stockholm: Center for History of Science at the Royal Swedish Academy of Sciences, 2010). Lambertus Visscher (1633–after 1690) was commissioned to engrave the frontispiece of Hevelius' *Firmamentum Sobiescianum, sive Uranographia* (Danzig: J.Z. Stoll, 1690) and Hevelius' *Cometographia Cometographia* (Danzig: Simon Reiniger, 1668), after designs of Andreas Stech. See *Nieuw Nederlandsch Biographisch Woordenboek* 10, kol. 117.

65 Heniger, *Hendrik Adriaan van Reede tot Drakenstein*, 129–130.

66 *The Cleveland herbal, botanical, and horticultural collections: a descriptive bibliography of pre-1830 works from the Libraries of the Holden Arboretum, the Cleveland Medical Library*

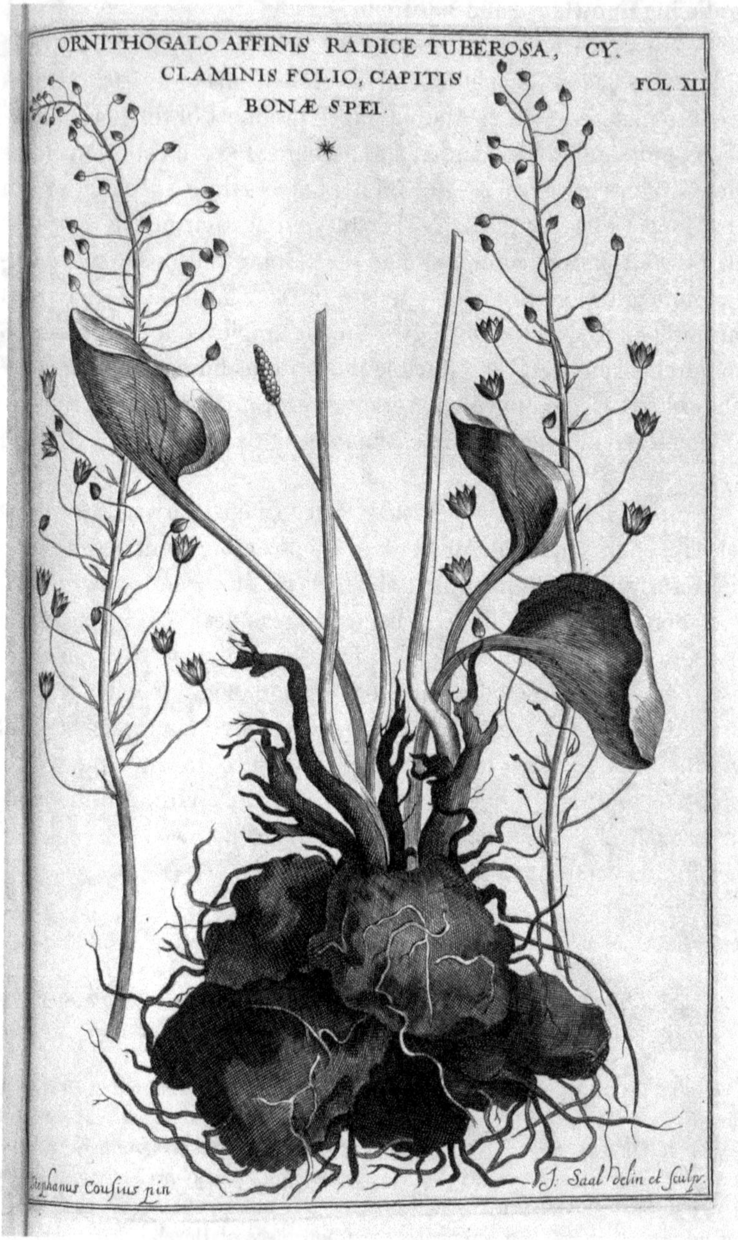

FIGURE 5.5 *Ornithogalo affinis radice tuberosa, cyclaminis folio, Capitis Bonae Spei* (Eriospermum capense (L.) Thunberg), folio 41, by Stephan Cousijns (painter) and Isaak Saal (draughtsman and engraver). From: Jacob Breyne, *Exoticarum aliarumque minus cognitarum Plantarum centuria prima* (Danzig, 1678)
COURTESY NATURALIS BIODIVERSITY CENTER, LEIDEN (BOTANY SECTION)

The *Centuria* showcased the combined efforts of Prussian and Dutch artists and patrons. The mindful work of skilled artists and artisans, together with the joint efforts of collectors and describers of botany in Holland, Poland, South Africa, and the Far East enabled the construction of botanical knowledge. An illustrated book can be considered a transportable parallel to an actual garden, as both are filled with (representations of) nature's creatures. Professor of botany Abraham Munting (1626–1683), an acquaintance of Breyne, propagated this notion in the introduction of his book on plants:

> Those who do not have gardens, and read this book will not regret this purchase, but enjoy the pleasure with heart's content in fostering the rarities of Nature regarding various plants; the printed images will provide for an optical garden in their homes.[67]

For producers of botanical books, it was a balancing act of rendering an image not too schematic and miss important elements or too detailed which made an illustration unclear.[68] A plant should be recognizable *and* an aide for both botanists and garden enthusiasts. A reader could enter a book with depictions as if he or she was walking in a garden or a field, encountering unfamiliar nature, page by page instead of step by step. As Breyne had seen and studied the actual plants, he could oversee the joint efforts of his artisans who transformed plants into comprehensible images of his chosen botanical specimens.

Association, and the Garden Center of Greater Cleveland, ed. Stanley H. Johnston, Jr. (Kent, OH: Kent State University Press, 1992) 236, on Stephan Cousijns and Jan Veenhuysen. For more on Andreas Stech and his flower paintings, see: Danuta Natalia Zaslawska, "The Floral Still Lifes of Andreas Stech," *Acta Historiae Artium Balticae* 1 (2005): 101–137.

67 Abraham Munting, *Nauwkeurige beschryving der aardgewassen, waar in de veelerley aarten bijzondere Eigenschappen der Boomen, Heesters, Kruyden, bloemen*, second edition (Leiden: Pieter van der Aa /Utrecht: François Halma, 1696). "De geene, welk geen hoven hebben, en echter dit boek zich eygen maken, zullen haar de koop niet misdanken, maar genieten de verlustiging haars gemoeds in 't betrachten van de zeldzaamheeden der Natuur ontrent allerley gewassen; en de printverbeeldingen zullen hunne oogen verstrekken voor een hof in haar huys."

68 Ogilvie, *The Science of Describing*. In chapter four "A Science of Describing," Ogilvie discusses illustrations were a working tool to make sense of the natural world. Images made in woodcuts, engravings or etchings could render nature schematic very elaborate. A botanist would seek a condensed picture, whereas an interested nature lover could prefer an ornate image, see 139–208. On the importance of "ad vivum" representations of nature, see: Sachiko Kusukawa, "Conrad Gessner on 'Ad Vivum' Image," in: *Ways of Making and Knowing. The Material Culture of Empirical Knowledge*, eds. Pamela H. Smith, Amy R.W. Meyers, and Harold J. Cook (Ann Arbor: Bard Graduate Center, 2014) 330–356.

In the hands of the Breyne family, local and exotic flora was changed into a visual and verbal language that allowed other botanists and amateurs to examine and appreciate nature. Breyne's publications allowed readers (then and now) to acquaint themselves with all types of plants through pictures, names, characteristics, habitat, and with references to publications from other botanists. Transforming nature became a means of communication amongst its producers and consumers to discuss taxonomy and type, so that botanists could create a common language in order to exchange nature and knowledge.[69]

Father and son Breyne changed knowledge and materials into objects that could be easily transported and reproduced, thus contributing to the establishment of shared plant idiom. To this growing body of natural history Jacob and Johann Philipp added plants and knowledge from the Baltic region. In their books, father and son combined these botanical findings from near and far into useful and well-received volumes. This location of knowledge allowed for the accumulation of plants, paintings, publications, and written papers that in the hands of skilled scholars and knowledgeable artisans was transformed into printed botanical knowledge.

5 Conclusion

For a long time, trade linked Danzig to Holland, Europe, Scandinavia, Russia, Africa, and Asia. The inhabitants of Danzig were a mixed group of Polish, German and Dutch speaking residents. Communities of scholars, traders, artisans, sailors, as well as circles of friends, acquaintances, business associates, and families formed the fabric of this bustling city. Astronomers, botanists, natural historians, painters, architects, and artisans jointly contributed to the formation of knowledge. Danzig may not have been considered a location of knowledge by historians of science as other European cities such as London, Paris, Amsterdam, Rome, or Berlin. The shipping routes allowed for speedy and regular transportation that brought people, plants, parcels, and books on botany and letters, drawings and manuscripts on exsiccates to and from Danzig. Because of its location, a busy harbour along the Baltic Sea, the Breyne family combined and contributed to a growing body of natural knowledge, which they actively discussed and exchanged with peers around the globe.

69 Dániel Margócsy, "'Refer to folio and number': Encyclopedias, the Exchange of Curiosities, and Practices of Identification before Linnaeus," *Journal of the History of Ideas* 71 (2010): 63–89.

This article focusses on the Breyne family, with their intricate web of family, merchants, artisans, and scholars around Europe and Asia. Their network became a location of botanical knowledge. It allowed the family to move upward on the scholarly ladder and it allowed Johann Philipp Breyne to tour through Europe and to become a member of the Royal Society. The Breyne family commodified – like their trade in commercial goods – their knowledge of local and exotic botany. Knowledge and material production and consumption can take place at any location within a smaller or larger network. Examining, accumulating, and publishing natural knowledge does not solely take place in one particular 'centre of accumulation.' It can happen anywhere – houses, gardens, fields – as long as the location is tied to a wider network. The Breyne family, because of its mercantile background, was in the position to create and to use its different networks to further natural knowledge.

CHAPTER 6

An Amphibious Science: Where Storms Took Shape

Azadeh Achbari

1 Introduction

In 1821 a hurricane, which became known as the 'Great September Gale,' hit the city of New York causing serious damage and destruction. The hurricane had such an impact on William C. Redfield, a former saddler who ran a steam transportation business, that the object of whirlwind storms never left his mind. He began studying wind patterns and paths of storms from the Atlantic. When he visited a part of the area over which the hurricane of 1821 had passed, he noticed that the toppled trees lay with their heads pointing to the northwest and others pointing to the southeast. A detailed study of eyewitness accounts led him to believe that storms rotated around a central point and that they moved in curved lines.[1] Redfield's theory, although supported by many respected men of science of the time, was in conflict with another theory, formulated by the American teacher, James Pollard Espy, whose studies of local storms suggested that air moved from different directions towards the centre of a storm forming a rising column of air and flowing out over the top of the storm.[2] The dispute about whether storms were rotatory or centripetal remained unresolved for a long time, until the Dutch physicist Christopher Buys Ballot offered an important piece of the puzzle: the relationship between pressure and winds.

Redfield, Espy and Buys Ballot were a few of a far larger group of nineteenth-century scientifically inclined men who were interested in understanding the movement of winds and storms. Pure curiosity about how atmospheric air circulates was one thing that motivated these men. With the growth of maritime trade and colonial expansion in the mid-nineteenth century, studies of storms obtained increasing social and practical relevance. The devastating effects of the September Gale and other major weather catastrophes at sea such as the wreck of the British *Royal Charter* on the northeast coast of Anglesey, or the French *Henry IV* at Sevastopol in Crimea gave a sense of urgency to

1 Harold L. Burstyn, "Redfield, William C.," in: *Complete Dictionary of Scientific Biography* 11 (Detroit, MI: Charles Scribner's Sons, 2008), 340–341.

2 James P. Espy, *The Philosophy of Storms* (Boston: Little and Brown, 1841).

 | DOI:10.1163/9789004264885_007

these studies.[3] Such unfortunate disasters propelled government-sponsored investigations of storms and the establishment of meteorological institutes in maritime countries in Europe and in North America. The main question of this essay is how location of observation and discovery influenced nineteenth-century investigations of storms. This essay consists of three sections. The first two discuss four settings where observations of winds were carried out, some at sea, others on land. It examines whether location was a factor of success in the establishment of networks of observation. In the third section the focus shifts to newly established scientific institutes around the mid-nineteenth century, where storm observations were studied, and examines the results of the investigations. It looks into how geography played a role in the discovery of general rules of wind flows.

2 The Storm Controversy

In the 1830s the science of the weather was widely popular, partly because of its practical relevance, but also because it was viewed as a field where major discoveries could be made.[4] However, meteorological investigations were not easy to carry out for they required discipline, standardised observation methods, a trained eye, sophisticated instruments and a large number of volunteers to carry out observations. Well-known organisers of meteorological research projects in the nineteenth century were often naval officers or men with close ties to naval and military offices, commercial shipping or government institutions, who could bring together large groups of men to carry out observations. Many of them were drawn to the field after having experienced devastating storms themselves like Redfield, who developed his theory of rotatory storms by studying eyewitness accounts of the notorious September Gale.[5] For his general theory on the nature of storms to be accepted, Redfield was in need of more proof. To this end, he set up a personal data collecting system with coastal stations and sea captains who sailed to and from New York. He gathered

---

3 James Burton, "History of the British Meteorological Office to 1905" (PhD dissertation, Open University, 1988), 41. Fabien Locher, *Le Savant et la Tempête. Étudier l'Atmosphère et Prévoir le Temps au XIXe Siècle* (Rennes: PU, 2008), 38, 47.

4 On meteorology as a new niche in science, see John F.W. Herschel, *A Preliminary Discourse on the Study of Natural Philosophy* (London: Longman, Rees, Orme, Brown, Green and Taylor, 1831), 328–329.

5 William C. Redfield, "The Prevailing Storms on the Atlantic Coast," *American Journal of Science* (1834). "Sketch of William C. Redfield," *Appleton's Popular Science Monthly* 50 (1897): 114–119.

information of wind patterns from logbooks that were kept aboard their ships. These logbooks, which contained information on the prevailing winds and precipitation on a daily basis, formed a good source to study the origins and paths of storms.[6]

Around the same time as Redfield's investigations, William Reid, a British lieutenant colonel of the Royal Engineers, started his own hurricane research project. He developed an interest in storms and hurricanes during his mission in the West Indies. Reid was sent to Barbados to help reconstruct government buildings, which were destroyed by a hurricane in August 1831.[7] Shortly afterwards he decided to write to Redfield when he learned that the American had already published a paper on revolving storms. Their correspondence grew into a friendship that lasted until Redfield's death in 1857.[8]

When Reid returned to England in 1834, he built on Redfield's studies by collecting weather data from logbooks of British ships. After four years he was ready to publish a treatise on *the Law of Storms*, which rested on an impressive amount of data.[9] But first he presented a paper on the subject to the Annual Meeting of the British Association on 20 August 1838. In this paper he supported Redfield's theory on the revolving nature of hurricanes and their motion in curved lines. He also provided guidelines for sailors as how to manoeuvre a ship away from storms (figure 6.1).[10] Reid's work was received well by fellow British weather investigators. In particular, the famous physicist John Herschel praised the work and publicly thanked Redfield for his efforts.[11]

In the US, however, Redfield's position was far from secure. As early as of 1834 he had been involved in a dispute over the nature of storms with an American teacher, James Espy. Whereas Redfield proposed gravity as the motive power for global wind patterns, Espy attributed the movement of winds to changes in pressure distribution, which he believed were caused by calorific heat. Espy rejected Redfield's notion of storms as revolving systems on the basis of contradicting evidence. He claimed that the centrifugal forces, which according to Redfield caused a hurricane to rotate, would soon come to an end by its outward motion. Instead, he supported another theory suggesting that

---

6 Burstyn, "Redfield," 340.

7 Olwyn Mary Blouet, "Sir William Reid, F.R.S., 1791–1858: Governor of Bermuda, Barbados and Malta," *Notes and Records of the Royal Society London* 40 (1986): 169–191.

8 James Rodger Fleming, *Meteorology in America, 1800–1870* (Baltimore: Johns Hopkins University Press, 1990), 37.

9 Blouet, "Reid," 174.

10 Pauline Halford, *Storm Warning. The Origins of the Weather Forecast* (Stroud: Sutton Publishers, 2004), 38–39.

11 Halford, *Storm Warning*, 39.

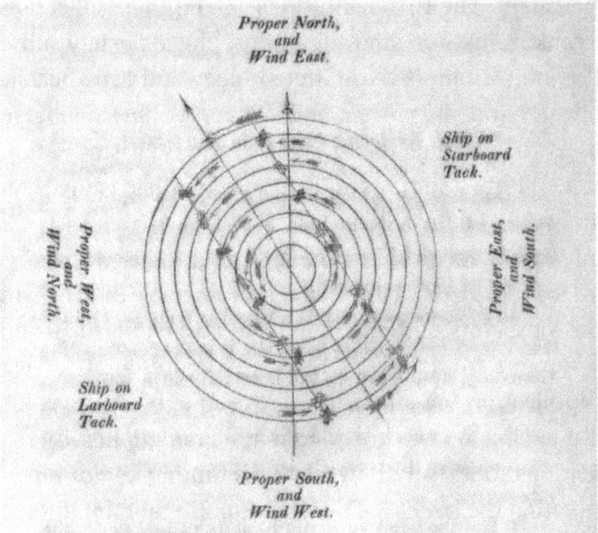

FIGURE 6.1 Reid's representation of a West Indian hurricane accompanying his guidelines for manoeuvring a ship away from it
SOURCE: W. REID, *AN ATTEMPT TO DEVELOP THE LAW OF STORMS* (LONDON, 1838), 426

air moved towards the centre of a storm forming a rising column of air and flowing out over the top of the storm.[12]

One way to resolve the dispute between Redfield and Espy was to test their contradicting views by comparing them with observations of a destructive tornado called the "Brunswick Spout" which occurred in 1835. However, there was no outcome to the problem as both Espy and Redfield saw completely different movement patterns in the traces of the spout, which as they claimed, only confirmed their own theories.[13] In fact, the controversy dragged on and triggered several initiatives to set up systems of observational data collection around the country.

In 1831 Espy began his own private collection of meteorological data, while he worked as a lecturer at the Franklin Institute in Philadelphia. Through the institute's journal he made a call for volunteer weather observers to send him their observations. In return, the *Journal of the Franklin Institute* started publishing monthly reports based on their observations. In 1834 Espy was offered support by the American Philosophical Society that worked together with the

12 Espy, *Philosophy of Storms*.
13 Halford, *Storm Warning*, 37.

US Army Medical Department. The two institutions shared their collections of weather data. The Franklin Institute and the Philosophical Society subsequently merged into the Joint Committee on Meteorology and Espy became its chairman. Meanwhile, a naval surgeon named William W. Ruschenberger offered to share the observations which were made at various coastal stations with the Joint Committee. Within a short period of time, Espy had gained control over an extensive network of data collection.[14]

In order to create unity in the records, Espy next turned to the government for support to purchase standardized instruments and to cover the costs involved in instructing his observers and compiling the observations. His application for financial support from the US House of Representatives was declined, but another application for support from the state of Pennsylvania was successful and a grant of $4,000 was provided to the Franklin Institute for nationwide weather investigations. Espy, however, encountered major practical difficulties in coordinating the system of observation. He found it difficult to find competent volunteer observers who could provide him with reliable observations. The costs to manufacture the instruments were higher than estimated and it took a long time to build standard instruments. County officials often came into conflict with the observers, because it was unclear who was responsible for what. Mismanagement caused the Joint Committee to dissolve in June 1838.[15]

In comparison with Espy, Redfield fared better. He obtained his data from logbooks kept by sea captains and clerks at coastal stations. These naval observers were far more disciplined and were used to making measurements as compared to Espy's volunteer observers in the country. Redfield also carried out observations himself and maintained a correspondence with other observers and men of science abroad. He received additional storm data from his partner, Reid, who had been collecting information from logbooks of British ships that sailed to the American east coast.[16] This close collaboration gave Redfield access to more observations. In 1838 Reid persuaded the British Colonial Office to enlarge the system of observations to the coasts of Britain's colonies overseas. Reid, furthermore, sought to recruit observers through his former superior, John Burgoyne, the Inspector-General of Fortifications. He agreed to organize observations at the overseas stations of the Royal Engineers.[17]

---

14 Fleming, *Meteorology*, 56–57, 62.

15 Fleming, *Meteorology*, 59–61.

16 Blouet, "Reid," 174.

17 Halford, *Storm Warning*, 43.

When comparing Espy's system of observation with that of Reid and Redfield, the question arises whether the sea as a location for observations offered advantages over land, to help advance the field of meteorology. In the opening address of the fifteenth meeting of the British Association in Cambridge in 1845, Herschel posed this question and he argued that the collection of large amounts of data could only be useful for meteorology if the observations were carried out in a uniform way. He wondered whether the best location for observations was on land, for mountains, lakes, forests, and other geographical factors distort the path of winds. The sea offered advantages, because the conditions for measurements were more consistent. In his words, "a ship is an itinerant observatory; and, in spite of its instability, one which enjoys several eminent advantages – in the uniform level and nature of the surface, which eliminate a multitude of causes of disturbance and uncertainty, to which land observations are liable."[18] The following section takes another network of observation at sea as its topic. Different from Redfield's network, this network, which did not only obtain observations from coastal stations, but also from locations far out on the ocean, is discussed so as to examine whether the sea as a location was a factor of significance for the progress of meteorology, as Herschel had argued.

3 A Sea of Observations

In the mid nineteenth century, Lieutenant Matthew Fontaine Maury obtained worldwide fame as the creator of a global network of observations that gave a boost to meteorology at sea. Maury was an able navigator in the navy before he became the superintendent of the Naval Observatory in Washington. At the observatory he performed routine astronomical tasks such as the determination of time, the examination and purchase of instruments, and the preservation of sea charts.[19] Under his direction, the observatory became involved in finding better sea routes for American navy ships. Maury found a new purpose for the great piles of logbooks of the American men-of-war that were kept at the Naval Observatory's Depot. These logbooks contained records of observations taken at regular intervals of the direction of the winds and their speed, air and water temperature, the direction and force of the currents, depth soundings,

18 John Herschel, "Address of the President," *Report of the Fifteenth Meeting of the British Association for the Advancement of Science* (1846): xxvii–xliv.

19 Frances Leigh Williams, *Matthew Fontaine Maury. Scientist of the Sea* (New Brunswick, NJ: Rutgers University Press, 1963), 101, 158.

and the appearance of tides. Realizing that these records contained valuable information on the path of winds and currents for different times of the year at different locations at sea, he wondered whether he could use this information to create charts depicting the best possible routes for his fellow navy captains to sail.[20]

By analysing and combining the information in the logbooks, Maury was able to create his first "Wind and Current Chart of the North Atlantic," which was issued in 1847. Shortly after, he produced eight more charts of the North and South Atlantic, the North and South Pacific and the Indian Ocean.[21] There was a profitable market for these charts. Sea captains, fishermen, insurers and bankers found these charts very useful, which were claimed to make sea journeys safer and shorter. Maury cleverly created a system of distributing free charts to navy and mercantile sea captains in return for more observations. He made a profit from the sale of his charts, which he could invest in his undertaking.[22] Soon he began to expand his system to increase its scale. He sought to unify sea observations by drawing up blank forms for all ship commanders to keep on their journeys. His superiors were easily convinced of the benefits of standard forms and agreed to distribute them among navy commanders. In the following years, Maury became engaged in reducing and analysing the logbooks that thousands of ships at sea sent to the observatory.[23] The result of the work was a renewed edition of his *Wind and Current Charts and Sailing Directions*.

The merit of Maury's work lay in the sea charts that helped sea captains to make their journeys shorter and safer. Compared to Espy, Reid, and Redfield, however, Maury did not succeed in formulating a general theory of wind and storm paths that nineteenth-century weather scientists were willing to accept. Although he very much hoped to make an impact on nineteenth-century science of global wind flows, his theories received strong criticism.[24] Nonetheless, his network of sea observations, which was extremely well organised, served as a model for other maritime countries.

Maury's charting work attracted the attention of the previously mentioned John Burgoyne, who sent him an invitation to participate in a British programme of meteorological observations. Instead of agreeing to cooperate with the British, Maury proposed an international conference in order to adopt a uniform system of observations of winds and currents, which the American

20 Williams, *Maury*, 179–183.
21 Ibidem, 180, 187.
22 Ibidem, 184. Robert V. Bruce, *The Launching of Modern American Science 1846–1876* (New York: Knopf, 1987), 177, 181.
23 Williams, *Maury*, 205.
24 Halford, *Storm Warning*, 102.

and British governments eventually approved. In 1852, when he was engaged with organising the conference, a Dutch naval officer, Marin Henri Jansen, came to visit him at his observatory. Jansen had obtained a copy of Maury's *Wind and current charts and Sailing Directions* before he started his sea journey and he decided to visit the author when he arrived in the United States.[25] During their meeting, Maury and Jansen appeared to have a lot in common. They were potential partners who could be of mutual assistance to each other. As a maritime nation, the Netherlands possessed the third largest fleet in the world.[26] Jansen proposed to help Maury by establishing connections with Dutch sea captains, whose observations Maury could use well. In return for their logbooks, Maury offered the Dutch sea captains a free copy of his charts.[27]

Jansen suggested proposing to the Dutch Minister of the Navy to cooperate with Maury's project, which had a clear practical benefit. Maury's charts could help Dutch captains to travel to the Dutch East Indies safer and faster. By June 1852 Jansen had spoken to the Minister and he had agreed to adopt Maury's system in the Netherlands.[28] The Dutch-American cooperation also helped to convince officials from other countries to participate in Maury's maritime conference.[29]

After meeting with Maury, Jansen adopted the plan to follow his example and use his assistance to create a similar system of exchange of observations for charts in the Netherlands. When Jansen returned to the Netherlands in June 1852, he met with the Dutch physics professor, Christopher Buys Ballot, and asked him whether he would be willing to cooperate. Buys Ballot, who had a great interest in meteorology, headed an observatory in Utrecht, where he carried out meteorological observations. He also received weather observations from five other stations around the country.[30] The Dutch professor accepted Jansen's proposal, for he could help him establish contact with ship captains, whose logbooks were useful for his studies of the weather.[31]

---

25 M.H. Jansen and S.P. L'Honoré Naber, *Het leven van een vloothouder. Gedenkschriften* (Utrecht: Kemink & Zoon, 1925), 278–279, 285.

26 Roger H. Charlier, "Fratres in Maribus 150 years ago. The First International Ocean-Science Conference," *Journal of Coastal Research* 20 (2004): 347–350, on 350.

27 E. van Everdingen, *C.H.D. Buys Ballot 1817–1890* (The Hague: Daamen, 1953), 61; Williams, *Maury*, 218.

28 Minister of the Navy to M.H. Jansen, 24 June 1853, Dutch National Archives, The Hague, Ministry of Navy, Entry 2.12.01, Inventory 2274.

29 Williams, *Maury*, 218.

30 Van Everdingen, *Buys Ballot*, 58–60.

31 Jansen and Naber, *Het leven*, 250. Van Everdingen, *Buys Ballot*, 60. On Jansen's role as mediator among men of science, the world of shipping and government officials, see also Achbari, "Building Networks for Science."

Buys Ballot hoped to transform his observatory into the centre of a European meteorological network. To this end, he hoped for his observatory to obtain the status of an official institute in the Netherlands. As an earlier study has shown, Jansen was a valuable ally for Buys Ballot, whose connection with Maury and his involvement in the organisation of an international maritime conference could help Buys Ballot to obtain governmental support for the transformation of his observatory into a state institute.[32]

In August 1853, Jansen participated in the maritime conference in Brussels as an official delegate from the Netherlands. At the meeting, several issues were discussed. The attendants conferred about ways to carry out accurate observations, verify instruments and unify the records. The participants agreed to adopt a standard form for records of observation. Maury suggested taking the standard form, which he had drawn up together with Jansen in Washington, as an example for registering maritime observations. He promised to distribute free charts to captains of all countries who kept the standard form aboard their ships.[33]

Another issue that Maury addressed was the desirability for each government to appoint a person to collect and process the logbooks. This person should also be responsible for the distribution of free charts to ship captains of the respective cooperating countries.[34] This last discussion point prompted several governments of maritime countries to establish state institutes for meteorological observations.[35]

As for the Netherlands, the conference attracted the attention of the government and helped Buys Ballot to confidently ask for state support to transform the Utrecht observatory into an official meteorological institute, where investigations that were practically relevant for Dutch seafaring were carried out. Thus the sea as a location of observations helped to increase the significance of Buys Ballot and Jansen's meteorological project. Within a month after the conference, the Utrecht observatory was transformed into the Royal Dutch Meteorological Institute (KNMI).[36] Buys Ballot was appointed director

---

32 Frans van Lunteren, "De oprichting van het Koninklijk Nederlands Meteorologisch Instituut: Humboldtiaanse wetenschap, internationale samenwerking en praktisch nut," *Gewina/Tijdschrift voor der Geschiedenis van de Geneeskunde, Natuurwetenschap, Wiskunde en Techniek* 21 (1998): 216–243, on 232–236.

33 Jansen and Naber, *Het leven*, 288–289.

34 Halford, *Storm Warning*, 65–66.

35 An official report of the meeting was published in 1853: *Maritime Conference held at Brussels for devising an uniform system of meteorological observations at sea. August and September 1853* (Brussels, 1853), 28–30. Halford, *Storm Warning*, 67.

36 Van Everdingen, *Buys Ballot*, 68. See also Van Lunteren, "De oprichting," 232–236.

of the institute, which was divided into two separate departments for land and maritime measurements. Jansen was happy to be head of the maritime department.[37] In the following years, there was a well operating system of maritime data collection in cooperation with the American naval observatory.

Returning to the main question whether the sea as a location played a role in the creation of networks of observations, Maury and Jansen's network of well-trained sea captains, who carried out masses of observations in a uniform, regular, and standardised way, certainly confirms the view that the sea as a location for observations offered advantages over land. Compared to Espy's land based network, Maury's network was far more efficiently organised. Disciplined naval officers kept accurate journals of the conditions at sea at regular times and offered a steady flow of maritime information for Maury to analyse. His network that got an official character through government support was much larger than Redfield's private undertaking. By taking a different approach than Reid, Maury lured naval commanders and captains of merchant ships to cooperate by offering them his charts in return. However, Herschel was in some respects right to doubt whether large numbers of data would help advance the field of meteorology. The observations that Maury and Jansen assembled were admittedly useful for charting the best sea routes. However, neither Maury nor Jansen was able to draw a meaningful and general theory of the nature of winds from their studies of the logbooks.

The storm controversy, with which the first section opened, dragged on for years without a solution. Redfield and Reid's theory of rotatory storms and Espy's centripetal theory divided nineteenth-century weather investigators into two rivalling camps. The next section shows how Buys Ballot's wind law put an end to the storm controversy while reconciling the two theories.

4 Locations of Discovery

As has been shown in the previous section, the international maritime conference played an important role in the establishment of meteorological institutes in maritime countries. This section takes a closer look at the investigations carried out at the Dutch institute and British meteorological department, both of which stood at the forefront of nineteenth-century storm forecasting techniques and meteorological research. A closer examination of the practical work carried out at these offices will show whether geography played a role in the discovery of a general rule of wind flows.

37 Van Everdingen, *Buys Ballot*, 60–63.

In the years after the establishment of the Dutch meteorological institute, Buys Ballot's daily work consisted of studying weather observations that were carried out at six stations spread across the country, Groningen and Den Helder in the north, Maastricht and Vlissingen in the south, and Utrecht and Nijmegen in the middle of the country. Each year he published the results of these observations in orderly arranged tables covering dozens of pages.[38] From the observations, he discovered a general relationship between the barometric pressure field and the subsequent force and direction of the wind.[39] In 1857 he presented his discovery to the members of the Dutch Academy of Science. As he pointed out,

> The difference between simultaneous absolute readings or deviations from the normal reading gives the most certain indication, even when one pays attention to places not too far apart. For out of a thousand times when, within the confines of the Netherlands, this difference at eight in the morning was less than two millimetres, the force of the wind never increased beyond thirty kilograms per square metre within twenty-four hours, and out of two hundred times when the difference was more than four millimetres, the wind attained a greater strength fifty times. The direction was or always became easterly (between NE and SE), when the reading was higher in Groningen and Den Helder than in Maastricht. And always west or south west, when higher at Maastricht than at Den Helder, almost without exception.[40]

Buys Ballot pointed out that this wind rule could be used as a means to predict strong winds that could not only be useful for Dutch seafaring, but also for the country in general. He proposed to use the wind rule for the creation of a system of storm warnings in the Netherlands. According to the rule, when the difference between the barometric readings at the northern and southern stations in the country was more than four millimetres, strong winds were expected. When a storm was approaching, he proposed to send telegraph messages

38 C.H.D. Buys Ballot, *Meteorologische waarnemingen in Nederland en afwijkingen der temperatuur en barometerhoogte ook op enkele plaatsen buiten Nederland* (Utrecht: Kemink en Zoon, 1852–1854).

39 On the transformation of Buys Ballot's wind rule into a law of nature, see: A. Achbari and F.H. van Lunteren, "Dutch Skies, Global Laws: the British creation of 'Buys Ballot's law,'" *Historical Studies in the Natural Sciences* 46 (2016): 1–43.

40 *Verslagen en Mededeelingen der Koninklijke Akademie van Wetenschappen. Afdeeling Natuurkunde* 7 (1857): 75–77.

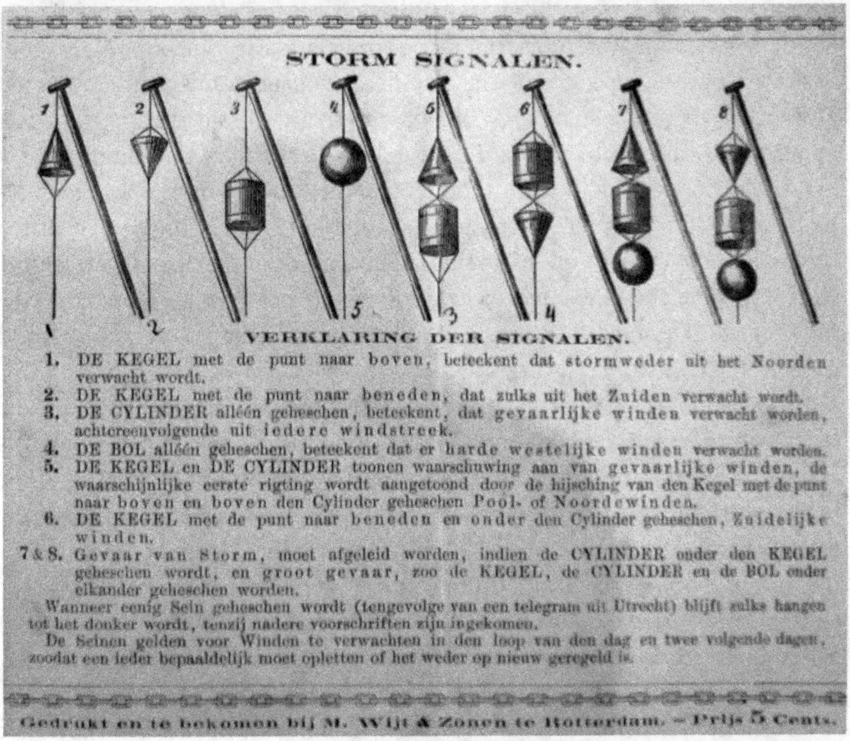

FIGURE 6.2 In 1864 the Dutch system of warning flags was replaced by storm signals. These signals were modelled after the British storm signals that were developed by Robert Fitzroy.
SOURCE: UTRECHT ARCHIVES, ENTRY 90, INVENTORY 1463

to the main Dutch ports containing barometric differences and the probable wind direction and force. In 1860 the minister of the interior, Schelto van Heemstra, accepted his proposal to establish a storm warning service, which soon was in operation in June 1860 (figure 6.2).[41]

Remarkably, the British meteorological department, which was established in the same year as the Dutch institute, began its own storm warning service in 1861. The department was headed by a naval officer named captain Robert Fitzroy. Fitzroy, who had been taught at the Royal Naval College at Portsmouth, was an experienced navigator. During the years between 1830 and 1836 he had

41 J. Schröder, "Excerpt from Resolution by the Minister of the Interior, dated 21 May 1860," in: *Eenige regelen voor aanstaande weersveranderingen in Nederland*, ed. C.H.D. Buys Ballot (Utrecht: Kemink en Zoon, 1860), 75–76.

commanded the *Beagle* through a surveying voyage along the coastline of South America. He was, furthermore, no stranger to carrying out observations at sea.[42] However, he would not be carrying out observations at his new office. Fitzroy was to head a small department with a staff of four assistants. He was appointed to distribute standard instruments to the navy and the merchant marine, collect logbooks from mariners, reduce their content, and tabulate this information in order to use them as material for preparing wind charts.[43] But he had far greater ambitions than to be a mere statistician. Once the statistical work at the office became routine, he turned to other more challenging projects. Among other things, Fitzroy designed a barometer and wrote a manual for its use. He also developed a network of weather stations, using ports as points where observations of rainfall, barometric pressure, wind, and temperature were carried out and he successfully proposed the creation of a system of weather reporting and storm warnings in Britain by the use of the telegraph.[44] He asked telegraph clerks to send their weather observations to the London office to be analysed. Fitzroy furthermore developed a set of signals in the shape of cones and drums indicating the direction and intensity of storms. In February 1861 he issued the first warning of an oncoming gale.[45]

It is interesting to know how the British warning and forecasting system differed from the Dutch system, as they were both established in the same period. A testimony written by Fitzroy's assistant, Thomas H. Babington, shows the method that Fitzroy used to forecast the weather in Britain:

> [T]here are two constant principal wind currents, north-east and south-west, of which the characteristics, especially with regard to temperature and degree of moisture or dryness, are totally distinct, all varieties of wind and weather in these latitudes may be traced to the operation of these two main currents.... Suppose a northerly (ENE-NNW) current to have been prevailing generally over this country with *fine* weather, the barometers at all the outports, during the continuance of such weather, will have been steady, or slowly rising, nearly uniform, or slightly higher at the northern than at the southern stations; there will have been much evaporation, and the sky will have been comparatively clear and free from cloud, while the temperatures, when free from the influence of

42 Halford, *Storm Warning*, 76–78, 82.

43 Katharine Anderson, *Predicting the Weather. Victorians and the Science of the Meteorology* (Chicago and London: University of Chicago Press, 2005), 108–109.

44 Ibidem, 109.

45 Ibidem, 112–114.

> radiation, will have been somewhat below the average. Suppose now the northern barometers rise rapidly above the average, while the temperature remains low and the sky clear. This is an indication of *more* wind from a northern quarter, but probably without rain. Should, however, the southern barometers fall at the same time, while the temperatures in the south are much higher than in the north, the wind will probably increase to a *gale* from a northern quarter, and a sudden chilling of the atmosphere in the south will ensue causing rain.[46]

As this fragment shows, the rules of forecasting at the British department were tacit and implicit. While barometric information was used in both Britain and the Netherlands, the British department lacked a specific rule to determine the force and direction of the winds, and thus of determining the occurrence of storms. It is, moreover, not hard to infer from Babington's explanation of the British method of forecasting that preparing daily weather reports was a time-consuming and tedious practice.

In his official reports Fitzroy mentioned that he received observations from thirteen stations around Britain and additional daily reports from five stations on the European mainland. From the synoptic analysis of this information he prepared daily forty-eight-hour forecasts for five weather regions in the British Isles. In 1863 he published a book entitled, *The Weather book*, in which he explained his methods of forecasting to a broad audience. He also offered a theory for weather change, which he named the *lunisolar* theory, from the supposed effects of the sun and the moon on the circulation of air masses.[47] The leading British weather scientists, however, were not willing to accept his theory and criticized his work.[48] Despite Fitzroy's implausible theories, the *Weather book* did contain interesting passages showing that he was close to discovering a rule for atmospheric circulation. He had observed, for instance, that "the entire mass of atmosphere in [the middle] latitude, has a constant, a perennial movement toward the east...."[49]

A comparison of methods used by Buys Ballot and Fitzroy to determine the wind, raises the question whether geographical location played a role in the preparation of forecasts and storm warnings. Could Buys Ballot's discovery of

---

46 Thomas H. Babington, "Report by Mr Babington on Forecasts and Storm warnings communicated with Mr. Farrer's Letter," *Proceedings of the Royal Society of London* 14 (1865): 309–310.

47 Robert Fitzroy, *The Weather Book. A Manual of Practical Meteorology* (London: Longman, Green, Longman, Roberts, & Green, 1863), 206, 244–256.

48 Anderson, *Predicting the Weather*, 119.

49 Fitzroy, *Weather Book*, 107, 173.

the relationship between the pressure gradient and the direction and force of the wind, for instance, have been influenced by the geographical position of the Netherlands? Did geography play a role in the discovery of general rules in meteorological research?

Recalling Herschel's recommendations for meteorology from 1845, the best location to carry out observations was at sea, where a uniform surface level offered the best conditions for measurements.[50] It is not unlikely that the shape of the Netherlands, which offered good conditions for carrying out observations, played a role in Buys Ballot's discovery. As a flat country, the Netherlands offered similar uniform conditions for observations as the sea. The lack of mountainous regions in the Netherlands allows unimpeded flow of winds from the Atlantic Ocean. Unlike on a boat, however, the stationary observer at a Dutch observation point had the advantage of a stable ground beneath his feet, where he could measure the air pressure and gauge the force and direction of the winds. Furthermore, the size of the country and the small number of stations that were necessary to estimate the state of the weather across the Netherlands offered another advantage. Buys Ballot had fewer data to analyse than Fitzroy, which may have helped him to find the relationship between the wind and the pressure field.

Once the Dutch system of storm warnings was established, four stations and a headquarters in Utrecht were sufficient to predict the weather in the Netherlands. In England, however, Fitzroy analysed data from twenty stations to prepare his daily weather forecasts in 1863. In that year, a merchant ship captain published a work that compared the forecasts of Fitzroy and Buys Ballot, and tested them against the actual state of the weather for a couple of months. The captain observed that the Dutch observations were as accurate as the British.[51]

In terms of forecasts, Buys Ballot could also benefit from the geographical position of the Netherlands, an advantage that Fitzroy did not have. As weather systems, in general, move from west/southwest to east/northeast across the Atlantic towards Europe, Britain is the first European country to experience depressions. The geographical location of the Netherlands allowed Buys Ballot to test his forecasts against the actual weather approaching from the west and southwest through an exchange of meteorological observations with Britain

---

50 John Herschel, "Address of the President," xxvii–xliv.

51 F.H. Klein, C.H.D. Buys Ballot and A. Adriani, *The foretelling of the weather in connection with meteorological observations by F.H. Klein together with a description of the telegraphic warning system introduced in The Netherlands, June, 1860, as proposed by the director of the Royal Netherlands Meteorological Institute Professor Dr. Buys-Ballot* (London: Benjamin Pardon, 1863), 25–29.

and France. Since the late 1863 he had been receiving weather reports from Paris and London.[52]

Fitzroy, however, could not benefit from weather information that he could receive beforehand, which made forecasting a lot more difficult for him. He also dealt with other obstacles because of the geographical position of Britain. Onshore ocean winds are much stronger than winds blowing on land due to friction, as land surface is much rougher than sea surface. This effect often caused for inaccurate forecasts of wind force. It is the more admirable that Fitzroy was able to issue accurate storm warnings. A present day assessment showed that 76 per cent of all warnings Fitzroy issued in the month of December 1863 were correct.[53]

As a final remark on the role of geography in nineteenth-century meteorological investigations it is interesting to point to Francis Galton's discovery of anticyclones. The European continent as an area of observation offered a wide scope for the study of larger weather systems than single countries. In 1863, Galton published a work entitled *Meteorographica*, which contained a series of maps depicting the state of the weather over Europe for the month of December 1861. The maps, which presented isobarometric curves connecting areas of equal pressure, gave important insights for the understanding of wind flows at the time. By the use of isobars Galton discovered two distinct centres of low and high pressure with winds spiralling inward in an anticlockwise direction into a pressure centre, or blowing outward in a clockwise direction away from the centre. The first phenomenon, named 'cyclones,' had been known since 1848, when an English captain of the East India Company, Henry Piddington, published his *Sailor's Horn-book*, in which he described the anti-clockwise movement of winds in tropical storms around a centre of low pressure. Piddington's model differed from Reid and Redfield's model in that it correctly described how winds curved inward to the centre of the depression. He coined the term 'cyclone' for these tropical storms from the Greek 'coil of a snake.'[54] As the opposite of 'cyclones' after Piddington's terminology, Galton named the areas of high pressure 'anti-cyclones.'[55]

---

52 Buys Ballot to the Minister of the Interior, 20 August 1863, Utrecht Archives, Entry 90, Inventory 683. Buys Ballot to the Minister of Interior, 19 January 1864, and Buys Ballot to the Board of Trade, 23 May 1865. Utrecht Archive, Entry 90, Inventory 684.

53 Burton, "History of the British Met. Office," 61.

54 Henry Piddington, *The Sailor's Horn-book for the Law of Storms: Being a Practical Exposition of the Theory of the Law of Storms* (London: Smith, Elder and Company / New York: Wiley and Putnam, 1848), 8.

55 Francis Galton, *Meteorographica: or Methods of Mapping the Weather* (London: MacMillan & Co, 1863), 7.

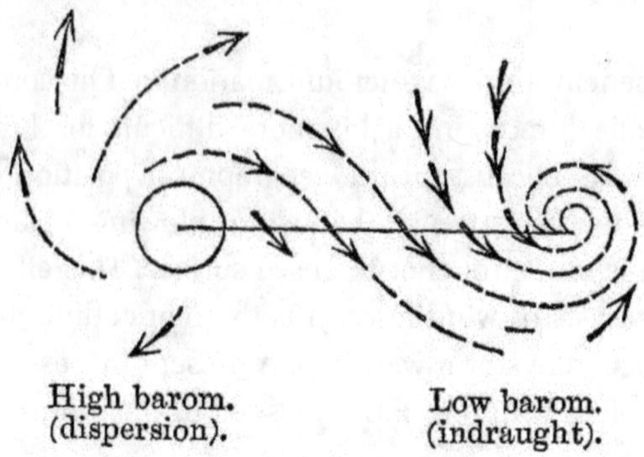

FIGURE 6.3 Francis Galton's systematic representation of atmospheric air circulation
SOURCE: FRANCIS GALTON, "A DEVELOPMENT OF THE THEORY OF CYCLONES," *PROCEEDINGS OF THE ROYAL SOCIETY* 12 (1863), 385–386

Galton's systematic maps of winds, in fact, offered a visual representation of Buys Ballot's wind law. The image shown in Figure 6.3 proves that Redfield's and Espy's rivalling theories of the shape of storms, and of winds in general, were both right. Wind flows rotate around a centre point and move at the same time toward the centre of a depression. After more than twenty years, the mechanical movement of wind flows, which fuelled the storm controversy, was finally figured out.

5 Conclusion

With the growth of international maritime trade, studies of wind patterns and storms attracted increasing attention in the early nineteenth century. Weather catastrophes formed a strong motivation to investigate the nature of storms and states were found willing to support large-scale investigations of storms and the weather in general. This essay examined the role of location and geography for the establishment of networks of observation and the discovery of general rules of wind patterns. Four nineteenth-century systems of observation, established by Redfield, Reid, Espy and Maury, were discussed to see whether location played a role of significance in the outcome of the projects. As has been shown, coordinating a system of observation at sea offered clear advantages over land. The Joint Committee on Meteorology, Espy's land-based network of observations, was confronted with many organisational difficulties. Espy had to establish his network from scratch. Among other things, he had to find and train volunteer observers himself, purchase standard instruments and coordinate a project that was loosely organised. In comparison, most of the groundwork for maritime observations was already in place. Naval and army officers like Reid, Maury and Jansen could build on existing organisational structure, instruments and large numbers of well trained and disciplined officers who were used to carrying out observations at fixed times and keeping accurate records of the state of the weather in logbooks on board ships.

Not only in terms of organisation, but also in practice, the sea as a location of observation had advantages over land. As Herschel pointed out, observations were only meaningful and helped advance meteorology if they were carried out in a uniform way. The sea as compared to land offered conditions that allowed uniform and standardised observations to be carried out. However, the larger the scale of the networks meant more observations to be analysed in practice, which made the formulation of general theories of winds a difficult task. As has been shown, the observations that Redfield obtained from his smaller scale observation project as compared to Maury and Jansen's global project were sufficient for developing a theory of local storms near the American east coast. Although Maury and Jansen were able to chart sea routes on the basis of their observations, they had to deal with a far larger number of observations than Redfield and Reid, who mainly carried out coastal observations. Despite his efforts, Maury did not succeed in formulating a global theory of winds that nineteenth-century men of science were willing to accept.

Scale was also an important factor in the practice of forecasting storms at the Dutch and British meteorological departments, as the last section demonstrated. Buys Ballot's network of six observation stations in the Netherlands made his daily practice of reducing data less time-consuming than Fitzroy's observation network of thirteen stations. The smaller size of the Dutch network, in fact, worked in Buys Ballot's favour in discovering the relationship between the pressure field and the direction and force of winds. Besides size, another factor that played a role in Buys Ballot's discovery of the wind rule was the geographical shape and location of the Netherlands. As a flat country, the Netherlands offered the same optimal conditions for uniform observations as the sea. Not directly bordering the ocean, the Dutch stations moreover, allowed for more accurate observations of the winds than the British offshore stations. The case studies of Dutch and British forecasting work have shown that less is indeed more. An even more important lesson to be learned from looking at nineteenth-century networks of weather investigations is the synergetic effect of competition and cooperation. The storm controversy, which divided nineteenth-century weather investigators into two camps supporting rivalling theories, encouraged the establishment of networks of observation and institutions for research. Cooperation strengthened the ties among investigators, allowed for the exchange of observations at sea and on land, and enabled the discovery of the dynamics of wind flows, which finally put an end to the storm debate.

CHAPTER 7

Science to Bring the Nation Together: The Formation of Nomadic Congresses in the Netherlands and Flanders

Ilja Nieuwland

1 Introduction

In the course of the nineteenth century, virtually every scientifically relevant nation, and those territories whose citizens want it to become one, saw the founding of annual or bi-annual gatherings of scholars of all scientific disciplines, in moving venues.[1] These "peripatetic" or "nomadic" organizations may be seen as a new kind of scientific class, but in their flexibility of changing locations they also added new aspects to the more traditional forms of scientific organization. This phenomenon has so far received incidental treatment, but its trans-national aspect has not been given much attention. What we see here is the confluence of two defining developments of nineteenth-century science and politics: the establishment of science and scientists as a separate, increasingly distinct professional community on the one hand, and the need for national (both political and cultural) self-assertion on the other.

These congresses with their changing venues had their root in the early to mid-nineteenth century, first in Switzerland and Germany, and later in other parts of Europe and North America. Among the latecomers were the Dutch-speaking territories: the Kingdom of the Netherlands and the northern part of the Kingdom of Belgium, Flanders. They form odd latecomers: firstly, because the size and infrastructure of this small part of Europe made travel much easier than was the case in, for instance, Germany; and secondly because in other areas the Dutch – and to a lesser extent also the Flemish – acted far earlier upon the need to establish initiatives that sought to rekindle a national sentiment for a number of reasons, such as the aftershocks of the Napoleonic period, the Belgian separation, and Flemish cultural and linguistic emancipation. After reviewing the development of these nomadic organizations,

1 Acknowledgements: I wish to thank Klaas van Berkel, Marieke van der Duin, Frans van Lunteren, Matthijs de Ridder and Huib Zuidervaart for taking the time to cast a critical eye on this text, or sections of it, in the various stages of its conception.

 | DOI:10.1163/9789004264885_008

I therefore want to address two questions: First, why did it take so long for Dutch scientists to organize a national traveling congress? And secondly, what was the impact of rotating venues on the Dutch and Flemish scientific communities – did it matter at all, and did it help these organisations in achieving their aim?

When we take a broader look at these meetings, it appears that the societies involved can be roughly divided into two categories, both of which are heavily coloured by local nationalist considerations. Where one type of nomadic conference sought to offer a showcase of scientific prestige to the nation, the other gathered scientists to create that nation, by establishing national awareness through the instrumentalization of a generally accepted and highly valued activity. The big difference is that the former tended to focus on scientific endeavour, whereas the latter used a general public esteem and respect for science and its participants to bring about a political transformation.

Below, I wish to illustrate these processes by analyzing the emergence of the Dutch and Flemish congresses for nature and medicine in the late nineteenth century. These organisations attempted, both in their own way, to rally the nation for the cause of science, and to employ science to serve the nation. Moreover, their cases demonstrate how, while the differences between both types of congress may sometimes appear trivial, they led to very different outcomes.

2 A New, Nationally Oriented Scholarly Society

In the nineteenth century many nations were to adopt the nomadic nature of scientific congresses to create a truly national canvas from which science could benefit, but which simultaneously benefited the 'nation,' even – or particularly – in cases where, in political terms, that 'nation' did not exist yet. The proliferation of these nomadic scientific congresses of can be traced back to central Europe, and to territories that were anything but homogeneous national entities.

The Swiss *Société Helvetique des sciences naturelles* or *Schweizerische Naturforschende Gesellschaft* was the very first. This organisation also largely determined the form that shaped future sister organisations. It had its origin in 'helvetism,' a political movement from the late eighteenth century that sought to strengthen the country by eradicating existing divisions of language, religion and class. The addition to the Swiss nation of Geneva, an esteemed centre of learning, had been one of the important consequences of the negotiations in Vienna after the Napoleonic Wars. The Geneva naturalist and pharmacist

Henri Albert Gosse, the main initiator of the *Société Helvetique*, expressly intended to use the new concept of a nomadic congress to strengthen 'national' science and forge a bond between the old *Eidgenossenschaft* and the new Swiss additions. The first meeting of Gosse's society took place in October 1816 in Bern – not coincidentally the capital of the old country. The congress became an instant success and was able to enlist 250 members for its second meeting. By moving the meeting venues, Gosse sought to bring science and scholars to all the major cities in the country. It was this element that proved to be instrumental, both in the conception and the success of the society's many imitators.

3 Science as a Tool for Unification and Education

The Swiss example was followed in many other nations that strove towards unification and the creation of a national identity. That most certainly was Lorenz Oken's aim when he succeeded in drawing the German naturalists together. As a veteran of the Helvetian society's founding meeting, and the nationalist *Wartburgfest* of 1817, Oken's credentials as both a German naturalist and nationalist were above suspicion.[2] At this time, Germany was still very much politically fragmented, and the definition of what was, or was not, 'German' remained largely oblique. Consequently, rulers from Italy and France to Denmark, but certainly those in Germany itself, had good cause to mistrust the outspoken nationalism for which Oken stood.

After 1816, Oken set out to organize something similar to the Swiss meeting. Although very clear in his ideas about what such a meeting ought to achieve, he was far less lucid when it came to the way in which the organization was to be structured. A child of the German *Klassik*, Oken was somewhat hesitant in abandoning its encyclopaedic ideals in favour of specialization, although most of his scholarly brethren favoured some sort of subdivision.

After the failure of another attempt to organize German scientists, carried out by the Halle chemist Johann Salomon Schweigger in 1820, Oken published his '*Ersten Aufruf zur Versammlung der deutschen Naturforscher*' ['First call for a

---

2 Hagen Schulze, *Der Weg Zum Nationalstaat.: Die Deutsche Nationalbewegung vom 18. Jahrhundert bis zur Reichsgründung* (München: C.H. Beck, 1985), 71–74. On the cultural history of nationalism, see Hannu Salmi, *Nineteenth-Century Europe. A Cultural History* (Cambridge: Wiley, 2013), 58–71, particularly 65*f.* For its meaning in the context of the scientific community, see Klaas van Berkel, *Citaten uit het boek der natuur. Opstellen over Nederlandse wetenschapsgeschiedenis* (Amsterdam: Bert Bakker, 1998) and Bert Theunissen, *'Nut en nog eens nut.' Wetenschapsbeelden van Nederlandse natuuronderzoekers 1800–1900* (Hilversum: Verloren, 2000), 168–171.

gathering of German scientists'] a year later. This provoked the Leipzig natural historians' association to extend an invitation to host the first meeting.[3] Thus, in 1822, the first session of the *Gesellschaft deutscher Naturforscher und Ärtzte* ['Society of German Natural Scientists and Physicians'] was opened. Oken's purpose was to unite all scientific endeavours and to come to a more homogeneous, but simultaneously all-encompassing kind of science. His aim was not to erect a formal structure to co-ordinate scientific research; rather, he wanted members of the *Gesellschaft* to share their knowledge in an informal setting. Evidently, one would meet each other as *German* scientists, and many authorities were well aware of the political ramifications of this self-definition. For instance, it would take three quarters of a century before the first meeting of the *Gesellschaft deutscher Naturforscher und Ärtzte* was tolerated on Austrian soil, and in the early years Austrian participants needed to clad themselves in anonymity if they were to have any chance of returning to their hometowns without problems.[4]

Another issue was the choice of venue for the meetings. Germany was a large territory, and although improvements in transport infrastructure began to shorten relative distances, it could still take considerable time (and effort) to travel from one end to the other. But because the meetings were intended to contribute to the unification of Germany as a nation as much as serving as a tool for bringing together the German scientific community, the practice of changing locations was never open for discussion. It was not unproblematic however, as it also made participation considerably more expensive and time-consuming than it might have been if the congresses had taken place in one of the main university towns. The success of the meetings of the *Gesellschaft deutscher Naturforscher und Ärtzte* seems to indicate that these troubles hardly dampened the visitor's spirits. At the first meeting in Leipzig in 1822, no fewer

---

3 Max Pfannenstiel, *Kleines Quellenbuch zur Geschichte der Gesellschaft Deutscher Naturforscher und Ärzte. Gedächnisschrift für die hunderdste Tagung der Gesellschaft. Im Auftrage des Vorstandes der Gesellschaft verfaßt von Max Pfannenstiel* (Berlin etc.: Springer, 1958), 18–23, 164. For the founding of the *Gesellschaft deutscher Naturforscher und Ärtzte*, also see Alexander Ecker, *Lorenz Oken. Eine Biographische Skizze. Gedächtnisrede zu dessen hundertjähriger Geburtstagsfeier, gesprochen in der zweiten öffentlichen Sitzung der 52. Versammlung deutscher Naturforscher und Ärzte zu Baden-Baden am 20. September 1879* (Stuttgart: Gesellschaft deutscher Naturforscher und Ärtzte, 1880), 220; Dietrich von Engelhardt, *Forschung und Fortschritt: Festschrift zum 175jährigen Jubiläum der Gesellschaft Deutscher Naturforscher und Ärzte* (Stuttgart: Wissenschaftliche Verlagsgesellschaft, 1997); Ilse Jahn, "Gründung und Spezialisierung der Gesellschaft Deutscher Naturforscher und Ärtzte im 19. Jahrhundert," in: *Zwei Jahrhunderte Wissenschaft und Forschung in Deutschland. Entwicklungen – Perspektiven*, ed. Dietrich von Engelhardt (Stuttgart: Wissenschaftliche Verlagsgesellschaft 1990), 101–121.

4 Melchior Treub, "Eene feest-vergadering," *De Gids*, 43–3 (1879): 128–156.

than 115 visitors attended, and by the time of the seventh meeting in Berlin, in 1828, turnout had risen to almost five hundred.[5]

While in one sense the meetings of the *Gesellschaft* were very successful, Oken's attempts to unite the scientific disciplines ended in failure. Not only did the chasm between medicine and other forms of scholarship remain firmly in place, these disciplines also became gradually more fragmented as time progressed, leading to what seemed like an inevitable separation of the field in its various sub-disciplines in 1828. Even if Oken mourned this specialization, he eventually felt compelled to accept it.

4 Nomadic Societies Elsewhere in Europe

The German congresses served as blueprints for similar organizations in other European and American countries. However, by completely adopting that example, a number of its idiosyncrasies were also copied. This was one of the elements that earned the *British Association for the Advancement of Science* much criticism. In a country in which the large majority of scientists were concentrated either in London and its surroundings or in Edinburgh, what point was there in organizing a congress in Wolverhampton, and even expressly stipulating that it would be held outside London?[6] Yet, in the case of the *British Association for the Advancement of Science* its nomadic nature served a very real purpose: to pull science from its ivory tower in Oxbridge and the English and Scottish capitals, and transport it to the newly established centres of industry in the country.

The failure of efforts to the Royal Society of London in 1830, and the stifling influence of the court's proximity, affirmed that an entirely new initiative was needed in order to galvanize scientific life in Britain.[7] The drive for greater specialization was the prime motivator that brought the *British Association*

5 Jahn, "Gründung und Spezialisierung," 108.

6 Jack Morrell, and Arnold Thackray, *Gentlemen of Science. Early Years of the British Association for the Advancement of Science.* (Oxford: Clarendon Press, 1981) describe the genesis of the BAAS. Also see Peter Briggs, "The British Association for the Advancement of Science (BAAS). From Scientist's Meetings to Annual Festivals of Science," in: *Zwei Jahrhunderte Wissenschaft und Forschung in Deutschland. Entwicklungen – Perspektiven* ed. Dietrich von Engelhardt (Stuttgart: Wissenschaftliche Verlagsgesellschaft, 1998), 183–189. Also see Treub's description of the *British Association for the Advancement of Science* in *The Times*: Treub, "Eene feest-vergadering."

7 Morell and Thackray, *Gentlemen of Science*, 57–60.

for the Advancement of Science together in 1833.[8] However, the promotion of 'science as a duty of statecraft,' as Sir Lyon Playfair put it in 1885, became an increasingly more important task for the organisation. This was a different type of nationalism from that which drove Oken, since it aimed to assert national prestige through the medium of a scientific meeting.[9] In other words, it was aimed outwards as much as inwards.

The Italians, inhabitants of another politically fragmented territory with a nascent unitary movement, had much more to gain from the German example. Ironically, it was the monarch of one of the smaller Italian countries that instituted the annual *Riunioni degli Scienziati* ['Meeting of Scientists'] in 1839 in an effort to exert control over a potentially dangerous gathering.[10] The conspicuous presence of government and army officials in the society discouraged political activity – but simultaneously once more pointed out the potential threat that such an organization could bring.

The United States followed suit, with the *American Association for the Advancement of Science* being founded in 1848: again, in different circumstances, but with largely the same aim. The *American Association* took the development towards a purely scientific organisation one step further: it can be argued that they started out as the organization Oken had tried to prevent the *Gesellschaft deutscher Naturforscher und Ärtzte* from becoming. Membership was still open to anyone, but offices within the association could only be occupied by professionals. Also, a committee was appointed to assess papers submitted for the association's meetings or publications. Especially this last measure was in direct conflict with the 'total freedom' so vehemently advocated by Oken.

Finally, the French *Association Française pour l'Avancement des Sciences* was founded in 1872, not coincidentally at the troubled beginning of the Second French Republic following the debacle of the Prussian-French war of 1871. That conflict had demonstrated how the state of French science – and scientific application – was less rosy than it had been at the beginning of the nineteenth century, but Krupp's guns also reminded everyone what the purpose of scientific endeavour could be. The *Association* should be seen as an attempt to bring

---

8 George Basalla, William Coleman, and Robert H. Kargon, *Victorian Science. A Self-Portrait From the Presidential Addresses of the British Association for the Advancement of Science* (Garden City & New York: Doubleday, 1970), 27–31.

9 Sir Lyon Playfair, "Address," in Basalla et al., *Victorian Science*, 64–83.

10 Giuliano Pancaldi, "The German model and the origins of the Italian 'Riunioni degli Scienziati,'" in: *Zwei Jahrhunderte Wissenschaft und Forschung in Deutschland. Entwicklungen – Perspektiven*, ed. Dietrich von Engelhardt (Stuttgart: Wissenschaftliche Verlagsgesellschaft, 1998), 171–178, esp. 174.

some form of unity, not only among scientists but also between the worlds of science, industry and the public. Although the element of national self-assertion was definitely present, it mostly served to gear the French nation for a more scientifically based outlook on life, in line with the progressivist and anti-clerical character of the Republic itself.[11]

Thus, by the late 1870s, the Western world's most influential nations all possessed their annual gathering of scientists, with the intention, explicit or implicit, to shape the nation's scientific 'identity' and sometimes reaching even further than that. There is a good case to be made for the idea that a nation that was to take itself seriously as a power – and be taken seriously by others – was required to own such a travelling scientific movement as part of a set of scientific paraphernalia that also included a national academy and a number of respectable universities and technological institutions.

5 Reception in the Netherlands

The developments that had sparked off the creation of national congresses took their time in reaching the Netherlands, and even then, little enthusiasm or initiative accompanied them. But once the *Nederlandsch Natuur- en Geneeskundig Congres* ['Dutch Congress for Science and Medicine' or NNGC] had been set up in 1887, it turned out to be an immediate success. It had taken a lot of time to arrive at that point, and several attempts had run aground before it was given the chance to become a valuable component of scientific culture in the Netherlands.

Following Jan Campert's account of the protohistory of the *Nederlandsch Natuur- en Geneeskundig Congres* – recorded at the founding meeting in 1887 – historians have traditionally credited the Amsterdam (and later Batavian) botanist Melchior Treub with bringing to attention the advantages of a 'nomadic' meeting for in-depth scientific discussion.[12] Treub had visited the 1877 edition

---

11 For further information, see *Diffuser et promouvoir les sciences (1872–1914)*, L'Association française pour l'avancement des sciences (AFAS), Rapport d'étape, GHDSO-Paris Sud XI (Paris, 1998). See also Joseph Ben-David, "German scientific hegemony and the emergence of organized science," in: Joseph Ben-David, *The Scientist's Role in Society. A Comparative Study* (Eaglewood Cliffs: Prentice Hall, 1971), 111–113.

12 J. Campert, "Verslag omtrent de geschiedenis en de inrichting van het Congres." *Handelingen van het Nederlandsch natuur- en geneeskundig congres* (1888): 31–34. See also R.P.W. Visser, "Het 'Nederlandsch Natuur- en Geneeskundig Congres' over de relatie natuurwetenschap en samenleving, 1887–1900," in: *Balans en Perspectief van de Nederlandse Cultuurgeschiedenis*, eds. J.J. Kloek and W.W. Mijnhardt (Amsterdam & Atlanta: Rodopi, 1991), 37–48; R. Aerts, *De letterheren: liberale cultuur in de negentiende eeuw: Het*

of the *Gesellschaft deutscher Naturforscher und Ärtzte* meeting in München, and he reported his experiences of that occasion to the readers of the influential literary journal *De Gids* in 1879. Although the majority of his article was devoted to impressions gained from the more prominent lectures held at the meeting, he also raised the question why such a platform did not exist in the Netherlands:

> Upon hearing about the good works of the British Association and the Association Française, one cannot but wonder whether such a society might not be viable and useful in the Netherlands as well; it should not be forgotten, that the aim of both these institutions is no less national than scientific. [...] It is difficult to see why such a goal could be useful in England and France, and not in the Netherlands.[13]

Although one should be careful not to underestimate the cultural and social importance, as well as the opinion-shaping force of an established journal such as *De Gids*, the link between Treub's article and the eventual establishment of a scientific congress seems to be less clear-cut than Campert suggested. Considerable time – no less than eight years – proved to be necessary to arrive at a consensus and come to the formation of a truly national congress in the shape of the *Nederlandsch Natuur- en Geneeskundig Congres*.

In his historical summary presented at the 25th anniversary of the *Nederlandsch Natuur- en Geneeskundig Congres* in 1912, the physician Hendrik Zwaardemaker pointed to a number of causes for this apparent lack of endeavour, but most of all to the 'international character of science.'[14] Zwaardemaker's inference (and later, his express claim) that the desire to communicate internationally somehow held back scientists from attending a congress in the Dutch language, seems to be strangely at odds with the twenty-five years of success he was celebrating at the time of his talk. Scientific internationalism certainly hadn't decreased during that period, yet few, if any, scientists seemed to have been inhibited by the use of the Dutch language at the meetings of the *Nederlandsch Natuur- en Geneeskundig Congres*. Another obstacle that Zwaardemaker perceived was the fear on the part of academic physicians to be surpassed by their scientific colleagues. But various attempts in the early

---

tijdschrift De Gids (Amsterdam: Meulenhoff, 1997), 397; Van Berkel, *Citaten uit het boek der natuur*, 233.

13 Treub, "Eene feest-vergadering," 133–134.

14 H. Zwaardemaker, *Voetstappen onzer wetenschap. Feestrede ter gelegenheid van het 25-jarig bestaan der Vereeniging 'Het Nederlandsch Natuur- en Geneeskundig Congres'* (Amsterdam: Vereeniging 'Het Nederlandsch Natuur- en Geneeskundig Congres, 1912), 6–8.

1870s by the medical profession to seek rapprochement to other disciplines appear to contradict that notion.[15] At his opening address for the first meeting, Barend Stokvis had also treated the reasons for the long-time lack of interest in establishing a national congress. His outlook was by and large more positive than Zwaardemaker's, and followed ideas set out by the Swiss botanist Alphonse de Candolle.[16] De Candolle had attempted to quantify the scientific merits and potential of various nations, and there the Netherlands turned out to be one of the highest-scoring countries, at least as far as the seventeenth and eighteenth centuries were concerned. As Stokvis observed, those nations that scored highly in De Candolle's statistics had one thing in common:

> They combine a number of physical and moral properties: diligent effort, persistent patience, steadfast perseverance, sound judgment, clarity of view, a glowing love of freedom, and an insatiable desire for knowledge and for the truth, that modern, delicious torment of Tantalus, that never grants less peace, than when she has just been satisfied by the indescribable joy of discovering a new fact, a new truth.[17]

On the other hand, the scientific crisis that had manifested itself in the Netherlands during the nineteenth century could hardly have escaped Stokvis. His unique explanation for this relegation of the Netherlands to the background of international science was that of a 'temporary withdrawal to recuperate,' a 'war stratagem,' from which the Dutch would strike back with renewed effort. In Stokvis' mind, the suitability of the Dutch nation for scientific activity was an inescapable fact because "all the important conditions for a flourishing of the sciences are present among us!"[18]

---

15 In 1870, the Amsterdam *Genootschap voor Natuur-, Genees- en Heelkunde* had started to allow representatives of other disciplines to visit the meetings of the section for medicine. One year later, the Amsterdam physician Barend Stokvis proposed the same for the Dutch Medical Association. Although the proposal was adopted, no effort was ever made to put it into effect. See *Weekblad bij het Nederlandsch Tijdschrift voor Geneeskunde* 19 (28 april 1883): 284–285. Also K. van Berkel, M.J. van Lieburg and H.A.M. Snelders, *Spiegelbeeld der Wetenschap. Het Genootschap ter bevordering van Natuur- Genees- en Heelkunde 1790–1990* (Rotterdam: Erasmus Publishing, 1991).

16 Alphonse de Candolle, *Histoire des Sciences et des Savants depuis Deux Siècles* (Geneva, 1873); B.J. Stokvis, *Nationaliteit en natuurwetenschap. Rede tot opening van het eerste Nederlandsch Natuur- en Geneeskundig Congres* (Haarlem: Vereeniging 'Het Nederlandsch Natuur- en Geneeskundig Congres,' 1887). See also Visser, "Het 'Nederlandsch Natuur- en Geneeskundig Congres,'" 41–42.

17 Stokvis, *Nationaliteit en natuurwetenschap*, 19.

18 Ibid., 28.

Stokvis' optimism was later reinforced by that of the chemist Jacobus Henricus van 't Hoff and the physicist Johannes Bosscha, both of whom observed even more admirable characteristics in the national spirit of the Dutch, making them eminently suitable for scientific endeavour.[19] These reflections upon the national temperament of the Dutch made the current lack of prowess so frustrating, because they perceived the national scientific potential as a constant: if the Dutch had been able to rise to such heights in the Golden Age, by necessity they should be capable of repeating that feat. The problem was that this analysis failed to consider changing international circumstances and their effect on the relative importance of the Dutch nation. That did not mean that no efforts were undertaken to restore the Netherlands to its supposedly natural prominent position among the scientific nations. Like in the world of politics, Dutchmen made efforts to come to greater international collaboration, always for the greater good of the nation.[20]

6 The Road towards the Founding of the Dutch Peripatetic Congress

Why then did it take relatively long for Dutch scientists to organize a national congress? First of all, the tone of particularly Zwaardemaker's account suggests that he considers such a congress to have been a historical necessity, something that each country that wished to assert itself in the arena of science ought to possess. That viewpoint, although possibly valid by 1880, ignores the historical contingencies which caused these organizations to come into being earlier in the century.

Obviously, the situation in the Netherlands around 1880 can hardly be compared to that in Germany around 1820. Even when an explicit nationalist element is recognised, it is of a radically different nature. The Germans were motivated by the desire to unite their territory, frequently against the wishes of their rulers, and in that sense the formation of the *Gesellschaft deutscher Naturforscher und Ärtzte* was at least as much a political statement as a scientific one. Nationalism of the brand carried out by the *Nederlandsch Natuur- en Geneeskundig Congres* could hardly be more different. Rather than portraying an inner struggle to achieve unity, it became the manifestation of an effort to

19 Theunissen, '*Nut en nog eens nut*,' 168–171.

20 Theunissen, '*Nut en nog eens nut*,' 171. With its growing national awareness, the Dutch began to see themselves as a 'guide' - a moral poltical judge whose authority came from the fact that it was so politically insignificant and therefore could claim no vested interest in major political conflicts. In science, basically the same idea was applied: Dutch science ought to be 'pure science.'

display it to the outer world. In that sense, it should be compared to the British rather than to the German example.

The need to improve scientific organization, which had been another main contributor to the formation of the *Gesellschaft deutscher Naturforscher und Ärtzte*, had largely disappeared in the Netherlands by 1880. The interests of sections of scientific activity were by and large taken care of by professional organisations, such as the *Nederlandsche Maatschappij tot bevordering der Geneeskunst* ['Dutch Society for the Advancement of Medicine'], founded in 1849, the *Nederlandsche Maatschappij tot bevordering der Pharmacie* ['Dutch Society for the advance of Pharmacy'], founded in 1847, and the *Nederlandsche Dierkundige Vereeniging* ['Dutch Zoological Association']. Rather than acting as a catalyst for scientific unity, these appeared to have functioned implicitly as something of an inhibitor. Furthermore, reorganizations in 1851 and 1855 gave the *Koninklijke Nederlandsche Akademie van Wetenschappen* ['Royal Netherlands Academy of Arts and Sciences'] a much more active role in the Dutch scientific community. However, the Academy was also a highly elitist organization, not well suited to showcase scientific activity to the outside world.

That is not to say that prior to the formation of the *Nederlandsch Natuur- en Geneeskundig Congres* there were no significant attempts to come to a truly national congress. However, even when supported, they seem to have met with little fortune. The first of the more serious initiatives was taken at a meeting of the medical *Nederlandsche Maatschappij* in July of 1883. A commission of this Society consisting of prominent members such as the aforementioned Stokvis and Anton Guye, brought a proposal to the vote which discussed the possibility of raising the level of its scientific discussions. A number of options were presented: first, alternating the annual meetings of the *Maatschappij* with a scientific congress; secondly, the organization of a scientific congress, in which the society suggested it would represent the medical sciences in a separate section.[21] The last option meant transforming the meetings of the *Maatschappij* in such a way that the emphasis would rest on scientific discussions. Although the members of the commission foresaw problems with the implementation of each proposal, they did attach great value to a more autonomous status of the Society's scientific discussions. Eventually, the commission agreed to propose that a scientific congress be held in a different university city each year.[22]

However, the members of the *Maatschappij* decided otherwise. The proposal was rejected in the general meeting by 51 votes to 18. Not so much because

21 *Weekblad bij het Nederlandsch Tijdschrift voor Geneeskunde* 19 (28 april 1883): 284–285.
22 Ibid., 286.

the need for such a gathering was not felt, but rather because it was not considered to be the Society's task to take the initiative in the matter. That privilege, it was suggested, should be granted to the Amsterdam *Genootschap voor Natuur-, Genees- en Heelkunde* ['Society for Natural, Medical and Surgical Sciences'], because of its prominence, and since it already incorporated all the branches of science. Moreover, some trepidation was expressed that such a congress might take over the Society and get in the way of its traditional activities.[23] It is a safe assumption that despite utterances to the contrary, the need for a cross-disciplinary organisation in the Netherlands was still not felt to be very acute. Certainly, it appears that in the opinion of members, the added value of a congress did not counterweigh the interest of hanging on to established ways of communicating information within the *Nederlandsche Maatschappij tot bevordering der Geneeskunst*.

A second, and decidedly more successful attempt was instigated by the *Nederlandsche Dierkundige Vereeniging* ['Dutch Zoological Association'] in 1886, specifically by one of the people who would eventually be at the cradle of the *Nederlandsch Natuur- en Geneeskundig Congres*, the Leiden zoologist Paulus Hoek (1851–1914).[24] Hoek expressly referenced developments abroad, before stating that:

> What we wish to propose is of a more modest nature: let us attempt to organise it that the different societies that are devoted to the practice of the natural sciences, from now on hold their summer meetings at the same time, and at the same place, in our fatherland. In that case, a cosy get-together of all the members could precede the meeting days of the individual societies, illuminated by a single, short talk, to be held alternately by a member of this or that society.[25]

This proposal was greeted enthusiastically by the meeting, which even went so far as to charge Hoek and the Dutch marine zoologist Max Weber with the task of further investigating the idea. Although this proposal did not intend

---

23 *Nederlandsch Tijdschrift voor Geneeskunde*, 19 (1883): 599–601.

24 For Hoek's career in zoology, see: Robert-Jan Wille, *Mannen van de microscoop. De laboratoriumbiologie op veldtocht in Nederland en Indië, 1840–1910* (Nijmegen: Van Tilt, 2018).

25 Noord-Hollands Archief, Haarlem (hereafter NHA): Archive Nederlandsch Natuur- en Geneeskundig Congres (hereafter NNGC), Box 10. Reprinted minutes of the meeting of the *Nederlandsche Dierkundige Vereeniging* [Dutch zoological association], without date. Due to the fact that the NNGC archives have not yet been inventoried, it is not possible to provide a more precise reference. Also see Visser, "Het 'Nederlandsch Natuur- en Geneeskundig Congres'."

to diminish the autonomy of the participating bodies, it can be argued that it might have led to a much closer integration of scientific organisations – and hence of the scientific community as a whole – than the *Nederlandsch Natuur- en Geneeskundig Congres* eventually accomplished. Moreover, it called for an association between purely professional organisations, whereas the participants of the congress would come from much more diverse backgrounds, as we will see.

It is significant that the zoologists' desire for a more collective approach originated in the fact that some of their activities were poorly attended; particularly interest in the summer meetings had been dwindling since the 1870s. By combining forces with other organisations, they hoped to reverse that trend. This represents a more defensive, and most of all practical attitude that contrasts with Treub's evangelism.

The execution of Hoek's plan immediately met with problems, mainly because the forms of these different summer meetings varied too much to adequately co-ordinate them. In addition, it turned out to be difficult to hold them at the same time. But it did spark the idea to found a new, separate organisation, which should be responsible for hosting a national scientific congress.[26] Hoek's efforts met with considerably more success than that of Stokvis, also because they did not hinge upon the sole efforts of the Zoological Association, but rather sought to broaden the basis for organizing a scientific congress. To this end, Hoek and Weber got in touch with representatives from the *Nederlandsche Maatschappij tot bevordering der Geneeskunst*, the *Nederlandsche Botanische Vereeniging* ['Dutch Botanical Society'], and the *Nederlandsche Entomologische Vereeniging* ['Dutch Entomological Society']. In order to assess the interest in such an organisation, it was decided that a circular should be sent to potential members. The circular was distributed widely, and a major part of the reactions came from non-professional circles.[27] The 133 replies demonstrated enough eagerness to go ahead with the project.

7 A New Scientific Organisation

The founding meeting of the *Vereniging het Nederlandsch Natuur- en Geneeskundig Congres* took place in the Amsterdam Krasnapolsky hotel on 11 April 1887. A temporary board, consisting of Paulus Hoek and Max Weber on behalf of the zoologists, the medical doctors Anton Guye and Jan

26 Minutes of the founding meeting of the NNGC. NHA, Archive NNGC, Box 3.
27 Reactions to Kerbert, 20 March 1887. NHA, Archive NNGC, Box 1.

Egens van Iterson (1842–1901) for the *Nederlandsche Maatschappij tot bevordering der Geneeskunst*, the botanist Jacob Gijsbert Boerlage (1849–1900) for the *Nederlandsche Botanische Vereeniging*, and the insect specialist Henricus W. Groll (d. 1900) for the *Nederlandsche Entomologische Vereeniging* needed no more than an hour and a half to decide on the form and organisation of the first congress.

That is not that surprising in itself. The Swiss had determined the form of these meetings in 1816 and it had subsequently been popularised by Oken's *Gesellschaft deutscher Naturforscher und Ärtzte*. It did mean, however, that some historical traits of these organisations were also duly copied. For instance, the idea to hold the congress in a different city each year made good sense in the context of German patriotism of the 1820s, which needed the injection of a national spirit, but it was far less obvious in the Dutch situation of the 1880s and 1890s, in which neither the political nor the logistic circumstances seemed to necessitate it. Also, in a country as small and easily traversable as the Netherlands, the reason of the *British Association for the Advancement of Science* for holding each congress as a scientific showcase in a different locale was far less obvious, given that the Netherlands, as a scientific nation, had never been focused on a single centre – despite a certain concentration in the western rim of the country – and the achievements of industry hardly merited it to be visited in the first place. Indeed, during its early existence, excursions undertaken as part of the meetings typically went to centres of research rather than industry.

Like the *Gesellschaft deutscher Naturforscher und Ärtzte*, the congress was divided up into thematic sections: one for medicine, one for natural history and chemistry, and another for mathematics and physics. At the time of the first congress, a fourth section, for geology and geography, was added. This division not only reflected the essence of Hoek's original proposal at the zoologists' meeting and the division that was generally maintained within the major societies in the country, it also indicated the increasing proliferation of scientific disciplines. Within time, the sections would operate like almost entirely separate entities, like sub-congresses – indeed a departure from the Okenian ideal of the early nineteenth century. As Zwaardemaker was forced to admit when speaking about the geographical section in 1912, he was a 'stranger in that place.'[28]

It had been decided that the first congress should be held in Amsterdam, and that it was to take place on a bi-annual rather than on a yearly basis, contrary to other national congresses in Europe. It was further agreed that

28 Zwaardemaker, *Voetstappen*, 20.

three resident members be given the responsibility of forming a seven-man board to organize the first congress. The choice for Barend Stokvis, Jan Willem Gunning and Hugo de Vries for this task is indicative of the nerve the plans had touched within the scientific community. This was truly the 'Premier League' of Dutch science, and its support would greatly enhance the reputation of the *Nederlandsch Natuur- en Geneeskundig Congres* and offer further testimony to the importance given to the new institution. The main issue was the date; the last thing the organisers wished was to give competition to existing efforts and organizations.

The Amsterdam *Genootschap voor Natuur-, Genees- en Heelkunde*, already seen as the most suitable candidate to set up a national congress by the meeting of the *Nederlandsche Maatschappij tot bevordering der Geneeskunst*, contributed the sizable sum of 300 guilders to the new organisation, but with the express condition that it only be spent on the scientific and not the social activities of the *Congres*.[29] This stipulation further emphasised the single-minded desire to resurrect Dutch science to its former glory and not be distracted by more mundane considerations.[30]

Membership was open to the professional scientific community, but its definition remained rather broad. In order to become a member of the *Nederlandsch Natuur- en Geneeskundig Congres* one had to possess an academic degree of some description, or be a member of a scientific society. However, a secondary education diploma was also recognised, and membership requirements for societies were not necessarily dictated by scientific qualifications. That makes it problematic to position the *NNGC* unequivocally within the increasing professionalization of the sciences in the Netherlands. Although one can say that it strove to represent the professional community, the doors were still wide open for anyone with a scientific interest, as the reactions to the circular had proved. There was some pressure to raise the threshold for occasional visitors, but a proposal to this effect was later withdrawn.[31] In effect, the stimulus to make this an exclusively professional organisation was no stronger than in most traditional learned societies, and less than in some, such as the *Provinciaal Utrechtsch Genootschap voor Kunsten en Wetenschappen*

---

29 Roughly equal to 8,000 euros in modern currency.

30 Tilanus sr. at the founding meeting of the NNGC. NHA, Archive NNGC, Box 3.

31 Minutes of the founding meeting of the NNGC. NHA, Archive NNGC, Box 3. Daams notes resistance to the lack of black-balling at the NNGC, but I can find no evidence to support this. J.H. Daams, "Uit de geschiedenis van het Nederlands Natuur- en Geneeskundig Congres," *Van vonk tot vlam. 100 Jaar natuurwetenschap in Nederland* (Amsterdam: Rodopi, 1989), 13–23.

['Provincial Utrecht Society for the Arts and Sciences'].[32] This does not imply that the *Nederlandsch Natuur- en Geneeskundig Congres* was an entirely egalitarian affair. The members of the board and important dignitaries at the congress were expected to have distinguished themselves to some degree in science, and the appointment of such board members was not taken lightly.

8 The Formative Years

The first *Nederlandsch Natuur- en Geneeskundig Congres* was opened on 30 September 1887 in Amsterdam, and lasted for two days. Great care had been taken to circumvent the activities of other organisations, such as the yearly meeting of the teachers in secondary education, the *Nederlandsche Maatschappij tot bevordering der Geneeskunst*, and so on, in order to prevent any notion of competition. The event was altogether less extensive than had been the intention of the organising committee. Two exhibits that were supposed to be part of the congress – one with scientific instruments, another with books – had to be cancelled. But visitors to the event would have enough left to fill their time. Apart from the official programme, there was a closing dinner, of which they could partake at the (hefty) price of five guilders per cover. Furthermore, they were given a brochure pointing them to other sources of 'entertainment' in the city. During later editions, this fringe programme became more extensive, with excursions as well as exhibits, and separate events for participants and their spouses. Visitors to the meeting of the *Nederlandsch Natuur- en Geneeskundig Congres* in The Hague in 1903, for instance, received tickets to a performance in the Royal Theatre, a conference dinner, excursions, displays, and experiments. Furthermore, a map pointing them to interesting places in the city and a hotel reservations card were sent along with the conference material.[33]

In total 654 members and visitors attended the first meeting in the *Gebouw der werkenden stand* (Building for the Working Classes) in Amsterdam; the ranks of the *NNGC* would soon grow to about 1,000 and stay around that number until they began to rise again in the early twentieth century. Within a few years, the congress could be said to represent virtually the entire Dutch scientific community. The meeting defined how subsequent congresses

32 N.J. Singels, *Gedenkboek van het Provinciaal Utrechtsch Genootschap voor Kunsten en Wetenschappen, 1773–1923* (Utrecht: Oosthoek, 1923), 71*ff.*

33 *Handelingen van het negende Nederlandsch Natuur- en Geneeskundig Congres, gehouden te Den Haag op 27, 28 en 29 maart 1903* (Den Haag, 1903). NHA, Archive NNGC, Box 8.

were organised. After hearing Stokvis' introductory lecture, a short historical overview by Jan Campert, and two plenary lectures about the Beri-Beri disease by Cornelis Pekelharing and Cornelis Winkler, the members withdrew into separate, thematic sessions. The next day, the same order was repeated, with plenary lectures of a general and philosophical character, followed by more specialised talks in the thematic sessions, which consisted of between four and six lectures. This pattern was to remain largely unaltered until the First World War.

Scientists who wanted to address the meeting had to submit the title of their talk some time beforehand, and were expected to hand in the text at the meeting or soon after. A committee would go through the contributions on offer, and select those that it considered to be of the highest merit. From the first congress onwards, the *Handelingen* [Transactions] were printed shortly after the congress, and always before the next one. However, the transactions were not the only printed matter the attendees could enjoy: the organisation would provide a daily leaflet (*Bulletin*) with the latest changes in the programme, brief descriptions of the subjects treated on the previous day, proposals for the members' meeting, and so on. This order established itself fairly quickly, and little changed after the second congress, which was held in Leiden in 1889.

During those first years, the meetings toured the Dutch towns with a university or 'hogeschool' (college): Utrecht in 1891, Groningen in 1893, Amsterdam again in 1895, Delft in 1897.[34] In 1901, the congress landed in Haarlem, not a university town but a venue with a rich scientific history.[35] More interesting are the next two choices of venue: Rotterdam, the country's main port and of great economic significance, but also arguably a city without substantial scientific importance at this time; and The Hague, the political capital of the Netherlands. Just as interesting is that those location choices are not in any way reflected in the subjects of papers at the conference: the 1901 Rotterdam congress did not contain a larger number of papers devoted to practical applications of science that benefited industry and trade, for example (*see figure 7.1*). The opening addresses, on the other hand, made clear that the organisers were well aware of the chances offered by these venues.[36] The choice

---

34 See various publications of the *Handelingen van het Nederlandsch Natuur- en Geneeskundig Congres*. The 1897 congress took place before, but in anticipation of, Delft's promotion to full university status in 1905.

35 Haarlem was the seat of the country's oldest scholarly society, the *Hollandsche Maatschappij der Wetenschappen*, founded in 1752.

36 See *Handelingen van het achtste Nederlandsch Natuur- en Geneeskundig Congres* (Rotterdam, 1902) and *Handelingen van het negende Nederlandsch Natuur- en Geneeskundig Congres* (Den Haag, 1904).

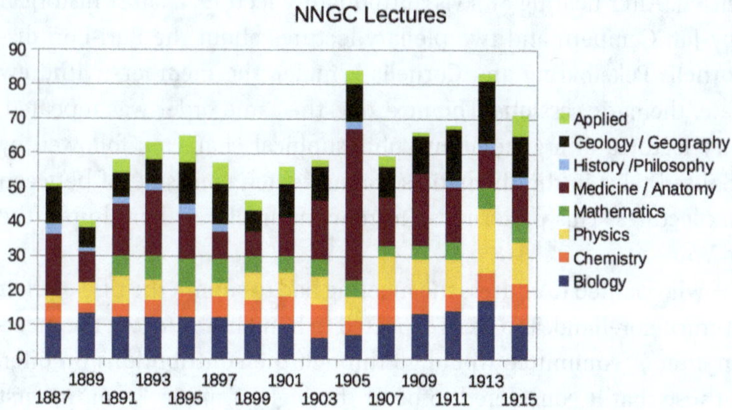

FIGURE 7.1 Thematic distribution of paper subjects at the meetings of the Nederlandsch Natuur- en Geneeskundig Congres, 1887 to 1915.
SOURCE: HANDELINGEN NNGC

for Arnhem in 1905 is more difficult to explain. It may have been a bridge too far for the Congress itself, because after this meeting we see the initial cycle of university towns return.[37]

9 Later Developments

The congress stayed true to itself when it came to its regulations. The first published set of regulations (1889) states that the aim of the organisation was 'to promote the flourishing of science and medicine in the Netherlands,' a phrase which was not altered until well into the twentieth century.[38] According to the regulations, the chosen means to do so was by organising meetings, and by 'promoting, encouraging, and funding of scientific research, preferably within the Netherlands and its colonies.' This granted the organisation wide powers for undertaking additional activities, something that was amply exploited. What *did* change was the content of the congresses themselves. Whereas the

37 Unfortunately, the deliberations of the organizing committee have not been preserved, which makes the rationale behind the choices of venue a matter of at least partial conjecture.

38 W.F.R. Suringar and A.W. Kroon, "Reglement van de vereeniging Het Nederlandsch Natuur- en Geneeskundig Congres," *Handelingen van het tweede Nederlandsch Natuur- en Geneeskundig Congres, gehouden in Leiden op den 26sten en 27sten april 1889* (Leiden: Vereeniging het Nederlandsch Natuur- en Geneeskundig Congres, 1889): XIII–XVIII.

physicians were responsible for about a third of the total number of contributions at the first congress, the prominence of medical subjects soon diminished. This happened to such a degree, in fact, that by the early twentieth century some concern was expressed about the way in which the medical disciplines were overshadowed by other disciplines. The board reacted by instituting a large number of medical subsections, much like the ones that were part of the structure of the *Gesellschaft deutscher Naturforscher und Ärtzte* meetings, which continued to be dominated by physicians. Consequently, at the 1905 meeting about half of all the contributions were of a medical character, where it had previously been about a quarter. On first inspection, this success appears to have been short-lived; at the next meeting, these sub-sections were mostly abandoned, and proportions returned to what they had been previously. The figures are however somewhat influenced by the fact that from 1907, the medical sections' meetings were organised differently from those of the other sections, with lengthier and more extensive contributions. A more realistic comparison may be the number of pages dedicated to each section in the transaction volumes, and here we see that the encouragement for medical contributions bore fruit for a longer time. The effect was nonetheless temporary, and overall the physicians' fears of being pushed aside by the other sciences appears to have been somewhat justified. Especially the rise in interest in sciences such as physics and mathematics (the first section) is conspicuous.

This situation is radically different from that in Germany. Within the *Gesellschaft deutscher Naturforscher und Ärtzte*, the medical disciplines never lost their grip on the situation (*see figure 7.2 for comparison*).

Consequently, those physicians and anatomists determined much of the discussion at the *Gesellschaft deutscher Naturforscher und Ärtzte*. Although the role of physicians such as Stokvis at the Dutch congresses should not be downplayed, their hold on the proceedings was not nearly what it had been in Germany. It could even happen that a board had no representatives from the medical profession – as was the case in 1903 – a situation that would have been hard to imagine at the *Gesellschaft deutscher Naturforscher und Ärtzte*. Another important difference was the relative lack of attention that was paid to applied science. Where military, tropical, veterinary and other practically applied forms of medicine flourished at the *Gesellschaft deutscher Naturforscher und Ärtzte*, they remained much less prominent at the meetings of the *Nederlandsch Natuur- en Geneeskundig Congres* – the exception being epidemiology.

Consequently, the *Handelingen* put much emphasis on 'pure' science. Out of the 854 lectures that were held at the meetings of the NNGC between 1887

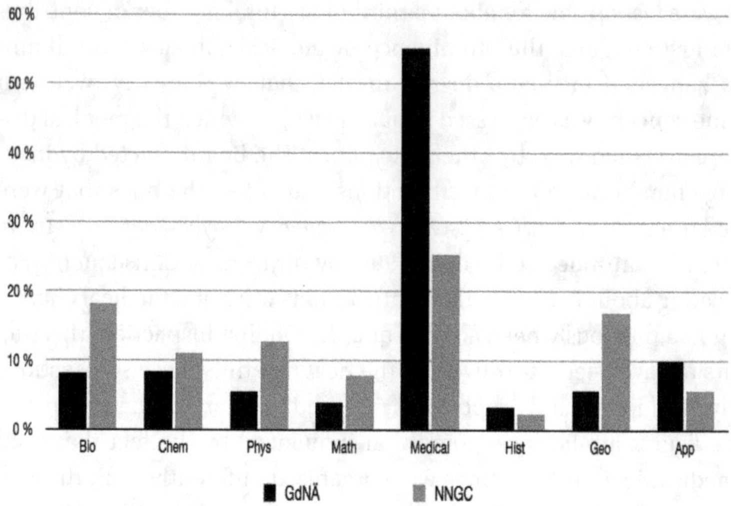

FIGURE 7.2 Cumulative relative thematic distribution of paper subjects at the Gesellschaft deutscher Naturforscher und Ärtzte (GdNÄ) and the meetings of the Nederlandsch Natuur- en Geneeskundig Congres (NNGC) compared, 1887 to 1915: Bio = Biology and natural history; Chem = Chemistry; Phys = Physics; Math = Mathematics; Medical = Medical and anatomical subjects; Hist = History (including history of science); Geo = Geology (including geography and palaeontology); App = Applied sciences.
SOURCES: TRANSACTIONS OF BOTH ASSOCIATIONS

and 1913, only 47 (or 5.5 percent) treated topics of applied science, usually in hygiene or agriculture (*figure 7.2*).[39] This number declined towards the end of the period of our investigation. This fact may seem noteworthy in view of the traditional emphasis on practicality and practicability within the history of Dutch science, but it is less surprising when one considers the missionary role of the congress in presenting the latest developments. Here, we see two phenomena that enhance one another. First of all, experimental research lent itself much better for showcasing than applied science. Secondly, the school of scientists that shaped this application-oriented tradition was already making way for a new generation of professionals by the time the Congress had become an established part of Dutch academic life.[40]

39 Statistics based on the *Handelingen*.
40 See also Theunissen, *'Nut En Nog Eens Nut'*; Ronald van Raak, *In naam van het volmaakte. Conservatisme in Nederland in de negentiende eeuw van Gerrit Jan Mulder tot Jan Heemskerk Azn.* (Amsterdam: Wereldbibliotheek, 2001).

10 Politics

Sometimes, a speaker at a congress might even broach a political subject. In 1895, Jacobus van 't Hoff, then in his second stint as chairman (he had already served in 1891), wondered which portion of the students in higher and secondary education was 'donated' by the Netherlands to its colonies. His conclusion – that the loss of about a third of the Dutch educated workforce to the colonies constituted a serious sacrifice – was certainly not politically uncontroversial. In any event, it was a break with the tamer and more mundane issues his predecessors had treated.[41] Rob Visser points to Johannes Bosscha's opportunism in his inaugural address of 1899 as though this were exceptional. But up to a certain point this type of lecture was an accepted forum to speak on behalf of the interest group one was representing. In addition to being technicians, Amsterdammers, or chemists, all these people had an interest in emphasising the achievements of science. In that respect, Bosscha did what was expected of him, much like Stokvis' and Van 't Hoff's celebrations of the inherent scientific talent of the Dutch, or Ambrosius Hubrecht's plea for higher standards in education. Interesting though they are as statements of faith, perhaps apart from Van 't Hoff's analysis of the colonial 'brain-drain' they contain few surprises, or, perhaps more importantly, convincing new insights.[42] On the other hand, that turns them into good indicators of the common opinion within the Dutch scientific community: these were representations of commonly held beliefs.

One modest surprise is in the number of contributions devoted to the link between science and medicine. Considering the origins of the Dutch scientific renaissance, such as it was, one might have expected this subject to be a central topic of discussion. On the other hand, the fact that it is treated in no less than 4 out of 14 addresses seems paradoxical in view of its apparent self-evidence. It is especially revealing that each of these explicitly treats the relevance of a scientific approach within the medical disciplines. There are more common factors, however, most of which evidently originate in this self-appointed spokesmanship on behalf of the scientific community. First is an emphasis on the importance of the nation as a beacon, also for the scientific community. Even when Van 't Hoff criticises the exodus of educated men to the colonies, he does so without misgivings towards those colonies; rather, he pleads for a

41 J.H. van 't Hoff, "Openingsrede," *Handelingen van het Nederlandsch Natuur- en Geneeskundig Congres* (1895), 1–9.

42 Visser, "Het 'Nederlandsch Natuur- en Geneeskundig Congres,'" 44–46. Also see Theunissen, *'Nut En Nog Eens Nut,'* 168–172.

more worked-out educational strategy to cope with the problem, in order not to further weaken the state of science in the fatherland.

11 Other Activities

The core activity of the *Nederlandsch Natuur- en Geneeskundig Congres* was, obviously, the organisation of the bi-annual congress and subsequent publication of its transactions. Nevertheless, from the beginning considerations were given to peripheral activities. The 'promotion, encouragement, and funding of scientific research' gradually began to assume more importance within the society. Since it had not been defined what such 'promotion' entailed, numerous possibilities were explored.

The first aim was the encouragement of scientific activity in the Netherlands and its colonies. The congress frequently donated major sums to friendly causes, and sometimes set up some of its own. The most famous (and probably most expensive) of these was the Siboga expedition. Sailing on the Dutch navy ship HMS Siboga, the marine zoologist Max Weber (one of the initiators of the congress and not to be confused with the sociologist of the same name) circumvented the Malayan archipelago for almost a year in search of deep-sea marine life, from 1899 to 1900. Although the expedition was largely funded by the *Maatschappij ter Bevordering van het Natuurkundig Onderzoek der Nederlandsche Koloniën* [Society for the Advancement of Scientific Research of the Netherlands' Colonies], the Ministry of Marine Affairs and the *Nederlandsch Natuur- en Geneeskundig Congres* also contributed substantial sums. This was mostly thanks to congress' treasurer Coenraad Kerbert, director of the Amsterdam Zoo Artis and one of Weber's closest friends.[43] In 1899, Weber sent some of his preparations to be published in the *Handelingen*, but he seems to have been unable to follow up on his promise to send more information at a later date. Nonetheless, the expedition was generally considered to have been a success, and Weber published prolifically afterwards, also in the congress's publications. But this was certainly not the only case in which research was partly or entirely funded by the congress.

The amount of Weber's subsidy (5,000 guilders) may have been somewhat higher than that of most, but sums varying between two hundred and a

43 The correspondence between Kerbert and Weber is held at the archives of the Zoological Society *Artis Natura Magistra*, Stadsarchief, Amsterdam. See also: Klaas van Berkel, "Het late begin van de Nederlandse wetenschappelijke expedities naar Indië," *Groniek* 186 (2010): 35–49, esp. 36.

thousand guilders were frequently awarded to research projects. Investments were also made to improve scientific infrastructure. In 1895, the congress supported the construction of a new zoological station in Den Helder by the *Nederlandsche Dierkundige Vereeniging* with the sum of a thousand guilders, thereby returning something to one of its founders.[44]

The *Nederlandsch Natuur- en Geneeskundig Congres* further attempted to raise interest in the history of various fields of science. Before the first congress, as we saw, time turned out to be too brief to organise a planned exhibition of scientific instruments and books. The historical commission of the NNGC, instituted in 1906, can probably best be regarded as a part of the wave of 'cultural nationalism' that swept over the Netherlands and indeed most of Europe at this time.[45] It was certainly involved with casting light upon the scientific past of the Dutch nation, and most of all that of its 'Golden Age' in the seventeenth century. Although history may have played a minor part in the contributions at the congresses themselves, it was a major concern in the 'extracurricular' activities of the society. Its most significant effort in this respect was the organisation of an historical exhibition at the occasion of the 1907 meeting in Leiden. The exhibition, held at the *Lakenhal* art museum, presented a somewhat eclectic collection of objects, mostly scientific instruments collected from all over the country.[46] Such was the success of this initiative that calls were heard for the founding of a national museum dedicated to the history of science, leading to the initiative for a *Vereeniging voor Geschiedenis der Genees-, Natuur-, en Wiskunde* [Society for the History of Medicine, Science and Mathematics] seven years later.[47] Another such initiative, although obviously influenced by other considerations, was a Darwin commemoration held at the Amsterdam zoo Artis in 1909. Generally speaking, the congress regularly contributed to projects that sought to enhance knowledge about the past of science in general and Dutch science in particular.

Another field in which the *Nederlandsch Natuur- en Geneeskundig Congres* made efforts to improve infrastructure was in providing access to scientific literature. First of all, this was done through the publication of a summary of 'that which has been written by Dutchmen in the field of physics' in the *Handelingen*, beginning in 1897. In the same year, a library commission was set up with the task to 'create co-operation between administrators, libraries, and

---

44 NHA, Archive NNGC, Box 13.

45 Van Berkel, "Dirk Huizinga."

46 *Catalogus van de geschiedkundige tentoonstelling van natuur- en geneeskunde, te houden te Leiden 27 maart-10 april 1907* [*etc.*] (Leiden: Sijthoff, 1907).

47 David Baneke, "De rol van een genootschap. Honderd jaar Gewina in het wetenschapshistorische landschap," *Studium* 6:3–4 (2013): 132–148.

those societies to which libraries are attached.'[48] Although the commission contained eminent scientists such as Barend Stokvis, Ambrosius A.W. Hubrecht, and Antoine P.N. Franchimont, it seems to have had little success at first. As its member Jan Frederik van Someren complained in 1898:

> The means, which have been chosen at [the] meeting for achieving our goal, are not very well chosen and ineffectual. [...] It seems to me that this way, we are putting the horse behind the cart.[49]

Van Someren's problem was that no funds were made available to the commission. He proposed to allocate the impressive sum of 3,000 guilders to an inventory of books and journals in society libraries, and to apply for a government grant in order to cover some of the cost. The problem was taken seriously, since the lack of clarity about the availability – or lack thereof – of scientific reading matter had been the main incentive to set up the commission in the first place. After the commission reported to the meeting, it was allocated a sizeable sum, but was also burdened with the responsibility to ask for a government grant, an undertaking whose potential success had earlier been questioned by Van Someren.[50] From 1898 onwards, the *NNGC* distributed a number of lists of society publications that were not available in public libraries, urging them to improve their scientific collections and to make them publicly accessible.[51] Like with the list in the *Handelingen*, it is significant that the library commission was focused on the purchase of specifically Dutch scientific material, following the regulations of the congress to the letter.

12 Offspring: Peripatetic Congresses in the Dutch East Indies and Flanders

Being the offspring of a host of European national congresses, the *Nederlandsch Natuur- en Geneeskundig Congres* itself fostered two more. The latter of these, the *Vereeniging Nederlandsch-Indisch Natuurwetenschappelijk Congres* ['Dutch East Indies congress of the Sciences'] was founded in 1918, but this organisation

48 Minutes of the library commission, 26 july 1897. NHA, Archive NNGC, Box 9.
49 Van Someren to the library commission (s.d. 1898). NHA, Archive NNGC, Box 9.
50 Rapport van de Bibliotheek-commissie, *Handelingen* 1899, 73–83.
51 Lijst van genootschapswerken en tijdschriften, welke op Ulto 1901 in de Openbare Bibliotheken in Nederland ontbraken (1902). NHA, Archive NNGC, Box 9.

remained a relatively small-scale operation.[52] The nomadic nature of the congress also remained limited to the main island of the Indonesian archipelago, Java. The first congress in 1919 was held in the capital Batavia (modern Jakarta), and until World War II further congresses were held in Bandung (1922 & 1931), Buitenzorg/Bogor (1924), Weltevreden (an outskirt of Jakarta, 1926), Soerabaja (1928 & 1938) and again Batavia (1935). Eventually, the calamitous events of the Japanese occupation, and then the Indonesian War of Independence, ended this colonial enterprise.

The *Vlaamsch Natuur- en Geneeskundig Congres* ['Flemish congress of Nature and Medicine' or VNGC] came into existence much earlier, in 1897, and found its origin in partisan rather than scientific motivations. The main initiator of the organisation, the Ghent physician Julius MacLeod, was also a well-known Flemish nationalist, whose intentions were only too evident:

> We ought to create a scientific movement with Dutch as its current language, a movement that could become a powerful factor in rising the masses outside of politics, not only because she were Flemish, but also because she was to remain strictly scientific in her ambitions.[53]

What MacLeod writes here is not only that scientific labour will help to uplift the repressed Flemish people by educating it, but also that the effect of that labour will unite the Flemish as a nation through the definition and discovery of a common goal. The nationalist agenda becomes even clearer when one looks at the background of some of the more prominent members of the VNGC.[54]

---

52 Compare Marius Jacobs, *Herman Johannes Lam (1892–1977). The Life and Work of a Dutch Botanist* (Amsterdam: Rodopi, 1984), 29.

53 MacLeod speaking to Staes and Vandevelde, according to the recollection of Vandevelde; A.J.J. Vandevelde, *Het Vlaamsch natuur- en geneeskundig congres van zijn oorsprong in 1897 tot in 1944, historische schets* (Brussels: Paleis der Academiën, 1944), 11. When opening the 16th edition of the Flemish congress, the linguist Gustaaf Verriest expressed similar sentiments: "… large part of the Flemish students, particularly those from the countryside, have a strange language as a mist around their minds, hindering acute reasoning and thinking…. Were we to conquer the Flemish University, a bright light will again ray in the young heads, and enable them to compete with their brethren in the North, whose scientific force is regarded so highly by all the civilised peoples of the world.". See G. Verriest, Opening address at the XVIth Flemish Congress, 1912. Letterenhuis Antwerpen (Archief en Museum voor het Vlaamse Cultuurleven), Archive Verriest, V483/H.

54 There can be some discussion about the way in which the Flemish movement was 'nationalist' at this time, since it was primarily aimed at the emancipation of the Flemish language and culture, rather than the establishment of an independent political entity. The open desire for a separate, Flemish state did not really become a widespread ideal until later, particularly during and after the First World War. See de Ridder, *Ouverture 1912*.

Julius MacLeod, Gustaaf Verriest, Jan de Windt, Herman Sabbe, and Cesar de Bruyker were all well-known activists for the Flemish cause.[55]

During the union with the Netherlands between 1815 and 1830, Flanders first witnessed a bilingual or even purely Dutch-language education, but this situation was reversed once the Dutch had been definitively ousted in 1831. Although a new secondary education bill in 1883 allowed for the use of Dutch in schools, the number of exemptions was such that the effect of the bill remained far below the *Flamingants'* expectations and desires, leading to an ever more embittered struggle with the almost exclusively francophone education authorities. Most importantly, the use of Dutch was still not permitted at universities, making them (and other institutes of higher learning) an obvious target for Flemish agitation.

Structurally, the *Vlaamsch Natuur- en Geneeskundig Congres* started out as a copy of the *Nederlandsch Natuur- en Geneeskundig Congres*, but it soon developed its own identity. An important distinction between the two meetings was the fact that the Flemish congresses took place annually instead of every two years. Moreover, the programme of peripheral activities around the meetings quickly grew into something decidedly lusher and more extensive than what took place at the Dutch gatherings. For instance, at the seventeenth meeting of the *VNGC* (held 1913 in Ghent), visitors could choose between the following activities: a 'musical feast' with the work of Flemish composers, a grand performance of Peter Benoit's huge patriotic oratorio *De Schelde* [Scheldt River], a historical guided tour through the city of Ghent, an excursion to (Flemish-language) schools and, not to be outdone, a banquet consisting of truly dionysian culinary excesses.[56] By comparison, the Dutch meetings could almost be described as spartan.

This difference in character can probably be traced partly to the origin of the participants. The Flemish meetings were decidedly less elitist than their northern counterparts. If we look at the membership lists for 1910, we see that out of a total 424, well over half (251 to be exact) list themselves as *geneesheer* (doctors) and twenty-one as pharmacists, whilst a mere twenty-two (about five

---

Literatuur en Vlaamse Beweging aan de vooravond van de Grote Oorlog (Antwerp: AMVC, 2008), 28 *et passim*.

55 See G. Lannoo, "Vlaanderen waarheen?: Een terugblik op 100 jaar Vlaamse Beweging," *Nieuwsbrief – tweemaandelijks ledenblad van de Orde van den Prince*, 24–3 (2005): 5–6. For the Flemish movement, also see De Ridder, *Ouverture 1912*.

56 *Programma van het XVIIe Vlaamsch Natuur- en Geneeskundig Congres. Gent, 20, 21 en 22 September 1913*. Letterenhuis Antwerpen, Archive Vlaams Natuur- en Geneeskundig Congres, inv. C338973/D1.

percent) cite various academic titles.[57] 'Geneesheer' in Flanders could mean anything from an old-school barbershop hacksaw to an academically-trained medical doctor, but it seems unlikely that the *medicinae doctores* would allow themselves to be lumped in with their less sophisticated counterparts.

The final important difference concerns the balance between scientific and political motivations. As evidenced by the comparatively limited scientific qualifications of the attendants of the *Vlaamsch Natuur- en Geneeskundig Congres*, their meeting (and the peripheral programme) was motivated by a desire to further Flemish identity and emancipation, preferably but not necessarily in combination with Flemish science. Despite Stokvis' referral to 'Dutch science' at the first congress, political subjects were rarely, if ever, treated at the Dutch gatherings.[58] Julius MacLeod, without question the most important force behind the Flemish congress and in more than one way a Flemish version of Lorenz Oken, had been involved with the *Nederlandsch Natuur- en Geneeskundig Congres* since its foundation, providing information to the Dutch about Flemish science and scientists.[59] In Flanders, politics was central, but implicitly rather than explicitly: in the choice of subject, cultural programmes, or the venues visited on field trips. During talks, off-the-cuff remarks would remind listeners of the speaker's ideological position.

Exchanging delegations to one another's meetings further enhanced the connections between the neighbouring organisations. From the outset, the Dutch congress has interest in involving the Flemish scientific community, and they kept close contact with the Dutch as well. This was reflected in the number of Dutch scientists who visited the meetings (around seven percent of attendance) as well as in their prominence. When the Groningen surgeon Hartog Jacob Hamburger spoke at the Flemish Congress of 1907, for instance, the event was very prominently advertised.[60] However, the number of Flemish scientists who visited meetings of the *Nederlandsch Natuur- en Geneeskundig Congres* remained almost negligible, suggesting once more that among the Flemish scientific community there was little interest in attending such meetings for their own (scientific) sake.[61]

---

57 Archive Vlaams Natuur- en Geneeskundig Congres, inv. 338973/II.

58 And patriotism in itself was hardly a politically contentious topic, of course. Historical and political themes make up just over two percent of all lectures held at the congresses between 1887 and 1915 (21 out of 925), and are virtually absent after the turn of the century. NHA, Archive NNGC, Boxes 6 and 8.

59 Minutes of the First NNGC, 2 May 1887. NHA, Archive NNGC, Box 5.

60 *Programma van het Elfde Vlaamsch Natuur en Geneeskundig Congres. Mechelen, 21, 22 en 23 september 1907*. Archive Vlaams Natuur- en Geneeskundig Congres, inv. C338973/D1.

61 Archive Vlaams Natuur- en Geneeskundig Congres, inv. 338973/II.

13 The Antwerp Congress of 1915

Mingling science and politics could have its disadvantages, however. Within the *Vlaamsch Natuur- en Geneeskundig Congres*, the priority of its goals was also not always expressed clearly: did it primarily serve a political or scientific goal? The problems created by this ambiguity showed themselves when the Dutch representatives Derk Coelingh (then *NNGC* secretary) and Hendrik Burger were approached at the Flemish congress in Sint Niklaas (1908) by Camille De Bruyne, Professor of Botany at the University of Gent. De Bruyne suggested they organise a future *Nederlandsch Natuur- en Geneeskundig Congres* meeting somewhere in Flanders – a Dutch meeting on Flemish territory. Coelingh further discussed the matter at the Groningen congress of 1911 and was able to report encouraging enthusiasm on the part of the Dutch to the two Flemish delegates, Albert J.J. Vandevelde from Gent and R. Wybauw from Spa. A meeting was arranged in Oostende to further discuss the possibility for a *NNGC* congress in Antwerp in 1915, but there the initiative came to a halt. Even though the Flemish, as we have seen, aspired to mimic the scientific prestige of the Dutch, their ulterior motive was more political, and would lead them into a vortex of misunderstanding with their northern counterparts.

Despite initial enthusiasm, the Flemish soon had to backpedal once it became apparent that within the *Vlaamsch Natuur- en Geneeskundig Congres* there was anything but unity with regard to the shape of such a meeting. At the last instance, the get-together in Oostende was cancelled because of the absence of Alphonse Van de Perre, who had previously been chairing a joint meeting of the *VNGC*, the Flemish lawyers' association, and the congress for Flemish language and history. Van de Perre, operating on an openly nationalist agenda, later spoke to Kerbert and Coelingh, suggesting a combined meeting of the *VNGC* and *NNGC* rather than just a *NNGC* session in Antwerp.[62] Among the Dutch, however, a Dutch congress was preferred over a joint effort, and soon little of the initial eagerness remained within the *Nederlandsch Natuur- en Geneeskundig Congres*. Vandevelde evidently felt somewhat embarrassed by the conduct of his compatriots, as he confessed to Coelingh in March 1912:

> Personally, I highly regret, as do several of my colleagues, that the magnificent prospect to hold the Congress in Antwerp may [...] not be realized. Because both Congresses do not have the precise same significance and organization, several among us also do not consider a merger as desirable.

62 Sixth meeting of the 13th congress, 30 December 1911. NHA, Archive NNGC, Box 5.

> [...] I cannot sufficiently express my regret about what has happened and I consider this as unfortunate for the Flemish cause.[63]

It is revealing that even though Vandevelde does not agree with the practices of his colleagues, he appears to be primarily motivated, as they are, by the positive result a congress could have for his political ideas, rather than by the virtues of the scientific meeting itself. A similar attitude regarding the use of the congress for political ends can be found in De Bruyker's letter to Coelingh from August 1912, in which he calls off the whole thing:

> Presently, we are confronted with a decision by our Antwerp friends, who [...] are unwilling – on grounds of principle – to initiate a purely Dutch Congress, at which the Flemish element would obviously remain in the background. [...] The decision in Antwerp and the reasons behind it should by themselves be good reason to ask you to abandon the plan at least for the time being.[64]

While one can be surprised about the blatant seizure of the meeting of the *Nederlandsch Natuur- en Geneeskundig Congres* by the Flemish for their political cause, the Dutch naiveté is also revealing. They appear to have been entirely blind to these motives.

14 Conclusion

It was no coincidence that, when he opened the first meeting of the newly-founded *Nederlandsch Natuur- en Geneeskundig Congres*, the Amsterdam physician Barend Stokvis spoke about "nationality and natural science."[65] Because as much as the 'classical' societies were an expression of eighteenth-century Enlightenment, it was the nineteenth-century prominence of the nation that may be regarded as the main inspiration for the establishment of the nomadic societies that sprung up around Europe. Of course, other aspects were as important, such as the opportunity to meet colleagues and to debate issues with peers, rather than with a collection of interested dilettantes. But when looking at the form of the nomadic congresses, it is apparent that their reason for

63 Vandevelde to Coelingh, 12 March 1912. NHA, Archive NNGC, Box 12.

64 De Bruyker to Coelingh, 5 August 1912. NHA, Archive NNGC, Box 12.

65 Stokvis, *Nationaliteit en natuurwetenschap*; also reprinted in the transactions of the first congress.

existence, at the outset at least, lies at least as much in nationalist sentiments as it does in considerations of scientific collaboration.

In the introduction, we pointed to the fact that the societies involved can be roughly divided into two categories, both of which are coloured by local nationalist considerations. The first one is most clearly defined by the *Gesellschaft deutscher Naturforscher und Ärtzte*: an organization primarily politically motivated, which at least initially used science as a tool to attain other ends. That does certainly not mean that scientific activity was irrelevant, but it does mean that many decisions about the form, location and audience of such meetings were made with other considerations in mind. The rotation of locations of the venues was crucial for its political aims. Movement was necessary to claim authority as nation of science. The second category was of the type of the *British Association for the Advancement of Science*, and encompassed those congresses that served to display of a nation's scientific prowess, and to forge a national community of scholars. Here the movement of the venues served primarily the aim of showcasing science to the nation. The Flemish congress can be put in the first category, the *Nederlandsch Natuur- en Geneeskundig Congres* clearly in the second group, which may explain something about the fact that despite ostensibly similar aims, the level of understanding between both organizations proved to be so unsatisfactory.

The general, immediate and enthusiastic acceptance of the *Nederlandsch Natuur- en Geneeskundig Congres* as a force that aspired to return Dutch science to its once glorious past is revealing, since it shows that a need for such an organization was clearly felt. On the other hand, so is the fact that it obviously was not needed for a considerable time. Yet, as soon as it came into existence, the *Nederlandsch Natuur- en Geneeskundig Congres* proved to be a powerful supplement to Dutch scientific life, and one that reshaped parts of it. It strove to be a truly national organization: both by representing the nation itself and by showing itself to it and thereby evangelizing what it stood for. And for a long time, it aimed to remain inclusive: as we have seen, it is difficult to categorize the practices of the congress as evidence for an increase in scientific professionalization without major reservations.

CHAPTER 8

Between Openness and Classification: Early Dutch Ultracentrifuge Development in International Context

Abel Streefland

1 Introduction

There probably was no time in history that people were more inclined to think in geographical terms than during the Cold War. If you were not from the West, you were from the East. These geographical indications were linked to political preferences. The Red Threat always came from the East and capitalism was Western. Having contact with Russians was similar to being a communist, to say it very bluntly. The term 'geopolitical' made this explicit: linking geography with politics. These sentiments were prominently felt in the United States and in Russia. In the Netherlands, they played a less prominent role, at least in the early Cold War period. The Dutch were more focused on rebuilding the country after the devastations of the Second World War than fighting a red threat. When this succeeded, and the Dutch slowly found their place in the geopolitical arena of the Cold War, Cold War sensitivities became factors to be taken into consideration.

After World War II, the locations of scientific hotspots had shifted. The US had become dominant in the world of nuclear physics, followed by Russia that had the will- and manpower to achieve complicated scientific goals (like building an atomic bomb) in a short span of time. For smaller countries in Europe, this was a time in which they could newly position themselves. Although the Dutch natural sciences had bloomed in the first decades of the Twentieth Century (the number of Nobel Prizes triggered the period to being named the 'Second Golden Age'), after the war not much was left. Some of the old institutes were able to resume successful research, but in general the field of physics could be redefined, leading to the foundation of several new governing institutions.

In the meantime, international contacts became more and more important. After all, knowledge was to be gained or bought from countries whose research was more advanced. Such efforts, in turn, did not always gain a favourable reception. Most countries, especially the US, were anxious about proliferation

 | DOI:10.1163/9789004264885_009

of nuclear knowledge and resorted to knowledge regulation and secrecy. After the war, the Dutch knew almost nothing of the advancements of the nuclear sciences in both the East and the West.

In this essay I want to take a look at the ways in which scientists dealt with these new kinds of problems. How could a small country like the Netherlands define its own scientific place in between the grand geopolitical forces of the US and Russia? And, on a much smaller scale, what was the influence of the early Cold War on scientists? What techniques were there to exchange scientific knowledge? What kind of scientific interaction was possible? In the first few sections I shall briefly sketch the field of nuclear science since the end of the Second World War. After that I will focus on the ultracentrifuge technique that could be used to separate uranium isotopes. This technique was developed in Amsterdam from 1954 onward. Focusing on an important symposium organized in Amsterdam in 1957 we will see that East-West sentiments started to play an important role in small-scale scientific practice. The end of this symposium was the stage of a dramatic and historically important exchange of knowledge, the implications of which still resonate in the present-day uranium enrichment branch.

2 Nuclear Science

Almost every account of Dutch nuclear science starts, perhaps surprisingly, before the dropping of the atomic bombs on Hiroshima and Nagasaki and even before the beginning of the Second World War.[1] An impulsive decision made by physicist Wander de Haas in 1939, just after Fritz Strassmann and Otto Hahn had discovered the fissionability of atoms, was the starting point of the Dutch atomic age. In the spring of 1939, Leiden-based De Haas went to The Hague to meet Prime Minister Hendrikus Colijn to talk about the acquisition of uranium ore from Belgium (mined in Belgian Congo). A few weeks later, 200 barrels filled with so-called yellowcake arrived from Antwerp in Leiden. During the war the uranium ore was kept hidden in Delft and was not discovered

1 J.A. Goedkoop, *Een kernreactor bouwen: geschiedenis van de Stichting Energieonderzoek Centrum Nederland Deel 1: periode 1945–1962* (Bergen: BetaText, 1995); Ad Maas and Hans Hooijmaijers, *Scientific research in World War II: what scientists did in the war* (London etc.: Routledge, 2009); Jaap van Splunter, *Kernsplijting en diplomatie: de Nederlan[d]se politiek ten aanzien van de vreedzame toepassing van kernenergie, 1939–1957* (PhD thesis, University of Amsterdam, 1993).

by the occupying forces.[2] After the war most of the yellowcake was traded with the British in exchange for uranium plates that were rejected for the purpose that the British had for them. This sample of uranium proved crucial for the position of Dutch nuclear science in the post-war scientific climate in Europe and beyond.

After the Second World War the world entered the nuclear age. Nuclear scientific plans popped up like mushrooms: plans to build a reactor, to enrich uranium, and in some countries to build atomic bombs. In the Netherlands, plans of this kind led to the founding of the Foundation for Fundamental Research on Matter (FOM). The name of the foundation is misleading, especially in English. There was nothing purely fundamental about the research proposed at FOM, where no foundational quantum mechanics or relativistic physics was done. On the contrary, the science that was done at FOM was what could probably best be described as normal science. Not purely applied, not purely foundational, but the kind of practical physics that always works towards a specific goal.[3]

One of the major early projects of FOM was the development of the Institute for Nuclear Research (IKO) led by Cornelis J. Bakker. Its main goal was to finish the construction of a particle accelerator, a so-called cyclotron. The Dutch electronics company Philips already started working on the cyclotron during the war and gave the designs to IKO under the condition that daughter company Philips-Duphar could use the finished machine overnight to produce mostly clinical radioactive isotopes. The machine was designed to accelerate light elements in a circular way by means of a strong magnetic field. During every revolution of the atoms, they would be accelerated to a slightly higher speed making the cyclotron much more powerful than the linear Cockcroft-Walton-machine that was built at the Vrije Universiteit (VU) at the other end of Amsterdam.[4]

Another major FOM project was the construction of machines that could separate isotopes. In the eastern part of Amsterdam, a few kilometres away from IKO, the Laboratory for Mass Spectroscopy emerged in an old electricity plant that belonged to the community of Amsterdam. In the great hall of the plant a huge electro-magnetic isotope separator was built by the physicist

---

2 See for example Dirk van Delft, "Preventing Theft: The Kamerlingh Onnes Laboratory in wartime," in: *Scientific research in World War II*, eds. Maas and Hooijmaijers.

3 See for the early history of FOM Friso Hoeneveld and Jeroen van Dongen, "Out of a Clear Blue Sky? FOM, The Bomb, and The Boost in Dutch Physics Funding after World War II," *Centaurus* 55:3 (2013): 264–293.

4 Goedkoop, *Een kernreactor bouwen*, 15.

Jaap Kistemaker and his co-workers.[5] Positive ions would be curved by a strong magnetic field into a semi-circular orbit. The radius of this circle depends on the mass of the ions. This principle makes the isotopes (which have slightly different masses) separable.[6]

In the first years, Kistemaker failed to inform his co-workers that the goal of the electro-magnetic machine was to separate uranium isotopes and thus produce enriched uranium.[7] Some of the first isotopes that were separated were neon-20 isotopes, which could be used for spectral research. Different isotopes were sold to various institutes and companies.[8] After some time, Kistemaker's co-workers came to realize that the actual goal of building the machine was to enrich uranium. On 6 November 1953, Kistemaker's lab was able to present the first 10 milligrams of enriched uranium to the board of FOM. It was approximately enriched with a factor of 10 (which means that there was 10 times as much U-235 in the sample as is present in natural uranium). The sample was sent to Norway.[9]

The reason it went to Norway was that another dream of Dutch post-war physicists was realized some 20 kilometres from Oslo in the village of Kjeller. As the Norwegians were in possession of heavy water and the Dutch had their uranium, this could well be combined to build a nuclear reactor. Heavy water could after all be used as a moderator in a uranium-based reactor. In 1951 the Joint Establishment for Nuclear Energy Research (JENER) was established between Dutch and Norwegian physicists and diplomats. The Norwegian scientists had already been working on a reactor since 1950. The Joint Establishment

---

5 Jaap Kistemaker (1917–2010) was a farmer's son from Kolhorn, a small village in the north of Holland. In 1935 he matriculated at the University of Leiden for a study in mathematics and physics. During the Second World War he was an assistant at the physics laboratory, founded by the Nobel laureate Heike Kamerlingh Onnes. Kistemaker obtained his PhD in November 1945 with a thesis on the thermodynamic properties of helium at very low temperatures. After working for half a year at the Niels Bohr Institute in Copenhagen to gain knowledge in the field of magnetic separation of uranium isotopes, he joined the Dutch Foundation for Fundamental Research on Matter (FOM) in 1947. Two years later Kistemaker became director of the FOM laboratory for Mass Spectrography (the current AMOLF), which institute he managed from 1949 until his retirement in 1982. Kistemaker was also curator of the Physics Cabinet of the Teylers Museum in Haarlem and from 1956 to 1984 professor at Leiden University.

6 Cees J. Zilverschoon and Cornelis J. Bakker, *An electromagnetic isotope separator* (Zilverschoon's PhD thesis, University of Amsterdam, 1954).

7 Interview by the author with Cees Zilverschoon, 21 March 2011.

8 Ibidem.

9 Jaap Kistemaker, *De Geschiedenis Van Het Nederlandse Ultracentrifuge Project* (Amsterdam: FOM-Instituut voor Atoom- en molecuulfysica, personally circulated by the author, 1991), 5.

Experimental Pile (JEEP) became critical for the first time on 30 July 1951. This was the first time that 'small' countries succeeded in building a nuclear reactor.[10]

The uranium that Kistemaker had sent to Norway in 1953 was used in the JEEP reactor to check the cross section of uranium for low energies.[11] The sample consisted of uranium oxide on an aluminium plate. After the uranium was purified it was used to determine the cross section of the atom as a function of the velocity of the neutrons originating from the reactor. This was highly important research. The United States Atomic Energy Commission (USAEC) had in 1952 only revealed the value of this cross section for one particular neutron velocity. The researchers at JEEP calculated the cross-section for a variety of velocities. Kjeller-based physicist Teun J. Barendrecht recalled: "On the first nuclear 'non-classified' symposium in Ann Arbor, Michigan (June 1954), Popovic presented the results. I remember the big consternation with the Americans; it was after all 'classified'."[12]

3 Atoms for Peace and the Geneva Conferences

Around the time that Kistemaker presented the first sample of enriched uranium to the board of FOM, the Americans announced an important change in policy. The monopoly that the US originally had on nuclear weapons was, after all, already lost (to Great Britain and Russia) and the US was still by far the biggest producer of isotopes in the Western world. These fissionable materials were not available outside the USAEC. The newly elected Republican president Dwight D. Eisenhower decided that the time was ripe to change this. Financial as well as political arguments guided this decision.[13] If the Dutch could produce pure isotopes and use them for economic purposes, the US should

---

10 The question which countries could be labeled 'small' is in my opinion not that useful. It is striking that nor the Dutch, nor the Norwegians could build a reactor on their own. The French had built their ZOE reactor around 1950. ZOE is an acronym for *Zéro, Oxide, Eau Lourde*. The reactor in Kjeller was regarded as superior to the French reactor, although the two had a lot of similarities. Goedkoop, *Een kernreactor bouwen*, 21.

11 Dragoslav Popović, "Energy Dependence of the U235 Fission Cross Section in the Low Energy Region: Preliminary Report," *Journal of Nuclear Energy (1954)* 1:1/2 (1954): 3–4.

12 Personal memory of Kjeller-based physicist Teun J. Barendrecht: "Op het eerste nucleaire 'non-classified' symposium in Ann Arbor, Michigan (juni 1954) heeft Popovic het resultaat medegedeeld. Ik herinner me de grote opschudding bij de Amerikanen; het was immers 'classified.' De bepaling geschiedde in Kjeller, eerder dan in Frankrijk," in: Kistemaker, *Geschiedenis van het Nederlandse UC Project*, 5.

13 See for example: Goedkoop, *Een kernreactor bouwen*, 35; John Krige, "Atoms for Peace, Scientific Internationalism, and Scientific Intelligence," *Osiris* 21:1 (2006): 161–81; Splunter, *Kernsplijting en diplomatie*.

also make a profit out of this. In December 1953 Eisenhower presented his famous Atoms for Peace program at the General Assembly of the United Nations in New York. An important part of the program was the permission to possess fissionable materials for (US) companies and institutions. Furthermore, the program also made it possible for other countries to set up bilateral contracts with the US to gain, under special regulations, both nuclear knowledge and materials. Atoms for Peace was meant to open up the nuclear market.[14]

The Atoms for Peace program made the atom a public symbol. In the Netherlands, for example, this development was strengthened by the 1955 American exhibition *The Atom, hope for the future*, which came to six Dutch cities to promote the usage of radioactive isotopes and nuclear power.[15] John Krige aptly contrasted the European interest to the overseas interest:

> In countries whose scientists and governments had an interest, not in preparing for a third world war but in rebuilding themselves on the ruins of the second, the atom was an opportunity, a symbol of modernity and a better world to come, nuclear power a promise for energy and independence.[16]

A direct result of Eisenhower's Atoms for Peace program was a conference that took place in Geneva in August 1955, under the auspices of the United Nations. The worldwide interest in this conference was huge: more than 1400 scientists from 73 countries occupied the halls of the Palais des Nations for two weeks. American firms were lining up to take orders to build reactors. The cross-pollination of knowledge and materials was encouraged in every front. The Dutch delegation consisted of 33 scientists and was led by Cornelis J. Gorter. All of the members of the delegation were in one way or another linked to the recently founded Reactor Centre of the Netherlands (RCN) or another branch of the Dutch nuclear effort. In Geneva, valuable contacts were laid with scientists and corporations from all over the world.[17]

4 Ultracentrifuges

In the following sections, we will focus on the ultracentrifuge technique for separating uranium isotopes. British physicists Frederick Lindeman and future

14 Splunter, *Kernsplijting en diplomatie*, 156.
15 Goedkoop, *Een kernreactor bouwen*, 46.
16 Krige, "Atoms for Peace," 170.
17 Goedkoop, *Een kernreactor bouwen*, 51.

Nobel-prize winner Francis Aston first proposed to use centrifuges for isotope separation in 1919.[18] The principle of the centrifuge is simple. A spinning tube (rotor) should spin fast enough to separate the isotopes in the gas inside this rotor. Heavy particles tend to go to the outside of the rotor; light particles stay in the centre. Experiments with isotopes in the early years nonetheless proved to be unsuccessful. Big temperature differences on the rotor wall due to friction with the surrounding air would cause convective currents in the gas which would result in a mixing of the gas. Any separation that would have occurred would then be undone. It took until 1934 before Jesse Beams in Virginia succeeded in separating the isotopes of chlorine by using a centrifuge placed in a vacuum, thus diminishing the problem of convection.

Within the American atomic bomb project during the Second World War, ultracentrifuges were seen as one of the candidates for enriching the uranium necessary for the development of the bombs. Beams became the leader of the centrifuge project. In August 1943 the first uranium separation tests using ultracentrifuges began. They ran for 93 days before the oil gland between the vacuum and the rotor developed a leak that made the centrifuge crash at 215 meters per second. A few weeks later Harold Urey, who headed the enrichment project, terminated the program, choosing other enrichment techniques above the centrifuge.[19]

After the war, a report on the Manhattan Project was distributed. Several copies of this so-called Smyth Report reached the Netherlands. Henry Smyth mentioned, in addition to some general aspects of the theory behind isotope separation (such as separation factor, efficiency, cost), almost all techniques imaginable for separating uranium isotopes. Although the reasons were kept vague, the diffusion method was chosen as the 'winner.' In the end this technique provided the enriched uranium necessary for the uranium bombs developed by the Manhattan Project. The centrifuge technique was mentioned in the report as a promising alternative to the diffusion technique and was "most likely to succeed with uranium."[20] It was, as we have seen above, nevertheless abandoned. In the report it was not made clear why the US researchers did not pursue and expand this technique further.

Kistemaker got the Smyth Report in October 1945 for a total of 24 hours from Hendrik A. Kramers, his supervisor. "The information in this booklet sparked a

18 Frederick A. Lindemann, and Francis W. Aston, "The Possibility of Separating Isotopes," *Philosophical Magazine Series 6*, 37:221 (1919): 523–34.

19 R. Scott Kemp, "Gas Centrifuge Theory and Development: A Review of U.S. Programs," *Science & Global Security* 17:1 (2009): 1–19, esp. 2.

20 Henry De Wolf Smyth, *Atomic energy for military purposes : the official report on the development of the atomic bomb under the auspices of the United States government, 1940–1945* (Princeton: Princeton University Press, 1945), 169.

chain reaction in the brain of many European physicists," he later wrote.[21] The Dutch ultracentrifuge program first started after Kistemaker successfully built the electromagnetic isotope separator. This separator was not meant to enrich large quantities of uranium. Ultracentrifuges on the other hand possibly could.

For most physicists outside of the US, the Smyth Report was the first time they learned of the possible success of this technique, although it remained puzzling why the US had terminated the centrifuge branch of the program. It was clear that the biggest problem of building centrifuges was the friction that would arise in combination with the extremely high velocities that had to be reached before the uranium isotopes separate. In the US, Beams stayed working on this problem until 1951, when the Committee on Isotope Separation of the Division of Research of the USAEC strongly advised against further centrifuge work and the program was terminated.[22] In the early fifties physicists and engineers on two other locations, however, continued the work on ultracentrifuges.

5 Russia

In the Soviet Union, work on the development of a gas centrifuge was started in 1946. A large group led by German scientist Max Steenbeck, based in Suchumi on the shores of the Black Sea, consisted of 60 to 65 both German and Russian workers. Most of the German physicists were taken as prisoners of war to the Soviet Union in the aftermath of World War II. The almost 3m-long centrifuges that they were working on were developed for final enrichment of already enriched uranium from Russian uranium diffusion plants. The leader of the experimental division, Austrian scientist Gernot Zippe, recalled in 1958:

> When the group was informed that this [long centrifuge] was no longer required, they devoted their efforts to the development of a device suitable as the basis for a complete plant starting with natural uranium. As a result of their experience with the long tube program, a very simple short tube gas ultracentrifuge with exceptional performance characteristics was built. Because of the inexpensive construction made possible by its simple design, the very low electric power consumption, and a novel method for handling the process gas this device is believed to be superior

21 "De informatie in dit boekje heeft een kettingreactie in het brein van diverse Europese fysici veroorzaakt." From: Kistemaker, *Geschiedenis*, 1.

22 Kemp, "Gas Centrifuge Theory," 2.

> for large scale separation of uranium isotopes to gaseous diffusion or other known types of gas centrifuges.[23]

This new and immensely lightweight technique was not meant to be used in addition to the diffusion technique, but stood on its own. A cascade of centrifuges could enrich uranium from its natural form to an enrichment degree that was enough to be used in nuclear reactors.

This simple technique did not leave the Soviet Union until Zippe himself left. But he was not allowed to leave just yet. For two years he was kept in a kind of mental quarantine. This was the Russian way of making German and Austrian scientists working in Suchumi forget what they had learned before returning them to the West. Zippe stopped working on centrifuges in 1954 and moved from Suchumi to Leningrad. Apparently by coincidence, this happened just at the time that the Soviet centrifuge development team also was transferred to Leningrad. In this way it was easy for Zippe to keep up with the newest developments, although he was not working on the most recent designs himself. When he got back to Vienna in 1956 he had therefore not forgotten how these lightweight centrifuges enriched uranium.[24]

6 Germany, and How This Sparked the Netherlands

In the late 1940s, early 1950s, German groups had made considerable progress with the Beams-type centrifuge from the 1930s. Centrifuge work in Germany had begun during World War II and was since 1945 concentrated in Hamburg, where Paul Harteck and Wilhelm Groth worked on different ways to separate isotopes (photo-chemistry, 'isotope locks' and gas centrifuges).[25] In 1950 Groth went to Bonn to become a professor and a few years later he became the first director of the Institute for Physical Chemistry. He continued his centrifuge work there. His designs were immensely complicated, very expensive, but successful in separating isotopes.[26]

---

23 Gernot Zippe, Jesse W. Beams, and A.R. Kuhlthau, *The Development of Short Bowl Ultracentrifuges* (Charlottesville: University of Virginia, 1958), 5.

24 Gernot Zippe, *Rasende Ofenrohre in stürmischen Zeiten. Ein Erfinderschicksal aus der Geschichte der Uranisotopentrennung im heissen und im kalten Krieg des 20. Jahrhunderts* (Vienna: Eigenverlag E. Kubasta, 2008), 181; R. Scott Kemp, "The End of Manhattan: How the Gas Centrifuge Changed the Quest for Nuclear Weapons," *Technology and Culture* 53:2 (2012), 272–305, 290.

25 Kistemaker, *Geschiedenis*, 6.

26 Kemp, "The End of Manhattan," 288.

In November 1954 Kistemaker went on a trip to Bremen. He went to the Atlas Werke Company to take a look at one of the industrial mass spectrometers that was developed there.[27] Although Kistemaker had succeeded in enriching uranium with the large electromagnetic isotope separator, there was no mass spectrometer available in Amsterdam that could be used for scientific research. According to Kistemaker:

> [t]he situation seems ridiculous now, but it was simply a consequence of the after-war shortage of everything: mass spectrometers were not available on the open market (they were also classified for special purposes). Kistemaker's group had not been allowed to develop such sophisticated instruments, next to their heavy separation program.[28]

For this reason he started to look abroad, also in Bremen. Perhaps a mass spectrometer could be bought there.

Of considerable importance to Kistemaker's centrifuge research was a little trip he undertook from Bremen to Hamburg on one afternoon. The accounts of the reason for this trip vary. One account claims that Kistemaker was bored in Bremen and went to Hamburg to check out the Institute for Physical Chemistry.[29] Another mentions that the mass spectrometer in Bremen was broken; therefore he had to change his plans. He decided to see how the famous institute in Hamburg was doing after the war.[30] According to Kistemaker, the institute was deserted when he arrived, but the front door was wide open. He walked around a bit and found the back door of the colloquium room open. Someone was giving a talk. This was Hermann Gerhard Hertz,[31] talking about

---

27 Jaap Kistemaker, "The Jakob Kistemaker Story," (June 2006), 6, located in the archive of AMOLF, Science Park, Amsterdam.

28 When writing about himself, Kistemaker in later life had the habit of doing that in the Third Person. Ibid., 5.

29 Henk Tolsma, "Prof.dr. J. Kistemaker (1917), initiator van het Nederlandse ultracentrifuge-project," *Technisch Weekblad*, January 16, 2004, http://www.technischweekblad.nl/prof-dr-j-kistemaker-(1917)-initiator-van-het.74317.lynkx [accessed 27 October 2014].

30 Kistemaker, "The Jakob Kistemaker Story," 6.

31 Finding out which Hertz was talking in Hamburg was quite a struggle. Kistemaker in Ibidem talked of Herz (without the 't'), the nephew of G. Herz, the Nobel-prize winner. Gustav Hertz (with 't') was indeed a Nobel-prize winner, who himself was a nephew of Heinrich Hertz. Kistemaker also mentioned a G. Herz from the university of Lund. This could only be the son of Gustav Hertz, who was professor at Lund University. Nonetheless, Hermann Gerhard Hertz is presumably the Hertz in question. He was indeed a nephew of Gustav Hertz and was at the time working at the University of Hamburg, on isotope separation

the separation of Argon-isotopes with gas centrifuges.[32] Kistemaker was impressed by the talk and its results: gas centrifuges could – in addition to what Smyth reported – be successful as an alternative to diffusion and electromagnetic isotope separation.

After returning to Amsterdam, he sent his co-worker Dieter Heymann to Groth in Bonn to take a look at the centrifuges that were built there.[33] In December of 1954, an evaluation was started to check the possibility to develop quickly rotating, horizontally placed centrifuges with plain bearings in the Netherlands. The evaluation was conducted in cooperation with the Dutch company Werkspoor, whose director A.W. Baars "spontaneously" freed fl. 50.000, – for this. The first experiments were done in an old war-shelter on the terrain of Werkspoor and were not very successful. But it was a start.[34]

In 1955, the Reactor Centre of the Netherlands was founded in The Hague. From its foundation it took over the funding of the ultracentrifuge project in Amsterdam. The execution of the project stayed in the hands of the FOM institute in Amsterdam together with the company Werkspoor.[35] The technical university in Delft provided a handful of young engineers and the Netherlands Organization for Applied Scientific Research (TNO) got contracts to do research on the special materials that were needed to make the centrifuges firm enough. The Municipal Energy Company of the city of Amsterdam (GEB) provided housing for the project. Every two weeks the whole group would meet to talk about the progress of the centrifuges. Every scientific staff member had to be present. Kistemaker recalled:

> Often also technicians took the floor. During the discussions the employees learned to listen to one another, and to act timely and disciplined. These progress meetings were the success formula of the UC team.[36]

---

with gas centrifuges. See for example: Hermann Gerhard Hertz and Erich Nann, "Neue Ergebnisse Bei Der Isotopenverschiebung Schwerer Gase in Der Gas-Ultra-Zentrifuge," *Zeitschrift Für Elektrochemie, Berichte Der Bunsengesellschaft Für Physikalische Chemie* 58:8 (1954): 612–716.

32 Kistemaker, *Geschiedenis*, 6.

33 Ibidem.

34 Ibidem.

35 Goedkoop, *Een kernreactor bouwen*, 115.

36 "Vaak voerden ook technici het woord. In de werkbesprekingen leerden de medewerkers naar elkaar te luisteren, en op tijd en gedisciplineerd te handelen. Deze voortgangsbespreking werd de succesformule van het UC team." Kistemaker, *Geschiedenis*, 12.

7 The 1957 Symposium on Isotope Separation

Parallel to the start of the Dutch ultracentrifuge program, Kistemaker started to organize a symposium. This was to provide a major boost to the ultracentrifuge project due to an unexpected visitor who showed up unannounced. But the symposium also merits a closer look on other accounts. Its organization provides an interesting view of the way science had to be organized in a time when on the one hand atomic knowledge had lost its innocence and on the other hand secrecy and geopolitical forces started to enter the scientific arena.

The symposium was organized under the auspices of the Dutch Physical Society (NNV). They organized a symposium every year. Every even year this would be a large international congress, and every odd year this would be a small-scale international symposium – a so-called 'specialist-symposium.' Kistemaker was asked to be the secretary of the committee for the 1957 specialist-symposium on isotope separation. Professor Jan A.A. Ketelaar from the University of Amsterdam was the chairman of the committee. The organization started in the spring of 1956. The symposium was planned for the days after Easter, from 23 to 27 April 1957 at the Vrije Universiteit in Amsterdam. In one of the first letters from the committee to the board of the NNV it became clear that Ketelaar's and Kistemaker's ambitions exceeded the small-scale that was prescribed by the NNV. They clearly aimed for another large international conference. At a special meeting on 30 June 1956 the board of the NNV made clear that this was not possible. The financial risks would be too large and too much would be asked of the NNV-members from Amsterdam. Therefore, the maximum number of participants for this symposium was kept at 200 persons: 150 scholars from abroad and 50 Dutchmen.[37]

In the fall, the committee really got started and began asking key-speakers to join the symposium. As special guests they planned to ask two Americans, one German, one Englishman and one Swiss. Next to that, Kistemaker sent around 70 personal invitations to physicists from all over the globe. As an inducement, he mentioned that many Russian scientists could be expected, as for example in this letter to a physicist in the GDR:

---

37 "Short Account of the Meeting of July 30 1956," 30 July 1956, Box 23, folder on the international symposium on isotope separation, NNV archive, Science Park, Amsterdam (hereafter NNV, 23, symposium).

> We also expect a lot of Russians and it is now also totally normal for an international congress that a lot of people from the Soviet-Union, Poland, Czechoslovakia and the East-Zone come here.[38]

As Kistemaker knew that much could be expected from Russian participation, he also invited a handful of Russian scientists. The Russians, however, had a reputation of not answering letters. During the previous NNV conference in the summer of 1956 on nuclear physics, some 30 Russian scientists showed up one day late. On top of that, they announced two weeks before the beginning of the conference their desire to read "about 40 lectures."[39] Therefore, Kistemaker already in September started to send them reminders via the Russian Embassy in The Hague. The letter to the Embassy contained five individual invitations meant for the Russian scientists in Moscow and Kiev and one for the president of the Russian academy of sciences.[40]

In October Kistemaker sent another letter to the President of the Russian Academy, Aleksandr Nesmejanov, containing the second preliminary program of the symposium.

> [...] we should appreciate the presence of our Russian colleagues most highly. We expect and know that many subjects to be treated at this symposium of specialists (the size will be only 200 people) are going to be declassified in the USA and England and we have good hope from the experience in Geneva that in the USSR the same will be done.[41]

Kistemaker was hoping that the Russian scientists would talk about yet unknown results of their work on isotope separation. Gustav Hertz, the Nobel Prize winner who had worked for a few years in Suchumi on the shores of the Black Sea on diffusion cascades, was also invited but declined to come to the conference. On the other hand, Wilhelm Groth, Jesse Beams and Harold Urey planned to come to Amsterdam in the spring. A few weeks after Urey had confirmed his attendance, he had to withdraw. His health was presumably not good enough to come to the other side of the ocean. Kistemaker found

---

38 "Wir erwarten auch viele Russen und es ist jetzt auch ganz normal dass für ein internationalen Kongress viele Leute aus der Sovjet-Union, Polen, Czechoslowakei und der Ost-Zone hierher kommen." Letter from Kistemaker to Dr. Bernhard, 27 August 1956, NNV, 23, symposium.

39 Kistemaker to A. Nesmejanov, Oktober 1956, NNV, 23, symposium.

40 Kistemaker to N. Jouravlev, USSR Embassy in The Hague, 25 September 1956, NNV, 23, symposium.

41 Kistemaker to Nesmejanov, loc. cit.

this disappointing and – by calling the withdrawal "a little strange" – expressed his doubts whether Urey's health was really the problem.[42] "Evidently, the Americans still consider the ultracentrifuge very important," he wrote a little sarcastically to Groth.[43]

Without Urey, the ultracentrifuge session that was planned for the Tuesday afternoon would be out of balance. Carl Cohen, also from the US, was to speak about theoretical aspects of the centrifuge. Not one American, however, would talk about experimental research.[44] Little did Kistemaker know that, after the Manhattan Project, the ultracentrifuge program in the US was fully terminated. In 1954 the program was nonetheless reopened, mainly because of the positive results reached in Germany. But, as R. Scott Kemp has recently written, "the revival was desultory: No machines were built because the AEC was confident that it would have full access to German technology, which was at that time the most advanced in the West."[45] The imbalance of the program of the symposium was therefore not caused by the secrecy implied by Kistemaker in his letter to Groth, but by the simple fact that no experimental work was done in the US at the time. Kistemaker's hopes of declassification from the other side of the Atlantic Ocean were therefore, in retrospect, in vain.

In December 1956, probably because of the Hungarian Uprising, a discussion started within the NNV and within the committee about the participation of Russian scholars. The general opinion about the USSR in the Netherlands was not only one of fear of a Soviet military attack on Western Europe. There existed a broadly felt resentment of communism.[46] The reactions to the Soviet intervention in Hungary were, therefore, quite strong and broadly felt – in contrast to, for example, the reaction in the United States. In Amsterdam and other cities, offices of the Dutch Communist Party (CPN) were targeted and demolished in the days following the Soviet intervention.[47] These sentiments, together with anti-Soviet sentiments that rose as a result of the Suez Crisis, found their way into the scientific world and started to influence Kistemaker's organization of the symposium.

---

42 "Ein wenig fremd," in: Kistemaker to W. Groth, 9 January 1957, NNV, 23, symposium.

43 "Offenbar halten die Amerikaner die Ultrazentrifuge noch immer für sehr wichtig," Ibidem.

44 Ibidem.

45 Kemp, "Gas Centrifuge Theory and Development," 3.

46 Duco Hellema, *Buitenlandse politiek van Nederland. De Nederlandse rol in de wereldpolitiek* (Utrecht: Spectrum, 2006), 153.

47 Duco Hellema, "The Relevance and Irrelevance of Dutch Anti-Communism: The Netherlands and the Hungarian Revolution, 1956–57," *Journal of Contemporary History* 30:1 (1995): 169–86, 174.

As a result of the protests, Prof. Hendrik Brinkman, chairman of the NNV, opposed any Russian participation at the symposium. Kistemaker, on the other hand, reminded the committee that the participation of the Russians was already announced in the second preliminary program. Many scientists from all over the world would come to Amsterdam because of the Soviet participation. In addition, the Russians had just been accepted as members of the International Union of Physics, so exclusion would be strange. After casting votes within the organizing committee, Russian participation was, nonetheless, not considered appropriate. A letter of that nature would be sent to the Russian Academy of Sciences and to the Union of Physics. But – and this was an intervention proposed by Kistemaker – not before the Dutch ministry of foreign affairs had been consulted.[48]

This ministry, in the person of the head of the department for international organizations J. Meyer, called Kistemaker in the evening of 29 January 1957. He, too, was against Russian participation to the symposium. A few days later, Kistemaker sent a long letter to Meyer, explaining in detail why Russian scientists were so important for the success of the conference.

> Very little is known of the practice of isotope separation in the USSR. This conference is a unique opportunity to hear something of this. Not using this chance means incalculable damage.[49]

The ministry responded positively to Kistemaker's plea, but only under the condition that a month before the start of the conference the ministry knew all the names and functions of the Russian participants. Furthermore, no official delegations from Russia were welcome, for this was meant to be a scientific meeting – for colleagues only.[50]

This way of dealing with problems was typical for Kistemaker: obstinate, dedicated and against strict regulations. Whenever something was decided that he did not agree with, he would go as far as he could to reverse the decision. This was also reflected in a letter he sent to a Dutch colleague at this time:

---

48 "Minutes of the Second Meeting of the Committee for the 1957 Symposium," December 6, 1956, 5, NNV, 23, symposium.

49 "Van de praktijken op het gebied van isotopenscheiding in de USSR is zeer weinig bekend. Deze conferentie is een unieke gelegenheid daar iets meer van te horen. Het niet gebruiken van deze kans betekent onberekenbare schade." In: Kistemaker to J. Meyer, 5 February 1957, NNV, 23, symposium.

50 Kistemaker to Mr. Kitaev, 12 February 1957, NNV, 23, symposium.

> I totally agree with your remark about the admission of Russian scientists. As a matter of fact I am the one organizing this symposium and, as I did not agree with the decision of the committee, nor did Ketelaar, it now seems that the Russians are nonetheless coming. The rest was dead set against it. By the way, I am far from a communist![51]

In the following weeks, it became more and more probable that the Russians would indeed show up. Their identities and the topics of their talks remained, nevertheless, unknown to the organizers of the symposium.

Meanwhile, the symposium became a kind of hype. From all over the world scientists started to write Kistemaker to ask if they could also join the conference. Most of these scientists had to be turned down. Only if they could show that it was a "matter of life and death" were they allowed to join.[52] The reason for this was not only that this was supposed to be a specialist symposium (its maximum of 200 participants was already exceeded) but also that most of the hotels in Amsterdam were full. It was the flower season and the city was flooded with tourists.[53] In the midst of all these requests, Kistemaker also received one from an unknown Austrian scientist. This Gernot Zippe too was kindly refused:

> [...] in reply to your letter of the 25th instant I regret to inform you that no more subscriptions to the symposium on Isotope separation which is to be held in Amsterdam in April next can be accepted, owing to the restricted number of seats available. Yours faithfully, Kistemaker.[54]

Kistemaker clearly had no clue that Zippe had, like the Nobel-prize winning Hertz, worked in Suchumi as the leader of the experimental group building a totally new kind of ultracentrifuge.

Zippe had finally returned from Leningrad to the other side of the Iron Curtain on 28 July 1956, more than eleven years after the end of the war. His

---

51 "Je opmerking over toelating van Russische onderzoekers deel ik volkomen. In feite organiseer ik het symposium en daar ik het met de beslissing van de commissie niet eens was, evenmin als Ketelaar, ziet het er naar uit dat de Russen thans toch komen. De rest was er vierkant tegen. Ik ben overigens verre van communistisch!" in: Kistemaker to J.H. De Boer, 18 February 1957, NNV, 23, symposium.

52 "Levensnoodzaak," in: Kistemaker to De Troyer, Brussels, 27 March 1957, NNV, 23, symposium.

53 Kistemaker sent a dozen letters refusing scholars to come. For example: Kistemaker to Juan McMillan, 8 March 1957, NNV, 23, symposium.

54 Kistemaker to G. Zippe, 28 March 1957, NNV, 23, symposium.

hometown, Vienna, had changed considerably since he had left it in 1939. It was hard for him to settle down, but he managed to find a little job. In the spring of 1957, two Americans who had the task to contact German physicists coming back from the USSR contacted Zippe. They were interested in his knowledge, and Zippe was interested in what had happened in the West concerning isotope separation since the end of the war. The Americans gave Zippe the tip to go visit the upcoming conference in Amsterdam.[55] The decline from Kistemaker did not discourage Zippe, on the contrary. He recalled:

> Someone who has eleven years of experience in Russian organization could of course not be impressed by such a letter. I timely placed myself on my Maicoletta [motorcycle] and nonetheless drove there; the journey through the spring, especially through the blooming gardens in Holland, was a pure delight. Finally free and master of my own decisions![56]

Meanwhile in Amsterdam Kistemaker was starting to get fed up with the duplicity of the Russians. In a final letter to the Embassy of the USSR, Kistemaker stressed that in case any Russian scientists would show up at the last moment, the Embassy would have to take care of their housing. Next to that, a place in the program for them could certainly not be provided, because the conference was full.[57]

At the same time, another problem arose. Kistemaker had planned an excursion to the Phillips Company in Eindhoven. But the company did not want visitors from the other side of the Iron curtain snooping around. Their access was denied. Kistemaker, on the other hand, made clear in all the letters to the "Ostzone" that, although the policy of Philips was to keep them out, they all were very welcome to visit his laboratory in Amsterdam.[58]

The Russians probably did not come to Amsterdam. In the NNV archive there is no mention of any Russian participation during the conference, and in the proceedings that were published in 1958 only two articles about

---

55 Kemp mentions the American contact from the perspective of the Intelligence Division of the USAEC. Zippe, in his memoir, from his own point of view. See: Kemp, "Gas Centrifuge Theory and Development"; Zippe, *Rasende Ofenrohre*, 189.

56 "Einer, der elf Jahre Rußlanderfahrung im Organisieren hatte, konnte natürlich so ein Brief nicht imponieren. Ich setzte mich zeitgerecht auf meine Maicoletta und fuhr trotzdem hin; die Fahrt durch den Frühling, insbesonders durch die blühende gärten in Holland, war ein reines Vergnügen. Endlich frei und Herr meiner Entschlüsse!," in: Zippe, *Rasende Ofenrohre*, 188.

57 Kistemaker to Mr. Kitaev, 10 April 1957, NNV, 23, symposium.

58 See for example: Kistemaker to Dr. Bernhard, J. Mühlenpferdt, Etc., 18 April 1957, NNV, 23, symposium.

ultracentrifuges were published: one by Konrad Beyerle together with Groth and the other by Kistemaker and Joop Los.[59]

Although probably no Russians visited the symposium, an important exchange of knowledge from the East to the West did occur. Kistemaker recalled it in the following way:

> At the end of this symposium, Friday afternoon at 4 o'clock, an – until then – silent Austrian physicist approached [me], requesting a private conversation. This was Gernot Zippe.[60]

Zippe had listened, probably without registering for the conference, to the lectures at the Free University and was surprised by the lack of knowledge on gas centrifuges. Kistemaker made an appointment with him on the next day, 28 April 1957 at 10 in the morning in Kistemaker's home in Amsterdam. They talked for two hours about the gas ultracentrifuges that Zippe had designed under Steenbeck in Suchumi.[61]

> His elementary communications led to an important change in our program. In our preliminary studies we of course considered both horizontal and vertical setups, by analogy with Groth's setup in Bonn. On Monday April 30 we decided to change from horizontal to vertical centrifuges.

---

59 J. Kistemaker, Jacob Bigeleisen, and Alfred Otto Carl Nier, *Proceedings of the international symposium on isotope separation held in Amsterdam, April 23–27, 1957* (Amsterdam: North-Holland, 1958).

60 "Aan het einde van dit Symposium, op vrijdagmiddag om 16 uur, kwam een tot dien zwijgende Oostenrijkse fysicus naar Kistemaker toe, met het verzoek of hij hem privé kon spreken. Dat was Gernot Zippe." In: Kistemaker, *Geschiedenis*, 10.

61 Every part of this meeting between Zippe and Kistemaker is, to say the least, quite strange: that Zippe, although he was not welcome, nonetheless came to Amsterdam; that Kistemaker was eager to hear new facts from the Soviet Union, but no Russian scientist came and then this guy popped up out of nowhere; that Zippe actually told Kistemaker sensitive and classified information. The scarce sources that I could find on the meeting are all quite heroic and are almost all based on memories. Kistemaker, *Geschiedenis* is based on the memories of Kistemaker and his coworkers in 1991 and some old possessions like agendas and memoranda. Zippe's memoirs, *Rasende Ofenrohre*, is based on interviews. An interview conducted by Alexander Lagaaij with Zippe is transcribed in C.D. Andriesse, *De republiek der kerngeleerden* (Bergen, 2000). Next to these sources, Kistemaker's ultracentrifuge archive that after his retirement was moved to the company URENCO and is presently managed by the daughter company Enrichment Technologies, Almelo, gives not much more insight into the episode. Nonetheless, a copy of the notes that Kistemaker made during the meeting was found there.

> This meant that we abandoned slide bearings and changed to pivotal bearings.[62]

The decision was allegedly taken within a day. And it was not only a matter of which bearing to use. Until then, the ultracentrifuge designs had been based on the German tradition of *Maschinenbau*.

> Everything had to be big, heavy and strong. They were traditionally enclosed in slide bearings. The *spinning top-centrifuge* of Steenbeck and Zippe was in every sense the contrary, an object for *precision engineering*: light, elastic and self-straightening, with both magnetic and pivotal bearings, almost floating![63]

The changes were thus far-reaching for the research. It implied a change of mind. Size, sturdiness and German machinery had to be forgotten. Lightness was the new goal, and simplicity. Three weeks later, Kistemaker went to Vienna to talk to Zippe once more. But in the meantime, the German company Degussa had approached Zippe to design a centrifuge for them. Zippe had accepted the offer and was subsequently not allowed to share information with others about the centrifuges. Kistemaker came home from Vienna empty-handed. In November 1958 Degussa succeeded in registering a patent for the gas centrifuge, only six weeks before the Dutch group tried the same. The Dutch patent was declined. But Kistemaker's collaborator Joop Los and the patent representative of the Reactor Centre of the Netherlands, A.F. Arnold, went to the patent judge. They had a case since the Dutch had started to work on this kind of centrifuge before Degussa did. In the end, the judge decided that the two patents were equivalent, which means that both the Dutch and the German patents were granted.[64]

Zippe, in the meantime, left Europe for the United States. As noted above, just after his return from the Soviet Union, two Americans had contacted him.

---

62 "Zijn elementaire mededelingen leidden tot een belangrijke omschakeling in ons programma. Bij onze voorstudies hadden wij uiteraard zowel horizontale als verticale opstellingen overwogen, naar analogie van Groth's opstelling in Bonn. Op maandag 30 april besloten wij van horizontale naar verticale centrifuges over te gaan. Dit betekende ook dat wij afstapten van glijlagers en overgingen op een taatslager." Cf. Kistemaker, *Geschiedenis*, 10.

63 "Alles moest groot, zwaar en sterk zijn. Ze waren traditioneel gevat in glijlagers. De *tol-centrifuge* van Steenbeck, Zippe was in alle opzichten het tegendeel, een object voor de *fijn-mechanische* industrie: licht, elastisch en zelfrichtend, met magnetische en taatslagers, bijna zwevend!," in: Ibidem.

64 Ibidem, 12.

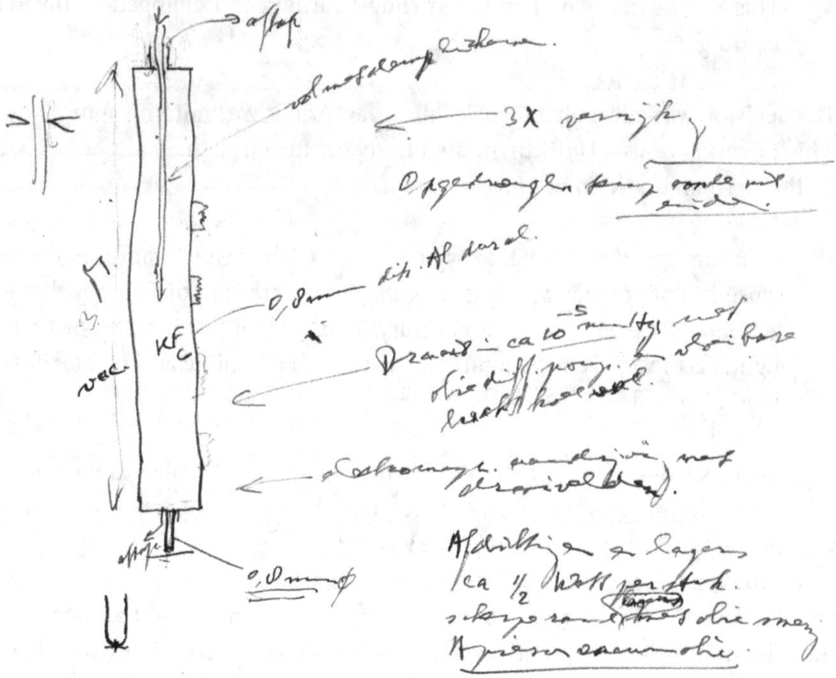

FIGURE 8.1 One of Kistemaker's notes from the meeting with Gernot Zippe the day after the symposium.
SOURCE: KISTEMAKER ARCHIVE, ET-NL, ALMELO

These first contacts sparked an important reaction in the United States. It was realized that gas centrifuges really could work successfully and could even be lucrative as a uranium enrichment method. The chief of the Physics and Mathematics branch of the AEC's Division of Research, George Kolstad, decided in 1957 that Zippe was needed in the US to re-create the Soviet centrifuge. In May 1958, a year after the symposium in Amsterdam, Zippe and Degussa agreed to a salary of $10,000. He went to the University of Virginia in August and replicated the Soviet apparatus within a year. Carl Cohen, who had assessed Zippe's work, was so impressed that he advised the AEC to expand work on the Soviet centrifuge. In 1960, such a program was started at Oak Ridge.[65]

Concluding this peculiar period of international cross-pollination of ideas and nuclear entrepreneurship was the 1961 decision to classify all Dutch centrifuge work as a state secret. This was decided after an American delegation led by Kolstad visited the Netherlands, Germany and the U.K. "They came to tell us

65 See: Kemp, "The End of Manhattan," 289; Kemp, "Gas Centrifuge Theory and Development," 3–4.

that we publish too much," Kistemaker recalled.[66] The visiting scientists and diplomats stressed the importance of secrecy and urged the European countries to follow the American decision to classify the whole program. The different Dutch ministries had long and fierce interdepartmental discussions about the way this should be done, but in the end gave in to the American plea.[67]

Reasons for the US request to classify the ultracentrifuge research existed on different levels. One was that proliferation of nuclear knowledge could lead to new, less trustable, countries possessing nuclear fissionable materials. This was a reflection of a more generally felt Cold War sentiment that started to spread after the launch of Sputnik, the Hungarian intervention and the Suez-crisis.[68] The relationship between the US and the Soviet Union lost its open, Geneva-like, character.[69] Next to that, it was no surprise that the delegation visited Europe shortly after France had exploded its first atomic bomb in the Sahara in 1960.[70] More and more countries were starting centrifuge programs or picking up existing programs after news reached them of the successes from Germany and the Netherlands.[71] Another reason, less outspoken by the US, but nonetheless discussed in diplomatic circles in Europe, was that the ultracentrifuge research in Europe would be slowed down by classification, which would grant the US time to catch up with their project, which was not as

---

66 "Ze kwamen ons vertellen dat wij te veel publiceerden in Jaarverslagen en Rapporten." Cf. Kistemaker, *Geschiedenis*, 16. Also found in Andriesse, *De republiek*, 63.

67 For a more complete account of this episode, see: Abel Streefland, "Putting a lid on the gas centrifuge: Classification of the Dutch ultracentrifuge Project, 1960–1961," in: *Cold War Science and the Transatlantic Circulation of Knowledge*, ed. Jeroen van Dongen (Leiden: Brill, 2016), 75–100; William Burr, "The 'Labors of Atlas, Sisyphus, or Hercules'? US Gas-Centrifuge Policy and Diplomacy, 1954–60," *The International History Review* (accessible online): 1–27; See also Elmar A.A. Hellendoorn, *Between the devil and the deep sea. The Netherlands and the struggle for European nuclear order, 1954–1966* (PhD dissertation, Utrecht University, 2016), whom I like to thank for the inspiring discussions.

68 Although recently the effect of Sputnik on Western European politics has been called into question by David Baneke. See his article "The Absence of the East: International Influences on Science Policy in Western Europe during the Cold War," in: *Cold War Science*, ed. Jeroen van Dongen, 163–183.

69 Hellema, *Buitenlandse politiek van Nederland*, 214–215.

70 See for an interesting reflection on the American delegation visiting the Dutch laboratories: Andriesse, *De republiek der kerngeleerden*, 63.

71 In the minutes of the visit by the American delegation (dated 21 June 1960) Kolstad mentioned this argument. As exampling countries he mentioned the United Kingdom and Russia restarting their program. Kistemaker mentioned Canada, which also started a centrifuge program. Nationaal Archief, The Hague, Ministerie van Buitenlandse Zaken: Code-archief 1955–1964, entry number 2.05.118, inventory number 13589.

advanced as the research done in Germany, the Netherlands and the U.K.[72] In this respect the argument has to be nuanced. Not only geopolitical sentiments played a role, but also economic arguments.[73]

The main result of the classification was that communication between the different international groups was hindered. For some of these groups the decision ruined their work. Other groups, on the other hand, proved to be strong enough to be able to manage the centrifugal developments on their own.[74] The Dutch program eventually led to the tripartite Treaty of Almelo in 1970, signed by Germany, the Netherlands and the UK. This treaty resulted in the installation of multiple large centrifuge factories within the company URENCO, founded in 1971.[75]

8 Conclusions

After the Second World War a program was started in the Netherlands to enrich uranium. At first, a large machine was built to separate isotopes using electromagnetic forces. The machine enriched its first uranium in 1953, a small but important sample. The Smyth Report had sparked the enrichment program, but the technological development was done without influences from outside the Netherlands. Almost no international contacts influenced the research in the early years. It was as though the Dutch program for atomic research was

---

72 Recently declassified documents made public by William Burr within the Wilson centre digital archive give a nice overview of American thoughts on European centrifuge projects. Accessible via http://digitalarchive.wilsoncenter.org. See also, Burr, "The 'Labors of Atlas, Sisyphus, or Hercules'?"

73 I have tried to develop this argument in Streefland, "Putting a lid on the gas centrifuge." Related 'soft power' arguments were recently prominently put forward by John Krige. See John Krige, "US Technological Superiority and the Special Nuclear Relationship: Contrasting British and US Policies for Controlling the Proliferation of Gas-Centrifuge Enrichment," *The International History Review* 36:2 (2014): 230–251; John Krige, "Hybrid Knowledge: the Transnational Co-Production of the Gas Centrifuge for Uranium Enrichment in the 1960s," *The British Journal for the History of Science* 45:3 (2012): 337–357; John Krige, "Maintaining America's Competitive Technological Advantage: Cold War Leadership and the Transnational Coproduction of Knowledge," *Humana.Mente* 16:4 (2011): 33–51; John Krige, "Building the Arsenal of Knowledge," *Centaurus* 52:4 (2010): 280–296; John Krige, "The Peaceful Atom as Political Weapon: Euratom and American Foreign Policy in the Late 1950s," *Historical Studies in the Natural Sciences* 38:1 (2008): 5–44; Krige, "Atoms for Peace"; John Krige, *American Hegemony and the Postwar Reconstruction of Science in Europe* (Cambridge, MA: MIT Press, 2006).

74 Kistemaker, *De Geschiedenis*, 16.

75 To read a version of this part of history, see for example: Kistemaker, *Geschiedenis van het Nederlandse UC Project*; R.B. Kehoe, *The Enriching Troika, a History of Urenco to the Year 2000* (Marlow, UK: Urenco, 2002).

aiming for a niche: to excel in one specific technological part of nuclear reactor science, namely uranium enrichment.

The first sample of enriched uranium did spark international reactions. Although any effect of the Dutch program on the American atomic policy is doubtful, it is plausible that the Atoms for Peace program was at least partly triggered by the Dutch achievements. If a small country like the Netherlands could enrich uranium, why would the US still need to keep all its atomic knowledge classified? And moreover, if in Amsterdam money could be made with separated isotopes, why shouldn't the US provide a framework to do the same?

The new American policy gave rise to a period of relative openness. On the one hand the atom became a public icon of the future, on the other hand the Americans were, by opening their policy up for (American) companies, able to control the proliferation of nuclear knowledge – also in Europe. In this respect the scientific hegemony that was built up by the US expanded in Europe during this period. Internationalism flourished, symbolized by the successful first Atoms for Peace conference in Geneva in 1955.

The Dutch endeavour nonetheless took another turn. By deciding to expand the uranium enrichment program with ultracentrifuges, a route was taken that was not already paved by the US. The American ultracentrifuge program had after all been terminated after the War. Because contacts with Western Germany could easily be established (there was no strict classification yet), the German tradition of ultracentrifuge building became the standard for the early Dutch program. This changed after Gernot Zippe met Jaap Kistemaker following the 1957 symposium on isotope separation in Amsterdam. The centrifuges would have to be designed in a totally different way in order to be lucrative for large-scale uranium enrichment. Cold war ultracentrifuge science thus depended on chance and knowledge-leaking physicists from the other side of the Iron Curtain, forces that were not controlled by the two geopolitical superpowers.

It was also in this time that geopolitical disputes started to enter the nuclear scientific arena in the Netherlands. The organization of the symposium by Kistemaker was troubled by political discussions after the Hungarian uprising of November 1956. The discussions that the committee organizing the symposium had with the ministry of foreign affairs also meant the starting point of direct political mingling with the scientific research on ultracentrifuges. The ministry would in the years to come keep contact with Kistemaker and the Reactor Centre of the Netherlands about international ultracentrifuge policy. This reached its first peak preceding the decision to fully classify the project in 1961. The decision was motivated by US policy. A bigger contrast to the US policy from the early 1950s, in which the government encouraged openness and nuclear entrepreneurship, cannot be imagined.

PART 3

Transformations

CHAPTER 9

The Disputation Hall in the Seventeenth-Century Dutch Republic: An Urban Location of Knowledge

Dirk Van Miert

> The ... Error ... is that the Seats of Knowledg, have been for the most part heretofore, not *Laboratories*, as they ought to be; but onely *Scholes*, where some have *taught*, and all the rest subscrib'ed. The consequences of this are very mischievous.
>
> THOMAS SPRAT, *History of the Royal Society*, 1667

∴

Thomas Sprat's condemnation of the learning of the schools might have been directed in the first place at English universities, but there was a general feeling outside of academia that learning at the academic schools was overall precisely that: scholastic. The prime pedagogical instrument of scholasticism was the disputation, and while disputations aimed at a critical assessment of the truth,[1] in practice they also served to inculcate the foundations of orthodox learning, accepted by the consensus of schoolmen.[2] Nowadays, disputations are in the process of being promoted to a major source of intellectual history.[3] It is the aim of this article to explain why disputations deserve more attention. I will do so by taking into account the spatial and social setting of the disputation as a liminal genre, straddling the divide between private teaching and public debate. The examples are drawn from the Dutch Republic, more particularly Amsterdam, Leiden and Utrecht, where (contrary to the stagnation of English universities Sprat complained about) major intellectual controversies

---

1 Olga Weijers, *In Search of the Truth. A History of Disputation Techniques from Antiquity to Early Modern Times* (Turnhout: Brepols, 2013), passim.

2 Sari Kivistö, *The Vices of Learning. Morality and Conduct at Early Modern Universities* (Leiden and Boston: Brill, 2014) 201.

3 The most recent contribution is Marion Gindhart, Hanspeter Marti and Robert Seidl, eds., *Frühneuzeitliche Disputationen. Polyvalente Produktionsapparate gelehrten Wissens* (Wien, Köln, Weimar: Böhlau, 2016).

 | DOI:10.1163/9789004264885_010

were spawned and fought out precisely by means of disputations. Professors at war used disputations as platforms for attacking colleagues in and philosophers outside the universities, and abused their students as infanterists to defend their own orthodox or heterodox theories.

1 Disputations: A Rediscovered Field

Disputationes usually took place in the lecture halls of universities and illustrious schools (academic gymnasia). In Utrecht there were various halls, such as the Great Auditorium, the Small Auditorium and the Hungarian Chapel on the Dom Square. In Leiden, the main lecture hall in the *Academiegebouw* on the Rapenburg canal served as a grand place to defend theses. Likewise in Amsterdam, the Chapel of St Agnes on the Oudezijds Voorburgwal was not only the place where the public lectures took place, but also the location for disputations.

Disputations are not to be confused with inaugural dissertations, with which students obtained a doctoral degree (see below). During a disputation, a student (*respondens*) orally defended several theses against opposition brought forward by his fellow students or by professors other than his own supervisor (the *praeses*) in order to practice his knowledge and skills. Before a disputation, the beadle made sure that the hall was in order, and during the ceremony the *praeses* was responsible for an orderly discussion. Most disputations lasted for about an hour. In the Dutch Republic, they usually took place on Wednesdays and Saturdays. A *disputatio* is also the title given to the written version of the theses. Printed disputations not only state the time, place and theme of the defence, but usually also feature dedications (which say something about the student's teachers, his family connections and social background) and are sometimes followed by laudatory poems composed by fellow students, friends or relatives. Students thus seized the opportunity to confirm and expand their professional, familial and social networks, very likely in order to secure future support for the advance of their own careers.

Disputations were the daily bread of early modern academic life. They constitute a rich source for both the organisation of teaching and the content of what was being taught. The shelves of libraries still testify to the *Disputierfreude* of sixteenth-, seventeenth- and eighteenth-century universities.[4] Historians of

4 I borrow the German term from Marion Gindhart and Ursula Kundert, "Einleitung," in: *Disputatio 1200–1800. Form, Funktion und Wirkung eines Leitmediums universitärer Wissenskultur*, eds. Eaedem (Berlin: De Gruyter, 2010) 1–18 (2, n. 5).

universities are increasingly turning to these disputations, and they are being supported by ever more bibliographical tools which facilitate the identification and location of the usually quite ephemeral pamphlet-like printed disputations. To give but a few examples: the historian of universities Laurence Brockliss has worked extensively with French disputations, demonstrating their value for an assessment of the substance of academic teaching.[5] For Germany, Hanspeter Marti in particular has testified to the growing popularity of disputations amongst intellectual historians.[6] A prime example is Sari Kivistö's wonderful study of learned moral conduct, which is largely based on disputations from German territories.[7] The thirteen thousand German disputations which have hitherto been catalogued may constitute only half of the material available,[8] because anyone who starts looking in the on-line databases of early modern books by searching with the keyword 'disputation' comes across thousands of titles.[9] Inspired by the dissertation of Margreet Ahsmann about Leiden juridical disputations, I myself used them to study the contents of academic teaching in Amsterdam in the seventeenth century.[10] The large

5 Laurence Brockliss, *French Higher Education in the Seventeenth and Eighteenth Centuries. A Cultural History* (Oxford: Oxford University Press, 1987).

6 Gindhart and Kundert, *Disputatio*. See for example Hanspeter Marti, 'Disputationswesen und Rhetorikgeschichte: Vorbemerkungen zu einer im Entstehen begriffenen auswahlbibliographie,' *Rhetorik* 1 (1980): 165–167; Idem, 'Der wissenschaftsgeschichtliche Dokumentationswert alter Dissertationen,' *Nouvelles de la République des Lettres* 1 (1981): 117–132. Disputations not only allow to analyse the contents of teaching, but also changes in styles of teaching and of communication: see Idem, "Kommunikationsnormen der Disputation. Die Universität Halle und Christian Thomasius als Paradigmen des Wandels," in: *Kultur der Kommunikation. Die europäische Gelehrtenrepublik im Zeitalter von Leibniz und Lessing*, ed. Ulrich Johannes Schneider (Wiesbaden: Harrassowitz, 2005) 317–344. See also the volume mentioned above in note 3. For the particular popularity of the disputation at German universities, see also Ku-ming (Kevin) Chang, "From Oral Disputation to Written Text: the Transformation of the Dissertation in Early Modern Europe," *History of Universities* 19–2 (2004): 129–187 (163).

7 Kivistö, *The Vices of Learning*.

8 Hanspeter Marti, "Philosophieunterricht und philosophische Dissertationen im 17. und 18. Jahrhundert," in: *Artisten und Philosophen. Wissenschafts- und Wirkungsgeschichte einer Fakultät vom 13. bis zum 19. Jahrhundert*, eds. Rainer Christoph Schwinges and Barbara Studer (Basel: Schwabe Verlag, 1999) 207–232 (209 and 232). For German disputations see also Manfred Komorowski, "Research on early German dissertations. A report on work in progress," in: *The German book 1450–1750. Studies presented to David L. Paisey in his retirement*, eds. John L. Flood and William A. Kelly (London: British Library, 1995) 259–268, and the literature mentioned there.

9 www.vd16.de and www.vd17.de. See also Gindhart and Kundert, "Einleitung," 2–3, n. 5, which lists other bibliographies.

10 Margreet Ahsmann, *Collegia en Colleges. Juridisch onderwijs aan de Leidse Universiteit 1575–1630 in het bijzonder het disputeren* (Groningen: Wolters Noordhoff, 1990); Dirk van

number of disputations allowed me to conclude that a period which previous historians had regarded a time of decline marked in fact a successful stage in the history of the Illustrious School of Amsterdam. The same method allowed Robert Evans to conclude, on a much larger scale, that in the second half of the seventeenth century, generally thought to be a time in which the German universities were in a dismal state, these universities were actually thriving.[11] For institutes in many countries, lists of disputations are now printed regularly, facilitating future comparisons between different institutes.[12] A survey of the

---

Miert, *Humanism in an Age of Science. The Amsterdam Athenaeum in the Golden Age, 1632–1704* (Leiden and Boston: Brill, 2009).

11 R.J.W. Evans, "German Universities after the Thirty Years War," *History of Universities* 1 (1981): 169–190.

12 See for example Michael Sharratt, "Theology and Philosophy at the English College, Douai: A Handlist of Sources," *History of Universities* 18 (2003): 197–225; Jaap Harskamp, *Dissertatio medica inauguralis …: Leyden medical dissertations in the British Library, 1593–1746. Catalogue of a Sloane-inspired Collection* (London: Wellcome Institute for the History of Medicine, 1997, i.e. 1998). Marius J. van Lieburg, "De medische faculteit te Groningen en de ontwikkeling van de medische wetenschap. Een studie van de Groningse academische geschriften uit de jaren 1614–1714," in: *Onderwijs en onderzoek. Studie en wetenschap aan de academie van Groningen in de 17e en 18e eeuw*, ed. A.H. Huussen Jr. (Hilversum: Verloren, 2003) 31–83 (76–83); B.S. Hempenius-van Dijk, "Antonius Matthaeus (I) en zijn juridisch onderwijs in Groningen 1625–1647," in: Huussen, *Onderwijs en onderzoek*, 85–126 (105–116); Ferenc Postma and Jacob van Sluis, *Auditorium Academiae Franekerensis. Bibliographie der Reden, Disputationen und Gelegenheitsdruckwerke der Universität und des Athenäums in Franeker 1585–1843* (Leeuwarden: Fryske Akademy, 1995); Jacob van Sluis, "Bibliografie van Deventer disputaties 1630–1815," in: *Deventer Denkers. De geschiedenis van het wijsgerig onderwijs te Deventer*, eds. H.W. Blom, H.A. Krop and M.R. Wielema (Hilversum: Verloren, 1993) 213–232; J. van Sluis, "Disputeren in Franeker en Groningen," in: *Zeer kundige professoren. Beoefening van de filosofie in Groningen van 1614 tot 1996*, eds. H.A. Krop, J.A. van Ruler and A.J. Vanderjagt (Hilversum: Verloren, 1997) 47–56; Van Miert, *Humanism in an Age of Science*, 372–392. Some disputations have been reprinted, e.g. Joachim Jungius, *Disputationes Hamburgenses. Krititische Edition*, ed. Clemens Müller-Glauser (Göttingen: Vanden Hoeck & Ruprecht, 1988); Keith D. Stanglin, *The Missing Public Disputations of Jacobus Arminius. Introduction, Text, and Notes* (Leiden and Boston: Brill, 2010). Stanglin on the flyleaf of this edition quotes Louis D. Petit (1894) and Paul Dibon (1954), who both deplore the neglect of the study of disputations. The Special Collections department of the University of Amsterdam holds an impressive card-file system, compiled by S. van der Woude, of all disputations, dissertations and orations, defended and delivered at Dutch institutes of higher education from 1575 to 1800. Many of the thousands of disputations catalogued in here are only to be found in one copy, and countless of these copies are not held by Dutch libraries. The card files of this system are organized per institute, in two ways: alphabetically by presiding professor, as well as chronologically. It would be a great help to early modern intellectual historians if this card file system would be digitised, or at least integrated into the *Short Title Catalogue of the Netherlands* (STCN), which gives the locations and shelf marks of material printed in the Netherlands.

transformation of the early modern disputation allows us to frame this important *genre* in a historical context.[13] Based on Dutch disputations, Rienk Vermij in his book on the Calvinist Copernicans and several other historians in smaller publications have been able to map the individual opinions of professors and differentiate between the many and various kinds of Aristotelian thinking.[14] The ecclesiastical historian Keith Stanglin based his study of Jacob Arminius, the notorious Leiden professor of theology, largely on his disputations.[15]

2 Types of Disputations

In the seventeenth-century Dutch Republic, there were two basic types of disputations: the *disputatio exercitii gratia*, i.e. for training purposes (often also called *exercitium academicum* or *exercitatio*), and *disputatio* (or *dissertatio*) *pro gradu* or *inauguralis*, with which the defendant obtained the doctoral degree.[16] Within the practice disputations, a further distinction can be made between serial disputations and individual disputations. The serial disputations were written by the presiding professor. Together, they often constituted an entire course. Professors sometimes collaborated in devising such a course, which could form a cycle of disputations taking place over a year or even two years. Many of these serial disputations were prepared during private instruction in so-called *collegia*. A collegium was a group of students who were taught at the professor's house, another location of knowledge about which we still need to learn more. Sometimes the students in such a group took turns defending theses which remained unpublished or which were published later. Disputations of this kind are sometimes called *disputationes privatae*, but hitherto little is known about them. Cycles, series or courses consisting of disputations were sometimes bound together and published under the title *Collegium*. Such *collegia* functioned as textbooks. Often, the disputations in these collegia were recycled by a new generation of students.

---

13 Chang, "From Oral Disputation to Written Text."

14 Rienk Vermij, *The Calvinist Copernicans. The Reception of the New Astronomy in the Dutch Republic, 1575–1750* (Amsterdam: Koninklijke Nederlandse Akademie van Wetenschappen, 2002); see also Van Sluis, "Disputeren," 47–56, plus the contributions in Huussen, *Onderwijs en onderzoek*, as well as those in *Franco Burgersdijk (1590–1635). Neo-Aristotelianism in Leiden*, eds. E.P. Bos and H.A. Krop (Amsterdam and Atlanta: Rodopi, 1993).

15 Keith D. Stanglin, *Arminius on the Assurance of Salvation. The Context, Roots, and Shape of the Leiden Debate, 1603–1609* (Leiden and Boston: Brill, 2007), and Id., *Missing Public Disputations*.

16 The following three paragraphs are condensed from my *Humanism in an Age of Science*, 149–159.

The second category of disputations are the *disputationes pro gradu*. They are never part of a series and were supposedly the work of an individual student in order to be promoted to doctor. 'Supposedly,' because there is evidence in a number of cases that the theses were written by tutors.[17] For getting insight into the contents of the day-to-day teaching at universities, the inaugural disputations are not as useful as the practice disputations because these inaugural disputations took place independently of the academic curriculum.

The opponents of a given thesis could formulate their arguments in advance, since the disputations were printed and published before the event. The print run was usually about a hundred, but of course, this depended on the size of the institution. In Leiden, a given disputation was printed in 100 copies during the sixteenth century; in 1598 this number was increased to 150, and later to 200. In the case of institutes which were financed by the cities, the printing costs were probably paid for by the city. This was certainly the case with the fledgling illustrious school of Middelburg.[18] The Amsterdam professor Johannes van den Broeck received permission in 1684 to teach classes and preside over disputations but "not at the expense of the city."[19] This explicit condition indicates that it was not unusual for Amsterdam to contribute to printing costs. The distribution and circulation of printed disputation-texts was in general the task of the beadle. The printed text helped to publicize the event, as well as help opponents to prepare themselves.

3 Rules of Engagement

In 1707, the German scholar Justus Böhmer (1674–1747) in a work *On literary wars* retold a story, often cited before and after him, about the sixteenth-century French scholar Dionysius Lambinus, who engaged in an aggressive disputation with the much read letter-writer Paulus Manutius (1512–1572) about the question of whether the Latin word *consumptus* should be written with or without the letter p. At one point during the ceremony Manutius broke Lambinus' nose

17 J. Roelevink, "'Gepromoveerde ezels.' Enkele kanttekeningen bij de origine en de originaliteit van dissertaties en disputaties," *Batavia Academica* 4–2 (1986): 25–37.

18 H.M. Kesteloo, "De Stadsrekeningen van Middelburg IV: 1550–1600," in: *Archief van het Zeeuwsch Genootschap der Wetenschappen* 7 (1894) eerste stuk, 87: "Diverse theses ter ordonantie van de gecomitteerde tot beleijt van de Latijnse Schole." Cited after Huib Zuidervaart, "Zeeuws preacademisch erfgoed. Een filosofische disputatie uit Middelburg uit de late zestiende eeuw," *Zeeland. Tijdschrift van het Koninklijk Zeeuwsch Genootschap der Wetenschappen* 18–2 (2009): 50–56 (56, n. 18).

19 Van Miert, *Humanism in an Age of Science*, 156.

with a stone which was inscribed with a p.[20] Academic lore had it that the medieval legal scholar Portius Azo during a disputation killed his opponent with a knife.[21] This was the type of conduct which horrified Böhmer's contemporary, the prolific literary historian Nicolaus Hieronymus Gundling (1671–1729). Dismissing the "disputing itch" (*disputandi pruritus*) he assumed there were less bellicose means of obtaining the truth.[22]

There were of course rules for drawing up disputations. We can deduce these rules by studying the disputations, but there are also seventeenth-century sources which dictate what a disputation should look like and how it should go about the theses.[23] The Bolsward school-rector Johannes Fungerus (1546–1612) warned his students not to lose control and start shouting during disputations. Using harsh words, biting, spitting and kicking were forbidden, and seeking pleasure in the applause of an audience was not done: the truth was more important than the assent of one's fellow students.[24] Arjen Dijkstra pointed out to me a set of rules for drawing up disputations, hidden in a chapter of the philosopher Johannes Phocylides' (1618–1651) book on elementary logic, printed in Franeker in 1648.[25] Most of these rules are rather obvious: one should not call into question obvious things (is snow white?) or propose heretical questions (is God just?), or absurd questions such as "What was God doing before he created the earth"?[26] More interesting is Phocylides's remark that the substance of a disputation should not be "nimis curiosa et nova": not too inquisitive and new.[27] This notion of curiosity as something pejorative appears somewhat old-fashioned, for 'curiosity' was one of the key words of the century: not only did it denote a constant openness to the unexpected situation and the willingness to spontaneously observe and record (crucial in a period in which the occurrence of phenomena was not always manipulated) but also the drive to question the natural world by setting up experiments.[28] Phocylides's

---

20 Kivistö, *Vices of Learning*, 147.

21 Ibidem, 196.

22 Ibidem, 190–191.

23 Donald Felip, "Ways of disputing and *principia* in 17th-century German disputation handbooks," in: Gindhart and Kundert, *Disputatio*, 33–62.

24 Kivistö, *Vices of Learning*, 148–149.

25 Johannes Phocylides Holwarda, *Elementale logicum* (Franeker, 1648) (Cambridge University Library, shelfmark Ven.8.64.9). This copy is not to be found in The Netherlands. Arjen Dijkstra was so kind as to send me his reading notes on this book.

26 Holwarda, *Elementale logicum*, 139.

27 Ibidem.

28 On the conflicting appreciations of *curiositas*, see Neil Kenny, *The Uses of Curiosity in Early Modern France and Germany* (Oxford: Oxford University Press, 2004). See also *Curiositas. Welterfahrung und ästhetische Neugierde in Mittelalter und früher Neuzeit*, ed. Klaus

rule would seem to imply that disputations were meant to rehearse received knowledge, not to present revolutionary or controversial matter, but the text modifies the injunction: new things are allowed if they are serious and seek to present a truth hitherto unknown. "This has been done in past ages by many and to good use."[29] In fact, disputations were often used for advancing new ideas. It is no coincidence that disputations gradually transformed from a didactic tool into something similar to a scientific article in the course of the nineteenth century.[30]

Holward's remarks remind us that it did not go without saying that public disputations were reserved for dogma.[31] I give two examples of disputations that sparked off far-reaching controversies. A case in point in the Dutch Republic is the start of the Arminian controversy, the theological debate that originated at the theological faculty of the University of Leiden in 1604, in the form of a debate between two professors of theology, Franciscus Gomarus and Jacobus Arminius. Arminius, who had previously been a minister in Amsterdam, had started to have his students defend theses which were deemed unorthodox by some of his former colleagues. They warned the professor of theology, Gomarus, of Arminius's inclinations. The anxiety was not just that Arminius had deviant opinions regarding predestination, but, above all, that he was a professor: a public teacher of the next generation of ministers, threatening to "infect" the students. Arminius doubted dogma in public. The matter was not much helped when Gomarus had one of his own students refute Arminius's disputation. Attempts to quench the nascent polemic by having the professors privately discuss the issue, or even publicly defend their positions – not in the lecture hall or disputation hall, but before the Court of Holland – were unsuccessful. The entire debate, which ultimately brought the Dutch Republic to the brink of civil war when church politics were drawn into the theological debate, had its origin in the disputation hall.

Krüger (Göttingen: Wallstein Verlag, 2002); Lorraine Daston and Katherine Park, *Wonders and the Order of Nature 1150–1750* (New York: Mitpress, 1998); Candice Delisle, "'The Spices of Our Art.' Medical Observation in Conrad Gessner's Letters," in: *Communicating Observations in Early Modern Letters (1500–1675). Epistolography and Epistemology in the Age of the Scientific Revolution*, ed. Dirk van Miert (London: The Warburg Institute / Turin: N. Aragno Editore, 2013) 27–42 (32). See now also Kivistö, *Vices of Learning*, 202–212 on good and bad curiosity.

29 Holwarda, *Elementale logicum*, 139.

30 Marti, "Philosophieunterricht," 223; Chang, "From Oral Disputation to Written Text," 156.

31 Gindhart and Kundert, "Einleitung," 3. Gindhart, Marti and Seidel, 'Einleitung,' in: Idem, eds., *Fruhneuzeitliche Disputationen*, 3–24 (20). Note that disputations acted as means to stabilize knowledge, but were equally used to launch new ideas and test the water.

A second major intellectual crisis in the Dutch Republic was likewise brought about by a disputation. In Utrecht in 1636, a curious incident occurred during a disputation supervised by the professor of philosophy, Arnold Senguerdius. The student Florentius Schuyl defended a number of metaphysical theses; cornered by a Cartesian opponent with non-syllogistic arguments, Senguerdius came to his defence. The medical professor Henricus Regius, indignant over Senguerdius's support, flew into a rage and assisted the Cartesian in turn.[32] The battle ended in a draw – the first confrontation in a long struggle, usually fought with sharp pen, acute tongue, and, once in a while, with bare fists and academic furniture. The First Cartesian War had broken out. In Utrecht, the senate of the university asked in April 1661 "that an iron grate or wooden fence would be erected around the seat of the repondents, so that these can freely defend their theses and for the audience to more easily understand every single word."[33] This bar around the disputational pulpit was apparently needed to protect the student against physical opposition from the floor. Scenes of physical confrontations were witnessed in the disputations hall in Leiden, where unrest was rampant.[34] As the German traveler Bernard Varenius reported to his friend Joachim Junius in 1647:

> It is, however, customary in this academy that the audience respond with scraping and trampling their feet on hearing something they disagree with. Thus loud yelling, huge turmoil![35]

Like the Arminian controversy, then, the Cartesian polemics entered public debate by means of disputations, and not only in the Dutch Republic, but also

32 G.A. Lindeboom, *Florentius Schuyl (1619–1669) en zijn betekenis voor het Cartesianisme in de geneeskunde* (Den Haag: Martinus Nijhoff, 1974) 15–16.

33 *Acta et decreta Senatus. Vroedschapsresolutiën en andere bescheiden betreffende de Utrechtsche Academie*, ed. G.W. Kernkamp (Utrecht: Kemink, 1936) 354–355: "ut clathrum aliquod ferreum, vel sepes aliqua ex ligno circa cathedram Respondentium formaretur, ut illi libere theses suas defendere, et auditores singula commodius intelligere possent."

34 H. Jamin, with M. Huisman, *Kennis als opdracht: de Universiteit Utrecht 1636–2001* (Utrecht: Matrijs, 2001) 45; Willem Otterspeer, *Groepsportret met Dame I. Het bolwerk van de vrijheid. De Leidse universiteit 1575–1672*, (Amsterdam: Bert Bakker, 2000) 375; Ahsmann, *Collegia en Colleges*, 285.

35 Bernard Varenius to Joachim Jungius, 17 May 1647 (in: Rob. C.G. Avé-Lallement, *Des Dr. Joachim Jungius aus Lübeck. Briefwechsel mit seinen Schülern und Freunden* (Lübeck: Aschenfeldt, 1863) 322): "Es ist aber Sitte in jener Academie, dass die Zuhörer mit Fussscharren und trampeln das begeleiten, was eben nicht nach ihrem Geschmack ist. Also grosses Geschrei, gewaltiger Tumult!"

at German schools.[36] It is therefore perhaps no wonder that the Amsterdam professor of theology Gerard van Leeuwen refused to organize disputations. As he explains in his inaugural address:

> I am thrilled that under the city's great authority, there should be no place for that tortuous and prickly genre of the disputation, from which all true, God-given knowledge, and even God himself, shrinks by its own nature. I have learned from long experience that a religion which, like ours, is full of mysteries, should only be steered and interpreted by a meticulous comparison of passages from Holy Scripture.[37]

One German author in 1659 satirized the disputants as pale, starving and sweating student-warriors quarelling over useless arguments such as the question of whether the hair of a goat was crispy. The litigious character of the disputation had already been the reason for Juan Luis Vives and Michel de Montaigne in the sixteenth, and Francis Bacon and René Descartes in the seventeenth century to reject it as method for seeking true knowledge, as it led students to think that winning over the audience was more important than striving for the truth.[38] This begs the question as to who composed the audience.

4 A Public Event and an Urban Theatre

Public disputations were in principle open to all, although we know very little about the attendance of people with no academic training, such as artisans and women. The disputations took place by the authority of the curators and were distinguished as ordinary or extraordinary disputations. The audience consisted of fellow students and professors of related disciplines, as well as the regents, family members, preachers, or former teachers honoured in the

---

36 Reimund B. Sdzuj, "Zum Cartesianismus am Bremer *Gymnasium Illustre*. Johann Eberhard Schwelings Dissertations *De anima brutorum* (1676)" in: Gindhart et al., eds., *Frühneuzeitliche Disputationen*, 179–198.

37 Gerbrandus van Leeuwen, *Oratio de Perpetuo Ecclesiae doctore Mose, habita in Illustri Amstelodamensium Athenaeo, cum SacroSanctae Theologiae Professionem auspicaretur, V. Calend. Mart.* (Amsterdam: J. Rieuwerts, 1686) 11: "Gaudeo quam maxime vestra autoritate ab hoc loco abesse debere disputationum illud genus tortuosum spinosumque, a quo cum ipse, tum omnis vera de Deo praecepta cognitio, sua natura tantopere abhorret. Longa enim factus sum certior experientia religionem quae, ut nostra, est mysteriis plena, sola et diligenti sacrarum literarum collatione esse moderandam atque explanandam."

38 Kivistö, *Vices of Learning*, 149–151.

disputations. Practitioners who had no academic position also attended, as is demonstrated by an incident in Amsterdam.[39]

The Amsterdam professor of medicine, Gerardus Blasius, complained that his colleague Johannes de Raei (a card-carrying Cartesian professor of philosophy who was bought away from Leiden University for a royal reward) was in the habit of boasting about his high salary and was constantly giving him orders. Blasius grew so weary of his pompous colleague that he provoked the Amsterdam physician Jan Baptiste van Lamzweerde, who had no formal relation to the institute whatsoever, to raise objections to De Raei's theses when De Raei presided over a student's disputation. When Lamzweerde rose to his feet to attack De Raei's student, the professor was taken completely by surprise. He afterwards complained to the curators that Blasius had passed his arguments on to Van Lamzweerde. In fact – according to De Raei – Blasius "asked, begged, and prayed" Van Lamzweerde to oppose De Raei.[40] This anecdote serves to illustrate that physicians with no formal binding to the pedagogical institute still had access to the disputations and that the disputation hall was a location in which learned debates were fought out at a serious level between professionals, and not merely for the sake of instructing students.

This was not the first time that Blasius used the disputation as an occasion to provoke confrontations. Blasius wrote a disputation, now lost, which criticized the work of the colourful Amsterdam physician Cornelis Bontekoe. Bontekoe was invited to the disputation hall, only to hear his own theories about jaundice being criticized in a disputation. Bontekoe felt that he was treated with contempt and that he had been denied sufficient time to defend himself.

Blasius's successor Pieter Bernagie also liked to denigrate Bontekoe. The latter had published a treatise on "the most excellent herb Tea" in 1678, against which Bernagie had taken up arms.[41] The second edition of Bontekoe's tea treatise was supplemented with others on coffee and on chocolate, as well as a less tasteful *Apology of the Author against his Slanderers*. In this piece of writing he

---

39 For Christian Thomasius' program of opening up the disputation to a wider public by switching from Latin to German, see Marti, "Kommunikationsnormen der Disputation," 329–332.

40 Luuc Kooijmans, *De doodskunstenaar. De anatomische lessen van Frederik Ruysch* (Amsterdam: Bert Bakker, 2004) 127.

41 Cornelis Bontekoe, *Tractaat van het excellenste kruyd thee: 't Welk vertoond het regte gebruyk, en de groote kragten van 't selve in gesondheyd, en siekten: benevens een kort discours op het leven, de siekte, en de dood: mitsgaders op de medicijne van dese tijd: ten dienste van die gene, die lust hebben, om langer, gesonder, en wijser te leven. Den tweeden druk vermeerdert, en vergroot met byvoeginge van noch twee korte verhandelingen. I. Van de coffi; II. Van de chocolate, mitsgaders van een Apologie van den autheur tegens sijne lasteraars* ('s Gravenhage: Pieter Hagen, 1679).

called Bernagie a "ground beetle," "full of wind, with envy in his heart, a tongue full of gall, and a pen full of lies," who criticized Bontekoe with "niggling, talking nonsense, distorting and finally spreading his filthy lies." Bernagie, or "silly Pete," is called names like "dumb cow," "woodlouse," "poisoned spider head" and many, many more.[42] Polemics like these, which were a familiar feature of the learned world,[43] more often than not originated with disputations that attempted to supply students with arguments against the enemy who lived and worked outside the university, but who did have access to the disputation hall.

As said, students usually dedicated their disputations to one or more persons: family members (fathers, uncles), city ministers, or professors or curators. It is almost unthinkable that those who found their names printed in the dedication would have been denied access to the disputation written in their honour.[44] This means that students' family members or even people with no direct or even indirect formal relationship with the university could still be present. Travelling students could witness disputations and seek out the best professors and most interesting courses.

Disputations were a public affair, and professors who often presided over disputations were well paid, because they increased the liveliness of the university. Disputations were a form almost of academic theatre.[45] The disputation hall was usually the same place as the lecture hall, and the whole setting must have given the impression of an interactive church service. Just as in a church, there was a *cathedra* from which professors delivered their *lectiones* and from which students defended their theses. In fact, most of the protestant universities in the Dutch Republic reconverted former Catholic churches to auditoriums: the Amsterdam auditorium or aula, for example, was a refurbished former convent chapel.[46] This must have reinforced the communal identity of the audience as being part of an academic parish. Contrary to the ecclesiastical cathedrae, however, the academic ones ("orateursstoelen") were usually lower (and not built up against a pillar) and they had a 'second,' smaller cathedra at

---

42 Van Miert, *Humanism in an Age of Science*, 104.

43 Ingrid De Smet, "*Calumnia dira pestis*: Slander, Memory and Collective Identity in the Republic of Letters, or How Polemics are Historicized," *Lias. Journal of Early Modern Intellectual Culture and its Sources* 38–2 (2011): 185–206.

44 Van Miert, *Humanism in an Age of Science*, 151–152.

45 Van Miert, "Retoriek in de Republiek. Vormen en functies van academische oraties in Amsterdam in de zeventiende eeuw," *De zeventiende eeuw* 19–1 (2003): 67–89.

46 Dirk van Miert, "Constructie en deconstructie van academische identiteit: de Agnietenkapel als *lieu de savoir*," in: *'Dingen die ergens toe dienen.' Verhalen over materiële cultuur van wetenschap*, eds. Esther van Gelder et al. (Hilversum: Verloren, 2017) 144–147.

the front, where the student had to stand (see figures 9.1, 9.2, 9.3 , 9.4 and 9.5).[47] In a depiction of a disputation in full swing at the illustrious school in Herborn, painted in 1751 in an Album Amicorum (fig. 9.6), we see that the professor sits in the pulpit, whereas the defending student stands before the front wooden panel, on which rests a book (presumably his printed disputation). An opponent stands opposite of him, facing him from behind a second wooden panel, and likewise holding a book. In this case there was no second cathedra, and the defendant and opponent were on the same level. The defendant awkwardly had his book declining top down, since the wooden panel, on top of which was a inclined reading board (for a psalter or bible?), was meant for an audience facing the pulpit. According to Marian Füssel, the spatial dimensions point at academic hierarchies, suggesting that the "epistemic choreography of the disputation was essentially supported by the social logic of the room."[48] This is all the more clear from figures 9.1–9.5, in which the defendant is placed midway between the opposing students down in the hall and the presiding professor breathing down his neck from above.

In figures 9.1a and 9.1b, the one on the pulpit may be Caspar Barlaeus, and the person sitting in front, may be the beadle. In my book *Humanism in an Age of Science*, I suggested that Barlaeus might also be the one sitting in front, and the person speaking (with the arm stretched out) a student, while the scene depicts a disputation. This is unlikely, however: Barlaeus as author of this collection of *Orationes* must take centre stage as speaker. Nevertheless, there clearly is a double cathedra, although the small chair seems to be a loose object (the Middleburg cathedra had its front small cathedra that could be detached). The image suggests that during lectures, the beadle sat in the small chair.

The beadle played a role not only during the *lectiones* of the professors, but also during disputations. Beadles were, as said, customarily responsible for the spread of the theses prior to the disputation and for the hall to be in order. They also headed the academic train of professors entering and exiting the lecture hall, and they might have done the same thing before a disputation.

---

47 I am grateful to Huib Zuidervaart for permission to reuse four images from Zuidervaart, "Het door George Sterlin vervaardigde academisch spreekgestoelte van de Middelburgse 'Illustre School,'" *Zeeland. Tijdschrift van het Koninklijk Zeeuwsch Genootschap der Wetenschappen* 27–1 (2018): 1–9.

48 Marian Füssel, "Die Praxis der Disputation. Heuristische Zugänge und theoretische Deutungsangebote," in: Gindhart et al., *Frühneuzeitliche Disputationen*, 27–48 (39): "Die Erhöhung des Präses, die räumliche Distanzierung durch die zum Teil aufwendig gestalteten Katheder und die Trennung von Akteuren und Publikum zeigen, dass die epistemische Choreographie der Disputation wesentlich von einer socialen Logik des Raumes getragen wurde."

FIGURES 9.1A–9.1B Details of the title pages of the two editions of Barlaeus's *Oratien en Blyde Inkomst van Maria de Medicis. Nieuwelijks met zijn gedichten verrijkt. Uit het Latyn vertolkt* (Amsterdam, 1662; Amsterdam, 1689). This is the Dutch translation of his work originally written in Latin, *Orationum liber, accesserunt alia nonnulla varii et amoenioris argumenti editio secunda* (Amsterdam, 1652). Both impressions are stylised: neither the auditorium of the Amsterdam Athenaeum nor Barlaeus are recognisable as such.
SOURCE: AMSTERDAM UNIVERSITY LIBRARY, SHELF MARKS BALIE 270: J AND OTM: O 63-6033

FIGURE 9.2
A double cathedra, constructed in the beginning of the seventeenth century for the auditorum of Leiden University, where it still is used today
PHOTO BY A. VAN HAASTER, ACADEMISCH MUSEUM, LEIDEN UNIVERSITY LIBRARY

FIGURE 9.3 The cathedra of the Middelburg Athenaeum Illustre, constructed in 1650 by carpenter George Sterlin
PHOTO BY CORNELIS HENNING (1906)

FIGURE 9.4 A double cathedra, built in 1729 for the University of Harderwijk
PHOTO BY R. ZIEGLER (1903), RIJKSDIENST VOOR CULTUREEL ERFGOED

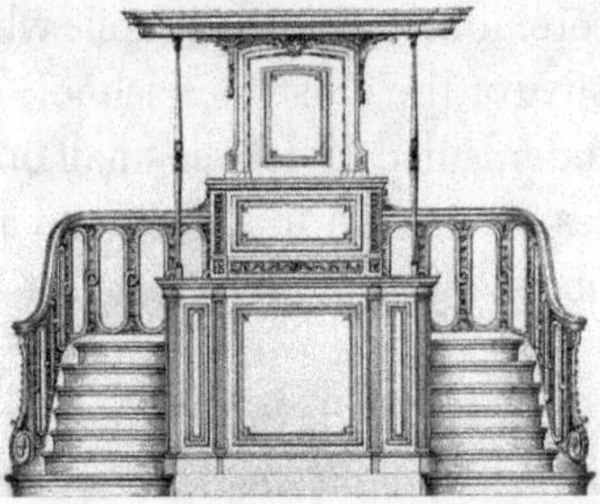

FIGURE 9.5 Design of a double cathedra made in 1777 for de Church of St John in Gouda, used since 1784 by teachers of the Latin School of Gouda

SOURCE: BIANCA VAN DEN BERG, *DE SINT-JANSKERK IN GOUDA: EEN OUDE STADSKERK VOLGENS EEN NIEUW RUIMTELIJK PLAN* (2008)

FIGURE 9.6 Disputation ceremony, painted entry by Johann Friedrich Winckel from Dillenburg, 2 April 1751, in the Album Amicorum of Johann Heinrich Achenbach, fol. 258r. Siegen City Archive, shelfmark Bestand Sammlungen, Sammlung 341-Stammbücher, A 92.

SOURCE: FÜSSEL, "DIE PRAXIS DER DISPUTATION," 39

From Middelburg we know that the double cathedra was also used by teachers and pupils from the Latin School during ceremonies at which pupils who finished school were promoted to the university (or the illustrious school) in an official ceremony during which the best students had to deliver small orations on being awarded prize-books.[49] Beadles made sure that everyone attending (professors, co-students, family members, occasional visitors) sat in their designated places and intervened when the audience disregarded the rules of conduct. The academic disputation was thus a public display in which the university showed the city their professors and students in action. As such, it was part of an urban landscape that hosted multiple platforms allowing for public performances, such as anatomical theatres, public theatres, markets, etc.[50] Teaching was a performance, and students had their share. Moreover, the disputation hall can be interpreted as a trading zone between university and city, between students and citizens, between professors and other scholars and between male academics and their (female?) relatives.

5 A Liminal Genre: Disputations and the Circulation of Knowledge

The disputation occupies territory between the private and the public sphere. Its mediating function is underscored by the fact that occasionally quotations from letters are found in the texts of the disputations. For instance, in a philosophical disputation defended in Amsterdam in 1657, the presiding professor of philosophy Arnold Senguerdius quoted an eye-witness account of the impact of a meteor from a letter from Andreas Colvius, thus bringing an unpublished letter to the attention of the academic community.[51]

Or take a more controversial example. In the early 1640s, many responses to the scandalous pre-Adamite theory of Isaac de La Peyrère were communicated by unpublished letters; later on, some letters were posthumously published; subsequently, those letters were cited in disputations. A Leiden disputation presided over by the professor of theology Johannes Hoornbeek in 1655 contained the earliest printed response to La Peyrère's printed *Systema theologicum*:

> Hence the origin of the pre-Adamites – so much for the not so recent *Dream of a nobleman from Aquitaine*, or of La Peyrère, an author indicated

49 Zuidervaart, 'Het door Sterlin vervaardigde spreekgestoelte,' 7.

50 Van Miert, 'Retoriek in de Republiek.'

51 Van Miert, *Humanism in an Age of Science*, 154 and 265.

> by Claude Sarrau in a letter to Saumaise from 1643, where he writes on p. 74: "Today La Peyrere came to see me and he said he wanted to publish this little work of his, in which he attempts to show that before Adam there existed other people."[52]

Such quotations point out how printed letters could immediately enter the public debate as near-contemporary witnesses.[53]

In a disputation of December 1655, too, Hoornbeek cited Hugo Grotius's opinion, who thought that La Peyrère's outrageous hypothesis posed a serious danger. Hoornbeek also refers to the public ban on the book, issued by the Court of Holland on 26 November 1655, i.e., less than a month before the defense of the disputation, according to which it was an "impious, scandalous work, meant to seduce people."[54] Indeed, this is precisely what the Court

---

52 Hoornbeek (praeses), Johannes Verscheepker (resp.), *Disputationum theologicarum ordinariarum repetitarum decima-octava De lege Dei*, 24 December (Leiden: J. Elzevier, 1655), thesis II, sig. Z2^v: "Inde natos Adamum et Eva. En originem Prae-Adamitarum. Adeo id non nuperum demum Nobilis Aquitani somnium, sive Peyrerii, quem auctorem indicat Cl. Sarravius ad Salmasium, epistola, a MDCXLIII ubi haec pag. 74 "convenit me hodie Peyrerius, dixitque velle se edere opusculum illud suum, quo docere conatur ante Adamum extitisse alios homines: consiliumque suum esse illud tibi inscribendi, sub hoc titulo, Somnium Nobilis Aquitani, de Prae-Adamitis, allusione scilicet facta ad scomma Grotii." The Latin has "non nuperum demum somnium," where "non nuper" and "demum" contradict each other. Perhaps Hoornbeek meant that the book, which had only (demum) come out that year in 1655, had already (non nuper) been conceived in 1643. The "dream of a nobleman" refers to a scathing remark by Hugo Grotius against La Peyrère: Hugo Grotius, *Dissertatio altera de origine gentium Americanarum* (Paris: Cramoisy, 1643) 15: "… aut aliquos ante Adamum fuisse conditos homines, ut nuper aliquis in Gallia somniavit. Si hae credantur, magnum video periculum pietati." For the quotation from Sarrau's letter to Saumaise, see the postscript in Sarrau to Saumaise, 12 November 1643, in: Claude Sarrau, *Epistolae. Opus posthumum ad Serenissimam Christinam Suediae Reginam* (Orange: s.n., 1654) 74–75.

53 This was not the first time Hoornbeek cited a recently published letter in a disputation: he had done so five years earlier, quoting – with apparent approval – Grotius's positive impression of the Catholic dissident Paolo Sarpi in Grotius's posthumously published *Epistolae ad Gallos*: Hoornbeek (praeses), Engelbertus Sloot (resp.), *Disputationum Theologicarum Ordinariarum repetitarum nona, de Praedestinatione*, 6 February (Leiden: J. & D. Elzevier, 1650), sig. N5^r . He also quotes a letter from Sarpi to Daniel Heinsius (sig. N5^r).

54 Hoornbeek (praeses), Johannes Verscheepker (resp.), *Disputationum theologicarum ordinariarum repetitarum decima-octava De lege Dei*, 24 December (Leiden: J. Elzevier, 1655), thesis II, sig. Z2^v: "Parum est, somnium dicere, quod Grotius, atque is addidit "magnum" inde "periculum" pietati: "impium, scandalosum, et ad seducendum homines natum" Curia Provincialis Hollandiae nuper judicavit, publico Decreto a MDCLV. d. xxvi Novembr."

of Holland and Zeeland said in its broadsheet which forbade the printing, reprinting, selling, translating, distributing or publishing of the work and which stipulated that all printers and booksellers must hand over their copies to the local authorities.[55]

Clearly, the disputation hall was not merely a location used for rehearsing received knowledge, but also for circulating and testing new knowledge or for warning against the dangers of new theories. On the threshold of the private and the public spheres (as far as these can be clearly distinguished at all), the disputation hall acted as an urban trading zone.

6 An Urban Location of Knowledge

Disputations enhanced the profile of a city. Having an academic institution within the city walls contributed to the city's reputation as an intellectual and cultural hub, which in turn strengthened its political power as well and added to its cosmopolitan character. Most universities in the United Provinces were paid for not by the cities but by the sovereign provincial states. The city hosting the university was keen on defending its privilege. Many cities that could not legally obtain a university (such as Amsterdam or Middelburg), came up with alternative ways of furnishing themselves with an academic image, by founding illustrious schools: university-like city academies, which sometimes competed with academic curricula elsewhere, but which more often than not were mere extensions of the municipal Latin Schools. Such illustrious schools did not have the right to grant degrees, but even there the practice of disputations, ideally a preparation for obtaining a doctorate, was held in high esteem. Practices differed, though. In Amsterdam there was a lively tradition of disputations, but in Middelburg repeated attempts to create an Illustrious School foundered time and again, and disputations were held only occasionally.[56]

---

55 *The Early Modern Pamphlets Online* (TEMPO), ed. W.P.C. Knuttel and G. van Alphen (Leiden: Brill) no. 7662: "schandaleus, Godtloos, ende tenderende omme de Ingesetenen van desen Staet ende allen anderen van de Christelijcke Religie af te leyden." See: https://primarysources.brillonline.com/browse/dutch-pamphlets-online.

56 Van Miert, *Humanism in an Age of Science*, passim; Willem Frijhoff, "Zeelands Universiteit: hoe vaak het mislukte en waarom," *Archief. Mededelingen van het Koninklijk Zeeuwsch Genootschap der Wetenschappen* (1986): 7–41. Zuidervaart, "Zeeuws preacademisch erfgoed; idem, 'Het door Sterlin vervaardigde spreekgestoelte,' 6–7. See also Zuidervaart, "The Middelburg Theatrum Anatomicum: A Location of Knowledge and Culture in an Early Urban Context," elsewhere in this volume.

To conclude, I suggest that historians should take disputations seriously as sources for the history of science and not dismiss them as mere educational repetition of received knowledge. Moreover, the disputations are important not only as a printed source of knowledge. If we want to understand the ways in which knowledge circulated in specific places, we should study the way in which knowledge was conveyed to the public: the occasions, the public, the timing, in short the entire social and spatial setting. This includes the acknowledgment that the disputation hall was fully integrated into the knowledge infrastructure of the early modern city.

Of course, much of this message applies to the academic auditorium not only as the scene for disputations, but also as the place where the academic community gathered for daily public lectures and the location where, every hour, students took notes. That, however, requires the analysis of orations and lecture notes. These sources are far fewer in number and rarely survive with the serial regularity of the printed disputations that patiently gather dust in our academic libraries.

CHAPTER 10

The Lorentz Transformation of a Museum

Martin P.M. Weiss

1 Introduction

Hendrik Antoon Lorentz' professional biography is testimony to a whole range of transitions in the history of science that occurred around the turn of the twentieth century.[1] In 1878, for instance, he was appointed to the very first chair of theoretical physics in Leiden in the Netherlands at the early age of 25. His subsequent work in this field proved pivotal to the emergence of both the theory of relativity and quantum theory, and he soon found himself in the midst of fundamental discussions on the implications of the ensuing break with what came to be known as "classical physics." Finally, he was one of the most prominent representatives of what has been described as the "Second Golden Age" of Dutch science, and one of the earliest beneficiaries of a new type of institute of higher education in the Netherlands, the so-called *Hogere Burgerscholen* (HBS). These transitions and Lorentz' role within the physics community of his day have been studied before, yet almost all of these studies focus on his work at Leiden; what is only addressed perfunctorily is that for the last 19 years of his life, Lorentz was in fact an employee of Teylers Foundation in Haarlem, having resigned his chair and accepted only an honorary professorship in Leiden.[2]

---

1 This article is an extended version of an account published in: Martin P.M. Weiss, *Showcasing Science: A History of Teylers Museum in the Nineteenth Century* (Amsterdam: Amsterdam University Press, 2019). The author would like to thank Frans van Lunteren, Marijn van Hoorn and the anonymous reviewer for many helpful comments on the subject of this paper.

2 On Lorentz and his position within the history of science, see: Anne J. Kox, "Hendrik A. Lorentz 1853–1928," in: *Van Stevin tot Lorentz: Portretten van achttien Nederlandse natuurwetenschappers*, ed. Anne J. Kox (Amsterdam: Bert Bakker, 1990) 226–242; Bastiaan Willink, *De Tweede Gouden Eeuw: Nederland en de Nobelprijzen voor Natuurwetenschappen 1870–1940* (Amsterdam: Bert Bakker, 1998) 105–109; Bert Theunissen, *'Nut en nog eens nut': Wetenschapsbeelden van Nederlandse natuuronderzoekers 1800–1900* (Hilversum: Verloren, 2000) 168–184. On Lorentz and Teyler, see: Marijn van Hoorn, "The Physics Laboratory of the Teyler Foundation (Haarlem) under Professor H.A. Lorentz, 1909–1928," in *Bulletin of the Scientific Instrument Society* 59 (1998): 14–20. Van Hoorn focuses on the research performed under Lorentz' purview and the instruments used. For a general overview of the history of

 | DOI:10.1163/9789004264885_011

FIGURE 10.1 Hendrik Antoon Lorentz in later years
COURTESY OF TEYLERS MUSEUM

This strong emphasis of his work at a university has to do both with the fact that Lorentz achieved his greatest successes on Leiden University's payroll, and that he remained on in Leiden as honorary professor even after having moved to Haarlem; yet it is also symptomatic of one of the most profound changes within the topography of Dutch science to occur around the turn of the century, which in turn is reflected by events as they unfolded around Lorentz' tenure in Haarlem: the transition of universities from mere institutes of higher education to places of both education and research. More importantly, by the 1920s universities had become synonymous with serious scientific research, certainly as far as physics was concerned, and especially if it was to be research of an unbiased nature.

This new-found role of universities as research centres was closely connected to two other developments: firstly, the idea that scientific research was worth pursuing for its own sake – this was typical of HBS graduates, many of whom went on to become professors of science at universities; and secondly the changing scale of research projects. As science became increasingly specialised and research involved ever more sensitive measurements, real progress could only be achieved through cooperation amongst scientists. Together with the increasingly elaborate equipment required by all these scientists, this cooperation made research ever more costly, with the result that ultimately

Dutch science during Lorentz' lifetime, see: *A History of Science in the Netherlands*, eds. Klaas van Berkel, Albert van Helden and Lodewijk Palm (Leiden: Brill, 1999) 130–210.

only governments, industry, or large philanthropic organisations were able to provide the required funds.

This brings us back to Teylers Foundation in Haarlem. Some 130 years before Lorentz came to Haarlem, this foundation had started providing his predecessors with both the premises and the equipment they required for their own research endeavours. Because these were unique circumstances, and because the Foundation's budget was notoriously high, the curatorship in Haarlem was a highly coveted position. Yet by the time Lorentz passed away, the Foundation was mired in financial difficulties, could no longer compete in terms of research with larger institutions such as universities, and was finding it difficult to attract well-qualified candidates to work at its research premises – they generally preferred the research environment provided by universities.

Sections three, four and five of this paper illustrate this changing role of Teylers Foundation within the Dutch research community. Questions will be addressed such as why Lorentz was approached by the Foundation in the first place, why he accepted the job of curator and resigned from his post at the University of Leiden, what kind of research was conducted at the laboratory during his tenure, and what effect the emergence of universities as centres of scientific research as well as the financial setbacks Teylers Foundation experienced had on the position of the laboratory.

The sixth, concluding section will then draw attention to another aspect of Lorentz' work for Teylers Foundation that, as will transpire, was in fact of little importance to Lorentz himself, yet can serve to illustrate the changing role and definition of another class of institution of importance to science besides universities, namely museums. In a nutshell, the first decades of the twentieth century saw the emergence of both science museums and museums for the history of science. Although there were many precursors to these institutions, they acquired a new quality over the course of Lorentz' tenure in the sense that they became overtly pedagogical, self-confidently advertising both the virtues of science and its seemingly noble history to the general populace.

The reason this bears any relevance to Lorentz' tenure in Haarlem is that as curator he was not only given responsibility for Teylers Foundation's laboratory, but also for all the instruments that his predecessors had acquired over the previous 130 years. Most if not all of them had been preserved, and were obviously gaining historical value. Crucially, most of these instruments were not stored in the laboratory itself, but in the adjacent Teylers Museum. This museum was home not only to these instruments but also to a collection of fine art, a geological collection, and a library. It had been founded with no pedagogical intentions whatsoever, and no instruments had ever been acquired purely for their historical value, but it did attract about 5000 visitors every year when Lorentz arrived and – given the emergence of museums for the history of science over the course of the decades after Lorentz' arrival – it is worth addressing the issue of how this museum compared to others, and how Lorentz regarded the collection.[3]

In preparation for this final section of the article, as well as in order to provide the reader with some more background information on the laboratory and the instrument collection Lorentz left Leiden for, the following section will provide a short overview of the history of both institutions.

2 A Brief History of Teylers

The history of Teylers Museum is intimately connected with that of Teylers Foundation's laboratory. For one, both were paid for by Teylers Foundation. More importantly, a distinction between the two only gradually arose and became increasingly pronounced as the nineteenth century progressed. One could even say that the museum had initially been conceived as a laboratory. Its initiator, Martinus van Marum, certainly used it as such after the museum had been completed in 1784, performing experiments in the original museum building – the so-called Oval Room – with his famous electrostatic generator. Yet once his interest turned from electricity to chemistry, he must soon have felt that the Oval Room had become inadequate for experiments: in 1791 the *Algemeene Konst- en Letterbode* proclaimed that Teylers Foundation had furbished "a spacious and extremely well equipped *Laboratory*" in a house near the museum.[4]

One of the reasons Van Marum would probably have felt the original museum was not suitable for chemical research was that it was so crowded. Not only did it house his instruments and a fossil collection and a collection of minerals he had acquired for the Foundation, but also books and a collection of fine art. The fact that the Oval Room was such a multifunctional room in turn had everything to do with Teylers Foundation's origins: it had been set up to look after the considerable fortune of Pieter Teyler van der Hulst after his death in 1778. More specifically, it was set up in order to ensure that his money would serve the causes his will had stated were close to his heart: alongside the study of theology and charitable causes, these included the arts and sciences. And in order to comply with his wish that both the arts and sciences were to

3 Visitor numbers for this period are given in the annual reports of the curator of the art department. Verslagen, 1908–1913: Haarlem, Teylers Museum, Archives of Teylers Foundation (hereafter ATS), inv. no. 147.

4 "... een ruim en ongemeen wel ingerigt *Laboratorium*": *Algemeene Konst- en Letterbode*, 7 (1791): nr. 182, 204. The author is grateful to Huib Zuidervaart for drawing his attention to this article. All translations are the author's except where indicated otherwise.

be encouraged with his endowment, the five trustees of the Foundation saw to it that a collection of prints and drawings was acquired alongside Van Marum's scientific collections for the museum carrying Teyler's name. Their most spectacular acquisition in this respect, in 1790, was a collection of drawings formerly belonging to Christina of Sweden. This included works of art by such famous artists as Michelangelo, Rembrandt, and Lorrain.[5] From the 1820s onwards, the Foundation also started acquiring paintings by contemporary Dutch artists alongside prints and drawings, and over the years two exhibition halls were added to display them, the first in 1839, the second in 1893. Before this second art gallery was added, another major annex was joined to the Oval Room. This gallery was completed in 1885, in neo-classical style, and housed the instrument collection and the collection of fossils as well as a lecture theatre and parts of the library.[6] It also provided the museum with an eye-catching monumental entrance façade. Visitor numbers to the museum subsequently rose about fourfold to about 3,000 per year.[7] The Oval Room was not affected by any of these extensions to the museum and remained intact and in its original state.

The laboratory that eventually fell under Lorentz' purview was constructed in 1884, in two houses adjacent to the 1885 extension to the museum which were joined. These premises were completely refurbished upon the initiative of Elisa van der Ven, Lorentz' direct predecessor, and had most likely already provided the premises for Van Marum's laboratory.[8] Van der Ven obviously felt that a smaller, one-room laboratory that had been built for his predecessor Volkert Simon Maarten van der Willigen between March 1866 and February 1867 (this was referred to as the "Observatory" because of a slit in its roof) was no longer sufficient.

In a way, by the time Lorentz arrived, the museum had therefore evolved from a place of experimental research into both an art museum that housed

---

5 On the history of this collection see Carel van Tuyll van Serooskerken, *The Italian Drawings of the Fifteenth and Sixteenth Centuries in the Teyler Museum* (Haarlem: Teylers Museum / Gent: Snoeck-Ducaju, 2000), 13.

6 On the features of this wing and the history of its construction see: T. van Gestel and A.W. Reinink, "Het 'Nieuwe Museum' van Teyler," in: *'Teyler' 1778–1978. Studies en bijdragen over Teylers Stichting naar aanleiding van het tweede eeuwfeest* (Haarlem/Antwerpen: Schuyt & Co, 1978), 223–321.

7 Geert-Jan Janse, "Uit nieuwsgierigheid en ter onderricht," in: *Teylers Museum: Een reis door de tijd*, ed. Marjan Scharloo (Haarlem: Teylers Museum, 2009), 11–29, 24.

8 On the decision to refurbish the laboratory, see Directienotulen, 4 January 1884 & 29 February 1884. ATS, inv. no. 11. The author is grateful to Anne Marie ten Cate for having drawn his attention to these passages of the minutes of the meetings of the trustees of Teylers Foundation.

scientific collections as well as a library on the one hand, and a scientific laboratory in an adjacent building on the other. Lorentz' colleagues in Haarlem included a curator of the art collections, a librarian, a curator of the palaeontological and mineralogical collections, and a range of caretakers, assistants and guards, including the laboratory attendant Jilles van Waveren who had already assisted Van der Ven.

3 An Offer He Couldn't Refuse

Elisa van der Ven passed away on 27 June 1909, after having served Teylers Foundation as keeper for more than 30 years. Even before his funeral, the first unsolicited application was sent off to the trustees, asking for details about the vacancy that had obviously just opened.[9] Although this applicant – a teacher at a local polytechnic – received a fairly curt reply, two others who wrote just days later were told that nothing had been decided yet with regard to this position, but applications could be sent to the trustees.[10] By the time these applications would have arrived (none have been preserved) towards the end of July, the trustees had, however, already set their sights on Lorentz. The minutes of the meeting of the board of trustees on 28 July 1909 read: "Regarding the vacancy for keeper of the Physical Cabinet advice will be sought from prof. Lorentz in Leiden, at which opportunity the professor will be asked if he is inclined to accept this position himself."[11] A letter inviting Lorentz to Haarlem was sent off that very same day – although no mention was yet made of the vacant post being offered to him.[12] Lorentz promptly replied he would be happy to come, and a meeting was scheduled for early August.[13] It was to take place at the home of one of the trustees, Jan Adriaan Fontein, and the two men were to be joined by another trustee of Teylers Foundation, Anthonie Wilhelm Thöne.[14]

---

9 Oosting to trustees of Teylers Foundation, 30 June 1909. ATS, inv. no. 36.

10 Meerburg to trustees of Teylers Foundation, 3 July 1909 & Stoel to trustees of Teylers Foundation, 4 July 1909. ATS, inv. no. 36. Copies of the replies they received: Trustees of Teylers Foundation to Meerburg, 10 July 1909. ATS, inv. no. 57, fol. 72 & Trustees of Teylers Foundation to Stoel, 10 July 1909. ATS, inv. no. 57, fol. 73. Further letters of inquiry have been preserved ATS 36, some of the replies they received in ATS 57.

11 "Over de vacature conservator Physisch Kabinet zal advies worden gevraagd aan prof. Lorentz te Leiden, bij welke gelegenheid dien hoogleraar zal worden gevraagd of hij genegen is zelf de betrekking te aanvaarden": Directienotulen, 28 July 1909: ATS, inv. no. 13.

12 Trustees of Teylers Foundation to Lorentz, 28 July 1909. ATS, inv. no. 57, fol. 82.

13 Lorentz to trustees of Teylers Foundation, c. 30 July 1909. ATS, inv. no. 36.

14 Trustees of Teylers Foundation to Lorentz, 2 August 1909. ATS, inv. no. 57, fol. 84.

Of course, the question arises why the trustees approached Lorentz, rather than anyone else. There was certainly no shortage of talented and internationally recognised physicists in the Netherlands at the time – the decades around 1900 have even summarily been described as the "Second Golden Age" of Dutch science – and Lorentz was one of the most prominent amongst them, respected not only as a brilliant physicist who had just won the Nobel prize, but also as a person.

In a way, Lorentz' high reputation was probably one of the reasons the trustees decided to approach him. They saw themselves as heirs to an institution with a long-standing tradition as one of the nerve centres of Dutch science. As the previous section showed, this was not entirely unjustified. What's more, because Van der Ven had been more of a populariser of science than a researcher, and because he had been frail for much of his last years, the trustees must have felt it was time to re-establish Teylers' reputation in scientific circles. In an account of his first meeting with Thöne and Fontein, Lorentz wrote that his task at Teylers would be "[t]o turn T.[eylers] into a place where work of considerable importance is performed and from which a certain influence is exerted."[15] These are certain to have been the trustees' words, as much as his own.

But circumstances suggest there was also a second aspect: Lorentz was deeply dissatisfied with his position in Leiden. The reason was, firstly, that he had no laboratory of his own – even though he had been appointed to the first chair of theoretical physics in the Netherlands, he would have liked to be able to perform the occasional experiment – and, secondly, that he was saddled with a high teaching burden which left hardly any time for his own research.[16] Both issues had a lot to do with his equally brilliant fellow physics professor (for experimental physics) in Leiden, Heike Kamerlingh Onnes. Kamerlingh Onnes was soon to receive his own Nobel prize for liquefying helium, which he had succeeded in doing less than a year before Van der Ven's death, on 10 July 1908. But Kamerlingh Onnes was also a gifted manager, the physical manifestation of which became the constant refurbishments and major extensions to the university physics laboratories in Leiden, for which he repeatedly succeeded to gain funding. It was not that Lorentz and Kamerlingh Onnes didn't get on – on the contrary – but somehow, despite the constant addition of work space, even a small laboratory exclusively for Lorentz' use never materialised. At one point two rooms in the building were earmarked for Lorentz so that he could use them for his experiments – however, they were ultimately – and possibly

15 "T.[eylers] tot een plaats te maken, waar werk van eenige beteekenis gedaan wordt en van waar een zekere invloed uitgaat": Lorentz to Bosscha, c. 17 August 1909. ATS, inv. no. 183.

16 See Dirk van Delft, *Freezing Physics: Heike Kamerlingh Onnes and the Quest for Cold* (Amsterdam: Aksant Academic Publishers, 2007), 352–357.

quite simply wrongly – assigned to someone else, again thwarting Lorentz' hopes of establishing a small laboratory of his own. When he was offered a professorship without teaching duties at the University of Munich in 1905, he only stayed in Leiden because promises had been made to lighten his teaching burden. Although he was indeed provided with an assistant, Lorentz still felt, in his own words, that "I may, it seems to me, regard matters in such a way that a change has only been brought about in the interest of the University, but in no way for me personally."[17] For Lorentz, these are unusually strong words. Ever conciliatory and in full control of his emotions, his daughter describes how the issue of the two rooms being wrongly assigned "was not discussed." She adds: "My father preferred, rightly or wrongly, to keep his peace of mind rather than to create a disturbance unless it were strictly necessary."[18]

Either way, these circumstances would have made Lorentz especially receptive to what the trustees of Teylers Foundation had to offer. But could they also have known about his situation? The evidence that indicates that in all likelihood they did is the intimate involvement of Johannes Bosscha Jr. in the negotiations preceding Lorentz' appointment. As Fontein later recalled, it was Bosscha "who showed the trustees Lorentz' path to Haarlem."[19] Bosscha was the general secretary of the Holland Society of the Sciences, and had already offered Lorentz to succeed him in this post in 1908, in an attempt at providing Lorentz with a graceful exit from Leiden and to relieve him of his teaching obligations.[20] The two men had known each other for a long time and respected each other deeply. Bosscha lived in Heemstede, near Haarlem. The Holland Society's offices were across the street from Teylers Museum. Fontein was a member of the Holland Society.[21] So multiple channels of communication were open, and as a letter from Bosscha to Lorentz reveals, they were used, too: just days after the first meeting between Thöne, Fontein and Lorentz, Johannes Bosscha writes to Lorentz reporting that he had met "his friend Fontein" during an evening stroll along the beach to enjoy "the beautiful sunset," and that Fontein had told him about the meeting at his home. Bosscha is joyous that "you are seriously considering Teylers' proposals," and then revealingly

---

17 "Ik mag het, dunkt mij, zoo beschouwen dat er alleen ten behoeve van de Universiteit en geenszins voor mij persoonlijk eene verandering is gekomen": Lorentz to Bosscha, c. 17 August 1909. ATS, inv. no. 183.

18 G.L. de Haas-Lorentz, *H.A. Lorentz: Impressions of his Life and Work* (Amsterdam: North-Holland Publishing Company, 1957), 98.

19 "... die voor Lorentz, den weg naar Haarlem aan Directeuren heeft gewezen": Directienotulen, 17 February 1928. ATS, inv. no. 15.

20 Lorentz to Bosscha, c. 17 August 1909. ATS, inv. no. 183.

21 J.A. Bierens de Haan, *De Hollandsche Maatschappij der Wetenschappen 1752–1952. Herdruk met Supplement 1952–1970* (Haarlem: H. D. Tjeenk Willink, 1970), 383.

adds – not without apologising – "I already knew of them, but was not allowed to speak about them, when I received your pleasant visit."[22] Unsurprisingly, Bosscha then strongly advises Lorentz to take up the position offered, praising the clean air in Haarlem as a great advantage not only for the family.

At any rate, Lorentz had obviously begun seriously considering accepting the offer to succeed Van der Ven, starting a round of four months negotiating with the trustees, even though as far as they were concerned, there was little to negotiate about. In their enthusiasm – probably Fontein's above all – they essentially agreed to everything Lorentz proposed. Yet not all of the trustees were equally enthusiastic: already during the very first meeting at which approaching Lorentz had been discussed, two of the trustees had voiced fears that the Foundation's financial clout was not sufficient to afford such a prestigious successor to Van der Ven – although they were quick to add that, otherwise, they were "very taken with the concept of associating prof. Lorentz with the Foundation."[23] These two were Pieter Loosjes and Louis Paul Zocher. Four months later, Loosjes's doubts had been alleviated, but Zocher still had it recorded in the minutes that he objected to Lorentz being taken on.[24] Again, he added, purely on financial grounds. To Lorentz and Bosscha, it was as yet inconceivable that Teylers Foundation's resources were anything but unlimited. When Lorentz confessed that he would feel uneasy accepting Fontein's offer of a continued full salary even in retirement, Bosscha replied that Teylers was "wealthy enough" to pay both him and a successor a full salary for some years.[25]

Zocher's reservations later proved not to be unfounded, but in 1909, Fontein's enthusiasm went unabated, and everything was done to entice Lorentz to come to Haarlem. Beside the problems irritating him in Leiden that were mentioned above, Lorentz was tempted for three reasons. Firstly, he repeatedly stated that he would be happy to work for a "Dutch" institution with such a rich heritage as Teylers Foundation.[26] Secondly, he told Bosscha that it was "tempting, before it is too late to do so, to take a slightly different path [professionally], on which I might possibly be able to work more fruitfully than in […] my current

---

22 "… gij Teylers voorstellen in ernstige overweging neemt" / "Ik wist er reeds van, maar mocht er niet van spreken, toen ik uw aangenaam bezoek ontving": Bosscha to Lorentz, 11 August 1909. ATS, inv. no. 183.

23 "… zeer ingenomen met het denkbeeld prof. Lorentz aan de Stichting te verbinden": Directienotulen, 28 July 1909. ATS, inv. no. 13.

24 Directienotulen, 10 December 1909. ATS, inv. no. 13.

25 "… vermogend genoeg": Bosscha to Lorentz, 17 August 1909. ATS, inv. no. 193.

26 He speaks of "eene zoo echt Nederlandsche instelling als T." In Lorentz to Bosscha, c. 17 August 1909. ATS, inv. no. 183; the same phrase is used in Lorentz to trustees of Teylers Foundation, 25 August 1909. ATS, inv. no. 36.

position."[27] And thirdly, he must have thought he might be more productive in Haarlem than he was in his "current position" because he had access to a laboratory of his own, could edit the Foundation's small scientific journal, the *Archives du Musée Teyler*, and would essentially be his own boss all the time, having to answer only to the trustees – who were obviously eager to support him wherever they could.

Yet he had some worries as well. The first was that he felt there was no appropriate successor for him in Leiden yet. He granted that "there are indeed some promising young folk, but it is precisely over the course of the next few years that it shall have to transpire what they are really capable of."[28] His second worry was that he could forfeit his rights to the considerable government pension which he would only receive if he worked at a university and remained in government service right up until his retirement age of 65, which would have been in 1918. Although he doubted that, after 31 years of service, both the university and the government would create unnecessary difficulties in this respect, he did want this to be resolved in an orderly fashion.

Having weighed all the pros and cons for about three weeks (Fontein had told him he could take all the time he wanted), Lorentz then wrote to the trustees.[29] As a solution to his worries he suggested a transitionary phase of three to four years in which he would already place himself at the Foundation's disposal, but stay on in Leiden as a full professor until the question of his succession had been resolved. After that, he would stay on as an honorary professor or a normal lecturer at the university in order to secure his pension, but move to Haarlem and consider the Foundation his primary employer. During this transitionary phase, the salary he received from the Foundation did not have to be a full one.

Lorentz then went on to explain how he would define his tasks at Teylers, should he be engaged by the Foundation. His first concern was the laboratory which he wrote "must not be neglected" but used for the preparation of lectures and, "to no lesser degree, [be] a place where scientific research is performed."[30]

---

27 "verlokkend, vóór het daartoe te laat wordt, nog eens een ietwat andere richting in te slaan, waarin ik misschien met meer vrucht zou kunnen werkzaam zijn dan in [...] mijne tegenwoordige positie." Lorentz to Bosscha, c. 17 August 1909. ATS, inv. no. 183.

28 "Er zijn wel een paar veelbelovende jongelui, maar juist in de eerstvolgende jaren zal moeten blijken wat zij kunnen presteren." Lorentz to Bosscha, c. 17 August 1909. ATS, inv. no. 183.

29 For Fontein's offer see: Lorentz to Bosscha, c. 17 August 1909. ATS, inv. no. 183.

30 "Het laboratorium van "Teyler" mag niet worden verwaarloosd." / "vooral niet minder, een plaats zijn, waar wetenschappelijk onderzoek verricht wordt." Lorentz to trustees of Teylers Foundation, 25 August 1909. ATS, inv. no. 36.

Interestingly, he then stated that the trustees were surely aware that, thusfar, he had concerned himself more with theoretical rather than with experimental work. Although he wanted to remain in overall charge of the activities at the laboratory and conduct the occasional experiment himself, he also stated that he could not oversee the day-to-day running of the laboratory, and therefore proposed that an assistant be appointed, preferably "a young, promising physicist [...] who has already earned his spurs, and who has shown that he can work on his own, so that I could, not only during the years I would still spend in Leiden, but also later, leave the laboratory almost entirely to him."[31] This young physicist could receive the title "conservator" (keeper), as assistant professors at universities then did, and also receive a comparative salary.

Together, Lorentz continued, they could then run Teylers like a Dutch equivalent to the Royal Institution in London. They would take care of the instrument collection (referred to as "the cabinet"), perform research in the laboratory and assist external researchers who might want to use the research facilities, give lectures and courses for the general public as well as science teachers, and edit the *Archives*.

Finally, Lorentz still pointed out that implementing his ideas could be costly, as the refurbishment of the laboratory as well as the keeper's salary would incur extra costs. But the trustees were unperturbed. They arranged for another meeting with Lorentz on 27 October 1909, and although no minutes of the meeting were taken or have been preserved, it is clear from their subsequent correspondence with Lorentz and Bosscha, that the trustees had gone along with all of Lorentz' proposals, and Lorentz had tentatively committed himself to working for the Foundation. All he still wanted to do was personally inform his colleagues in Leiden and the board of the faculty of his plans to leave them. He reassured the trustees that their reaction would, however, have no impact on his decision to come and work for the Foundation. One interesting point he still brings up is a suggestion by Herman Haga to turn the laboratory at Teylers into a Dutch equivalent of the *Physikalisch Technische Reichsanstalt* in Berlin, which would have meant it would calibrate all instruments in the Netherlands. Lorentz himself was only lukewarm about the idea, and it never materialised.[32]

---

31 "... een jong, veelbelovend physicus [...] die zijne sporen reeds verdiend heeft, en die getoond heeft, zelfstandig te kunnen werken, zoodat ik hem, niet alleen in de jaren die ik nog te Leiden zou doorbrengen, maar ook later, het laboratorium zo goed als geheel zou kunnen overlaten."

32 Lorentz to trustees of Teylers Foundation, 4 November 1909. ATS, inv. no. 36. Together with Kamerlingh Onnes and Bosscha, Haga had earlier suggested that this task was to be performed by HBS-teachers across the country. See: Frans van Lunteren, "Van meten tot weten," *Gewina* 18 (1995): 102–138.

A final meeting was scheduled for 18 November in Haarlem to agree on contractual details. Lorentz gave his assessment of the laboratory he was to be made the head of (he had inspected it some days earlier), and reported that he had come to the conclusion that although a refurbishment was necessary, a new building was not – at least not for another twenty to twenty-five years. Once that period of time had passed, the new laboratory space could be created in an adjacent building. As for his title, "professor," "consultant" and "advisor" were floated, but the title "curator" was eventually agreed upon. So Lorentz was to be the "curator" of the Foundation's laboratory, and his assistant the keeper. Lorentz agreed to draw up a list of possible assistants as soon as possible.

Lorentz set to work doing so over the course of the following weeks, and on 9 December 1909 he suggested Gerhard Johan Elias, who at that point was working in Berlin with Henri du Bois, assisting him in his magneto-optical research.[33] Lorentz wrote that he was an excellent candidate because he had gained a lot of experience in running a laboratory in Berlin, and because magneto-optical research was what Lorentz intended to focus on in Haarlem. Just as interesting, however, is the list of candidates who either refused the job or whom Lorentz rejected. His first choice, the director of the Royal Magnetic and Meteorological Observatory in Batavia, Willem van Bemmelen, refused to leave the Dutch Colonies unless he could get a significant pay rise. Chemists, on the other hand, Lorentz did not take into consideration because he wanted to run a physics laboratory. He did seriously consider choosing an assistant from a long list of science teachers at schools around Haarlem – most of them well educated and with ample experience in running their institutions' laboratories. And finally, he decided not to take on Wander Johannes de Haas, even though he was a promising candidate, because he had only just completed his studies and had not obtained his PhD yet.

The trustees did not doubt Lorentz' recommendation, and on 10 December 1909, the decision was taken to appoint both Lorentz and Elias as curator and keeper of the laboratory of Teylers Foundation respectively.[34] Their tenure was to begin on 1 January 1910, although it soon transpired that Elias could not leave Berlin so quickly.

Until he gave up his full professorship in Leiden, Lorentz was now to receive an annual salary of *f* 2,000. Once he had come to Haarlem, this was to be raised to *f* 5,000. An additional *f* 10,000 per year were reserved for what was referred to as the "cabinet," which in this case meant everything that fell under

33 Lorentz to trustees of Teylers Foundation, 9 December 1909. ATS, inv. no. 36.

34 Directienotulen, 10 December 1909. ATS, inv. no. 13.

Lorentz' purview. This included the keeper's annual salary, which was gradually to rise to *f* 3,000, as well as an extra *f* 3,000 for any equipment for the laboratory, *f* 1,000 for the laboratory attendant (*amanuensis*) Van Waveren who had already worked with Van der Ven, *f* 2,000 for the publication of the *Archives*, and finally *f* 1,000 for lectures and courses. It was at this meeting that Zocher objected to Lorentz' appointment for a second time, because of the costs involved. Van der Ven had received an annual salary of about *f* 3,000 and a far smaller budget to cover costs. And not only was Lorentz significantly better-paid than his predecessor, but the costs were even higher than the initial estimate Zocher and Loosjes had objected to four months earlier – mainly because the initial estimate had not included a well-educated assistant.

4 Refurbishment of the Laboratory and Subsequent Research

The first indication that the Foundation's budget might be stretched to the limit came even before the new curator and keeper were officially appointed, on 24 December 1909. Lorentz had asked whether he could participate in a meeting of the trustees in order to relay some points he had discussed with his new assistant.[35] First of all, Elias could not leave Berlin as quickly as the trustees had hoped, because he wanted to complete a series of experiments he had been performing with Henri du Bois. With the trustees' blessing, he eventually arrived in Haarlem in April 1910. But in addition to this, Elias had drawn up some plans for changes to the laboratory, and, as was to be expected, they came at a cost. The estimate Elias and Lorentz gave the trustees was about *f* 9,000 spread out over a couple of years. Although Elias considered the laboratory "very suitable," he did want to have "a few minor things" installed, as Lorentz put it, such as running water, gas, and a fume cupboard. The trustees "considered such reasonable," and Lorentz then suggested ways of economising in order to obtain the *f* 9,000 required. The trustees were quick to remark that "the calculation of prof. Lorentz does not add up entirely," for the simple reason that they did not want to save *f* 2,000 in one year by not publishing a volume of the *Archives*, but Elias was given the green light to go ahead with his plans for refurbishing the laboratory.[36]

By April 1911, Van der Ven would probably have hardly recognised his former work environment. While Lorentz was still in Leiden, Elias and the laboratory attendant Van Waveren had overseen the laboratory's refurbishment. Electric

---

35 Lorentz to trustees of Teylers Foundation, 22 December 1909. ATS, inv. no. 36.

36 "zeer bruikbaar" / "eenige kleinigheden" / "achtten zulks billijk" / "de berekening van Prof. Lorentz niet geheel opgaat": Directienotulen, 24 December 1909. ATS, inv. no. 13.

lights had been installed everywhere, and most rooms were provided with running water and gas.[37] All the walls had been plastered and painted, and fire doors installed. In order to prevent any possibility of fire spreading to the museum, the door connecting the museum building and the laboratory had been sealed off. A whole range of instruments and machines as well as laboratory equipment had been acquired as well.

Lorentz and Elias now had a total of nine rooms at their disposal, six of which were spread over the three storeys of two formerly separate houses adjacent to the museum that had already been conjoined to form the laboratory building during Van der Ven's times. The ground floor consisted of two rooms, the so-called "large downstairs workroom" and a workshop for Van Waveren. A one-metre thick concrete slab had been installed in the large workroom under some removable floorboards, so that instruments could be set up on it in order to minimise any interference of vibrations with measurements. The fume cupboard was installed in this room as well, and all of the laboratory's electricity was controlled through a large switchboard set up here. The workshop was equipped with a lathe and a workbench, amongst other equipment and tools. By 1912, Lorentz proudly proclaimed that "only for harder jobs it is occasionally necessary to turn to other workshops."[38] Over the years, Van Waveren built or improved upon a wide range of instruments with the tools available in his workshop.

The second floor consisted of three rooms: first an office for Lorentz, which for some reason he always referred to in inverted commas; secondly another workroom, referred to as "the large upstairs workroom"; and thirdly a small room that was referred to as the "small upstairs chamber." This chamber was used as storage space, at least initially. On the third floor, the top story of one of the two houses that had been conjoined was converted into a darkroom for developing photographs.

Finally, there were two – or, depending on how one counts even three – other buildings associated with the laboratory, providing the remaining three out of the total of nine rooms. The first of these buildings was Van der Willigen's "Observatory," essentially consisting of one room. It was not used for scientific purposes during Lorentz' tenure, but must at some point have been converted into a studio to be used by the curator of the art collections. The last of the instruments that had still been stored there were transferred to the "Museum"

---

37 Unless otherwise indicated, the information presented in the following section is taken from Lorentz' annual reports to the trustees of Teylers Foundation: Verslagen, 1885–1944. ATS, inv. no. 191.

38 "... alleen voor zwaarder werk is het in sommige gevallen noodig zich tot andere werkplaatsen te wenden." Jaarverslag 1911–1912, 6 April 1912. ATS, inv. no. 191.

by 1912.[39] Lorentz suggested that same year that, should it eventually prove necessary, additional space for laboratory work could be created by extending the Observatory.

The second building was referred to as the "acids house" (*zuurhuisje*). It is not clear whether this was indeed a separate building, or part of an adjacent building. It consisted of two sections or rooms, one of which housed a 12hp (9 kW) engine to convert electricity (this indicates that it can't have been too far away from the laboratory itself), while the other was furnished as a chemical laboratory. Not all of these rooms were used exclusively by those permanently employed by the Foundation (Van Waveren later took on two of his sons as apprentices). On the contrary, over the years a whole range of guests made use of the facilities, some even for a period of several years, for their own research. This is also why one could argue that the last building in which research was performed did not necessarily belong to the laboratory itself: in 1912 what is referred to as the "garden house" (*tuinhuisje*) was equipped so that Anton Hendrik Blaauw could conduct a series of experiments to find out more "about the influence of radiation on the growth phenomena of plants."[40] No record remains which indicates that any of the keepers ever worked there as well, although no resources or efforts were spared by both Elias and Van Waveren to assist Blaauw. Especially Van Waveren spent a lot of time on Blaauw's equipment, prompting the author of Blaauw's obituary some thirty years later still to refer to at least one "beautiful instrument" that Blaauw had been able to use in Haarlem.[41]

The *zuurhuisje*, too, was placed at the disposal of a guest researcher between 1913 and 1916, the geologist Christoph Georg Sigismund Sandberg, for some work on "questions of dynamic geology."[42] The trustees only agreed to this under the condition that "suchlike does not entail any significant costs."[43] Lorentz saw to it that he was provided with all necessary equipment, assuring the trustees that he had made it clear that Sandberg could only expect financial support for everything that "[can] be considered to form part of the equipment of a scientific [*natuurkundig*] laboratory."[44] Sandberg's research

---

39 Jaarverslag 1911–1912, 6 April 1912. ATS, inv. no. 191.

40 "Over den invloed van bestraling op de groeiverschijnselen bij planten": Jaarverslag 1913–1914, 8 April 1914. ATS, inv. no. 191.

41 "Fraai instrument": W.H. Arisz, "Levensbericht A.H. Blaauw," *Jaarboek KNAW* (1942–1943): 228–249, esp. 231.

42 "Vraagstukken der dynamische geologie": Jaarverslag 1913–1914, 8 April 1914. ATS, inv. no. 191.

43 "... zulks geen kosten van beteekenis mede brengt": Directienotulen, 6 June 1913. ATS, inv. no. 13.

44 "... gerekend [kan] worden tot de uitrusting van een natuurkundig laboratorium de behooren": Jaarverslag 1913–1914, 8 April 1914. ATS, inv. no. 191.

in Haarlem doesn't appear to have resulted in any publications.[45] The same year that Sandberg arrived, the darkroom on the top floor of the laboratory was converted into a room for the linguistic research of a Mrs. Bakker, later referred to as Mrs. Bakker-Bezemer. She needed a place to analyse recordings she had made in Utrecht, and was provided with the necessary equipment from Teylers' instrument collection. She stayed on until 1916, after which she was evidently able to work from home.

From 1919 onwards, the former darkroom was occupied for two afternoons every week by Dr. M.J. Huizinga. He was a mathematics teacher interested in "questions that arise from current conduction through flames."[46] Before he left in 1920 he might still have met the engineer Gijsbert Frans van Dissel, who wanted to learn more about electromagnetic waves because he wanted to specialise in wireless telegraphy. An announcement in the *Indische Courant* of 1929 seems to suggest that he successfully pursued a career along these lines.[47] But even though guests who only passed through for assistance have not been included in the list above, and Huizinga's presence even prompted Lorentz to start holding monthly colloquia attended by other mathematics and science teachers from the Haarlem area, the laboratory was still first and foremost his and the keeper's.

Elias avidly set to work once the laboratory had been refurbished and all his equipment had arrived and been installed. As was Lorentz' intention, most of Elias's work revolved around magneto-optic effects. De Haas assisted him in 1913 in performing a series of experiments on "the structure of absorption lines of sodium vapour."[48] After Lorentz had been succeeded by Paul Ehrenfest in Leiden and subsequently moved to Haarlem in 1912, Elias assisted the curator in a series of experiments on hysteresis, a problem that the Nobel prize laureate would repeatedly return to over the coming years.

But just as Elias's own efforts were gaining momentum after all the instruments had been delivered, assembled, and calibrated to his satisfaction, that momentum was checked through events far beyond his control: the outbreak of the First World War. And even though the Netherlands remained neutral, the effects were of course felt in Haarlem too, as everyone everywhere was forced

45 He gained some notoriety later for his autobiographical account of his times in the Dutch Colonies, and as an anti-semite and supporter of the German Nazi party.

46 "... vragen, die zich bij de electriciteitsgeleiding door vlammen voordoen." Jaarverslag 1919–1920, 7 April 1920. ATS, inv. no. 191.

47 "Een onderscheiding," *De Indische Courant 76*, 12 December 1929. See also: Suzanne Lommers, *Europe – On Air: Interwar Projects for Radio Broadcasting* (Amsterdam: Amsterdam University Press, 2012), 116–118.

48 "De structuur der absorptielijnen van natriumdamp": Jaarverslag 1913–1914, 8 April 1914. ATS, inv. no. 191.

to economise. Worst of all, Elias was called up for military service in 1914. He was allowed to return after a few months, but it didn't take long for economising to take its toll. By the end of the War, Lorentz had stopped giving public lectures that were traditionally held in the winter, as the lecture hall could not be heated because fuel was scarce. No new instruments were acquired either.

Elias left Teylers on 8 August 1916, after he was offered and accepted a position as an honorary professor at the University of Delft (*Technische Hogeschool*). He would most likely have changed jobs even if the war had not been raging. Not only was the entire country, including the University of Delft, affected by the hostilities, but as early as March 1911 Elias had already accepted – with the trustees' blessing after Lorentz assured them that this would not detract from his work at Teyler – a teaching position in Utrecht, standing in for the frail Cornelis Harm Wind for two hours a week. The following year, Elias had been made an assistant professor (*privaatdocent*) in Utrecht.

Elias exchanging his position at the laboratory in Haarlem for a position at a university was the first indication of a pattern that was to emerge over the course of the following years: all of Elias's successors as keeper lamented the absence of students in Haarlem or left once they were offered a position at a university. And the list of successors was not a short one: most of the subsequent keepers did not remain in Haarlem for very long.

Elias was succeeded by De Haas, who had already been considered for the position in 1909, and in the meantime had received his doctorate. Although he left just a year later to join Elias as a professor of physics in Delft, he managed to recreate and describe more accurately than anyone else previously a physical phenomenon that has since been called the Einstein-De Haas effect. De Haas and the increasingly famous Einstein had first discovered this phenomenon while working together in Berlin. These credentials also go to show that De Haas's appointment had little, perhaps even nothing, to do with the fact that he had also become Lorentz' son-in-law in 1910.

After De Haas left, Lorentz employed Johannes Martinus Burgers, a recent graduate of Ehrenfest with strong Marxist sympathies, on 8 January 1918.[49] Lorentz set him to work on a set of experiments related to his research on hysteresis. Burgers was only to complete these experiments "up to a certain level" though, as he was simultaneously granted time to edit his PhD thesis and, like his predecessors in Haarlem, left for Delft after only ten months to accept a professorship of engineering.[50]

---

49 On Burgers, see A.J.Q. Alkemade, lemma "Johannes Martinus Burgers (1895–1981)," in *Biografisch Woordenboek van Nederland* 5 (Den Haag: SDU, 2002).

50 "… tot op zekere hoogte": Jaarverslag 1918–1919, 07 April 1919. ATS, inv. no. 191.

Lorentz then chose Balthasar van der Pol as the new keeper. At the time, Van der Pol was working in Cambridge at the Cavendish Laboratory as an assistant to J.J. Thomson. Lorentz allowed him to continue with his research into the propagation of electromagnetic waves, which he had already begun in Cambridge. Some of the instruments he required in Haarlem, Van der Pol ordered in England. Interestingly, in his report to the trustees on Van der Pol, Lorentz emphasised that his previous research had been conducted not with an eye to improving wireless telegraphy, but "the study of physical phenomena."[51] Until he left in late 1922 – for a position at the newly created industrial laboratory of the *Philips Gloeilampenfabriek* – Van der Pol experimented a lot with recently developed triodes, focusing especially on cases of hysteresis. Meanwhile, Lorentz himself had succeeded in measuring a "curious double refraction in crystals of the regular system" which had been theoretically predicted, but never observed yet.[52]

Van der Pol was succeeded by Dirk Coster, who stayed for a little more than a year before leaving in 1924 for a professorship in Groningen. While in Haarlem, he experimented with X-rays. He was succeeded by Adriaan Fokker in July 1926.[53] Fokker had previously worked with Ernest Rutherford in Manchester and with Einstein in Zurich. Lorentz agreed with the trustees that Fokker could succeed Lorentz as curator. Together they turned towards questions pertaining to quantum mechanics, specifically the existence of light quanta, but their efforts were cut short when Lorentz passed away on 4 February 1928.

5 "The Isolation of Haarlem"

As has already been pointed out, all of the keepers – all of them obviously budding young physicists with great potential and future members of the elite of Dutch science – evidently considered universities a more challenging work environment with better career prospects than Teylers' Laboratory. The question why naturally arises. Two points were certainly irrelevant, and a third was of little importance. First, their leaving cannot have had anything to do with Lorentz' scientific credentials or his role as their superior. No incidence is known of any of the keepers speaking ill of Lorentz. Secondly, the trustees

---

51 "De bestudering der physische verschijnselen." Jaarverslag 1918–1919, 07 April 1919. ATS, inv. no. 191.

52 "eigenaardige dubbele breking bij kristallen van het regulaire stelsel": Jaarverslag 1918–1919, 7 April 1919. ATS, inv. no. 191.

53 Directienotulen, 9 July 1926. ATS, inv. no. 15.

were as supportive as they could be, and certainly didn't interfere with work at the laboratory. Their demands – that a number of lectures should be held annually – were minimal, and certainly not comparable with the burden of teaching at a university. Finally, the salary the Foundation paid was competitive as well, albeit considerably lower than a university professor's.[54]

Two other points however were of crucial importance: the changing nature of science, and, eventually, the Foundation's finances. As for the first of these points, research was increasingly being conducted in large scale projects, requiring teams of scientists. Because of the larger scale of these projects, funding could increasingly only be provided by the government, industry, or perhaps major philanthropic organisations such as the Rockefeller and Carnegie Foundations. Eventually, a two-person, privately funded laboratory in a town such as Haarlem without a university could no longer compete. Perhaps the best example of the changing dimensions and character of scientific research is the immense expansion of the facilities at Lorentz' own university, Leiden. As was already mentioned above, Kamerlingh Onnes had immediately set about refurbishing the laboratory he was provided with upon his appointment in 1882; but during the years Lorentz spent in Haarlem, the physics department's premises were significantly extended upon Kamerlingh Onnes's initiative, with a major new wing being added to the original building between 1920 and 1926.[55] Lorentz' laboratory in Haarlem paled in comparison. And even if Teylers Foundation would have had the resources to expand its own premises, it would have faced another obstacle: Pieter Teyler's old town house was situated in what had now become the middle of town, and the space around it was obviously limited, all the more so because tearing down the museum was out of the question.

As scientific research became large-scale, research networks became increasingly important as well, especially as specialisation amongst experts increased too. As Ehrenfest demonstrated through his famous colloquia in Leiden, students could provide a valuable stimulus to research as well as a reservoir of future research talent, all the more so as their number increased significantly over the first decades of the twentieth century. This is where another of Lorentz' character traits becomes significant: he was not a natural-born charismatic leader with instincts of power. This in turn meant he was not the type to establish a scientific "school" of his own. Where he displayed leadership, it was more on a rational level, out of a sense of duty when jobs had

54 The salary for a full professorship was about *f* 6,000 p.a., the "conservator" earned about *f* 3,000 p.a.

55 On this see Van Delft, *Freezing Physics*, 530ff.

to be done or problems solved. It should be emphasised that this is not to say that his word did not carry immense weight, or that he was not respected, that he did not stimulate exchanges amongst scientists, or that he did not nurture students' talents. On the contrary, as his pivotal involvement in – and presidency of – the Solvay conferences shows, as well as his commitment to his own students and matters of education in general. But the contrast with Ehrenfest, for instance, did not escape contemporaries. As Lorentz' daughter put it:

> How great a difference between Ehrenfest and Lorentz as regards their relationship to their students and the influence which they exercised upon them! Whenever Ehrenfest met a young man in whom he saw great possibilities as far as physics was concerned, but who spent too much time and energy on other things, he would spare time nor effort to influence him to take up the study of physics seriously. Lorentz, facing the same situation, would regret the fact, but would come to the conclusion that, for better or for worse, this young man was more interested in other matters than in physics, and that this was his own business. Only when a student asked for help, or was in need of his assistance, on account of illness or similar circumstances, would Lorentz be ready to offer help.[56]

Or, as Fokker later remembered during an interview:

> [Lorentz] was very, very kind, and his conversation was never dull although he was quiet. But he didn't make an appeal on your soul, so to speak, you see; he just left you to your own intentions and to your own desires. And that's not the way to make a school; if you are making a school, you must just shake people, you see. The same thing I noticed later when I talked with Rutherford – I was some weeks with Rutherford. There's a great difference; it's a pity that Lorentz did not have a school around him just like Ehrenfest.[57]

So Lorentz' style did not help establish Haarlem as one of the nerve centres of Dutch science in the sense that Lorentz brilliance and high reputation was not enough to compensate what was beginning to emerge as a structural disadvantage, i.e. the relatively small size of Teylers' Laboratory and Haarlem's lack of a university. This was amplified by two factors: firstly that Lorentz stayed

---

56 De Haas-Lorentz, *H.A. Lorentz. Impressions*, 107. Original in English.

57 Transcript of an interview John L. Heilbron, held with Fokker at Beekbergen, 1 April 1963: http://www.aip.org/history/ohilist/4607.html (accessed 7 Febr. 2014).

on in Leiden as an honorary professor, giving his legendary "Monday morning lectures" even after he had officially reached retirement age; and secondly his time-consuming involvement with – and later presidency of – the so-called *Staatscommissie Zuiderzee* ('Zuiderzee committee'). This body was established to calculate what effect the construction of the *Afsluitdijk*, an enormous enclosure dam in the province of Friesland, would have on the water level and pressure on the surrounding dams. Lorentz developed a theoretical model to simulate the hydrodynamics involved and oversaw the immense calculations this required.[58]

As some of their correspondence shows, the keepers were beginning to feel this isolation. In 1922, for instance, Van der Pol wrote to Lorentz, who was abroad at the time,

> At Teylers it is quiet and lonely. Only now that you are gone I realise, actually for the first time, what it means to work in a town in which there is no university life.[59]

A year later, Van der Pol left for Eindhoven. And when Lorentz turned to Fokker to ask him whether he might want to take up the post of keeper, Fokker had grave doubts about his own qualifications, citing Haarlem's isolated position as the main reason he would prefer to take up an alternative offer he had received from the University of Delft:

> Without question there is very much that is great about a private Foundation [the Teylers Foundation] that wants its resources to serve science, and to ensure that these great intentions are done justice and developed even further is a tempting task. I also understand very well that many things could perhaps be done for the general cause of all physicists by someone who is not burdened by other duties of his position, with regards to cooperation on a national level as well as for international relations. But this is an ambition that aims high, to attract those members of the scientific circle of friends for regular interaction. In order to break the isolation [*isolement*] of Haarlem a lot of friendship will be necessary,

---

58 See Anne J. Kox, "Uit de hand gelopen onderzoek in opdracht: H.A. Lorentz' werk in de Zuiderzeecommissie," in *Onderzoek in opdracht: De publieke functie van het universitaire onderzoek in Nederland sedert 1876*, eds. L.J. Dorsman and P.J. Knegtmans (Hilversum: Verloren, 2007), 39–52.

59 "Op Teyler is het stil en eenzaam. Nu U weg is besef ik pas, eigenlijk voor het eerst, wat het zeggen wil, in een stad te werken waar geen universitair leven is." Van der Pol to Lorentz, 15 March 1922: Haarlem, Noord-Hollands Archief (hereafter NHA), inv. no. 364–62.

> and a certain amount of authority. Following this line of thought it then seems again as if I needed to earn my spurs in Delft first, and become familiar with the nature of and [... unreadable] of relations.[60]

In other words, "the isolation of Haarlem" weighed more heavily than the absence of any teaching obligations at Teylers and the possibility to focus entirely on research. By this time, the second aspect could no longer be ignored either: the Foundation's finances. The changes in the global economy were beginning to take their toll on the Foundation's budget, which of course had already been stretched to the limit when Lorentz was taken on. The war, inflation, the Russian Revolution, and the introduction of personal taxes all brought about change in the financial world during this time. When the trustees were looking for a successor to Van der Pol, they therefore openly told Lorentz that they were "of the opinion that they would have to be prepared for a possible cave-in and significant loss of value of stocks."[61] They had already lost a lot of money as a result of the Russian Revolution, when the Russian government bonds they held became worthless. Even though the trustees were able to compensate for this loss by investing in Dutch companies and bonds, the financial markets were still far from stable.[62] The year Coster was taken on, 1923, was the year in which hyperinflation brought the young Weimar Republic to its knees. Lorentz accordingly warned Coster that he had to consider his new position a temporary one, as there was a possibility that the entire physics department would have to economise – small wonder Coster soon left for Groningen.[63]

---

60 "Ontegenzeggelijk is er erg veel moois in een particuliere Stichting [Teylers], die haar middelen aan de wetenschap wil dienstbaar maken, en om te zorgen dat dit mooie der bedoeling tot zijn recht komt en nog verder ontwikkeld wordt is een aantrekkelijke taak. Ook zie ik heel wel in, dat er vele dingen wellicht te doen vallen voor de gemeene zaak der physici juist voor iemand, die geen drukke ambtsplichten heeft, zoowel voor de binnenlandsche vriendschappen als voor buitenlandsche verstandhouding. Maar het is een ambitie, die nogal hoog mikt, deze van een wetenschappelijken vriendenkring aan te trekken tot geregeld verkeer. Om het isolement van Haarlem te breken zal veel vriendschap noodig zijn, en een zeker gezag. In dezen gedachtengang lijkt het dan weer alsof ik in Delft eerst mijn sporen daartoe moet verdienen, en bekendheid verwerven met karakter en den binnenkant [?] van verhoudingen." Fokker to Lorentz, 3 January 1923: NHA, inv. no. 364–24.

61 "… van oordeel zijn dat zij rekening moeten houden met een mogelijke ineenstorting en waardevermindering van rentegevende papieren." This is the way Fokker quotes Lorentz. Fokker to Lorentz, 3 January 1923: NHA, inv. no. 364–24.

62 See the accountant's reports covering this period: Rekeningen met Toelichting, 1889–1945. ATS, inv. no. 609. The worthless Russian bonds were kept: Coupons. ATS, inv. no. 2414.

63 Jaarverslag 1922–1923, 7 April 1923. ATS, inv. no. 191.

Interestingly, even well into the 1930s the Foundation had in fact not actually lost any of its capital, which, at about *f* 2,000,000 was still considerable.[64] But, crucially, the trustees were no longer able to gain much of a return on the assets they owned. The accountant's annual reports confirm what Fontein, by this time president of the board of trustees, pointed out in 1926: "that one needs to be very frugal in order to retain *f* 10m [*f* 10,000] at the end of each year, which is quite necessary."[65] By this time, he and his fellow members of the board were no less concerned than they had been after Van der Pol left. It is worth recalling that the Foundation had a lot more to pay for than Lorentz' laboratory: all the other departments at the museum – the most costly of which for many years was the library –, the museum itself, the *Hofje*, the two learned societies, and a wide range of charities all required regular payments.

After lengthy discussions on how to economise, during which the idea to close down either the laboratory or the art department was floated several times by different members of the board, it was finally decided that all the prints and drawings that the Foundation owned in duplicate were to be sold off.[66] Sensing what else might still be coming, the curator of the art department, H. van Borssum Buisman, subsequently agreed to select prints and drawings that could be put on the market, but repeatedly protested heavily against any paintings being sold.[67] Lorentz himself also went on record protesting against other departments economising in order to maintain his own department's budget. But as Fokker soon noticed, all departments felt the squeeze, with the laboratory's annual budget being set at *f* 14,000 rather than the *f* 17,000 both Fokker and Lorentz considered necessary and had asked for.

So, in a way, it looked as if Zocher's premonitions about the Foundation's finances had been correct. Zocher himself had passed away in 1915. Yet Fontein's enthusiasm was never broken; he never showed any signs of regretting having employed Lorentz. On the contrary, when Lorentz passed away, Fontein gave a small eulogy at the following meeting of the board of trustees, in which he spoke of Lorentz' tenure and personality in nothing but glowing terms. Acknowledging that Lorentz had not been able to perform as much research at Teylers as he had wanted to, Fontein nevertheless drew attention to the "successes" of the keepers, emphasised Lorentz' impeccable character and fine intellect, and proudly quoted Kamerlingh Onnes's statement that the country

---

64 Rekeningen met Toelichting, 1889–1945. ATS, inv. no. 609.

65 "... dat men heel en heel zuinig moet zijn om jaarlijks f10m [f 10,000] te kunnen overhouden wat toch noodzakelijk is." Directienotulen, 14 May 1926. ATS, inv. no. 15.

66 Directienotulen, 30 April 1926–11 June 1926. ATS, inv. no. 15.

67 Directienotulen, 14 May 1926. ATS, inv. no. 15, fol. 5 & fol. 9.

should be eternally grateful to the Teylers Foundation because it had created a position for Lorentz in which he was free to focus exclusively on his research. But, Fontein continued, far more than the Foundation had been able to assist Lorentz, it had benefitted from his worldwide fame:

> On his great flight to world fame, which he never actively sought, but which simply came to him, he inadvertently took Teylers' Laboratory and all that accompanies it (including the entire Foundation) under his wing so that all would share in his world fame.[68]

Fontein's case can indeed be argued. Renowned colleagues of Lorentz such as Einstein or Du Bois came to Haarlem and to Teylers because Lorentz worked there. As Lorentz noted in his 1917 annual report, "Prof. A. Einstein from Berlin and Prof. H. Du Bois showed their interest in the laboratory on their visits."[69] Other well-known scientists include Van der Pol's visitor Appleton.

Yet at the same time, Kamerlingh Onnes's remark about Teylers Foundation that Fontein referred to was the only reference made to this Haarlem institution in a total of six speeches held in Lorentz' honour during the festivities to mark the 50th anniversary of his doctorate in 1925.[70] This might have had something to do with the fact that these festivities were held at the University of Leiden, but that in itself is not insignificant: they were not held in Haarlem. The scientific community obviously still associated Lorentz with Leiden more than with Haarlem.

Lorentz himself in fact regretted he had not been able to do more for Teylers Foundation. At the height of the discussions on the Foundation's finances in May 1926, he summarised his own view of the previous one and a half decades. As the minutes of the meeting read:

> With regards to himself Prof. Lorentz states that his initial plans have not been fully implemented. Speaker [Lorentz] hasn't done for Teyler what

68 "Op zijne groote vlucht naar wereldberoemdheid, welke hij nooit gezocht heeft, doch die hem eenvoudig te gemoet is komen waaien, heeft hij onwillekeurig Teyler's Laboratorium met al zijn aanhang (waaronder de geheele Stichting) onder zijn arm medegenomen op dat dit alles in zijne wereldvermaardheid zoude deelen." Directienotulen, 17 February 1928. ATS, inv. no. 15.

69 "Prof. A. Einstein te Berlijn en Prof. H. Du Bois toonden bij bezoeken hunne belangstelling in het laboratorium." Jaarverslag 1916–1917, 7 April 1917. ATS, inv. no. 191. This was one of at least two confirmed instances in which Einstein came to Leiden. The first was in 1911.

70 At least this was the only time its name appears in the published versions of these speeches. *Physica* 6–1 (1926): 1–21.

> he had wanted to. Numerous circumstances prevented him from doing so. Leiden did not let him go, the War claimed a lot of his time, as did peace. Then came the calculations for the Zuiderzee, etc. etc. Yet there is also reason to be satisfied.[71]

Before summarising Lorentz' ensuing account of everything that he thought did go well – above all Lorentz was proud of what the keepers had achieved – the trustee taking the minutes of the meeting still recorded verbatim Lorentz' very brief overall verdict of his tenure at Teylers, which, considering that Lorentz passed away less than two years later, has a final ring to it: "Het is wel aardig gegaan," which can be translated roughly as "It didn't go badly."

6 And the Rest is History

Let us briefly summarise some points that so far have transpired from the description of Lorentz' activities as curator of Teylers Foundation's laboratory. Firstly, he was taken on largely because of the enthusiasm of one trustee, Jan Adriaan Fontein, and the mediation of another eminent scientist, Johannes Bosscha. Secondly, Lorentz accepted the job in Haarlem at least in part because he was dissatisfied with his situation at the University of Leiden. Thirdly, the Foundation's budget was stretched to the limit from the very beginning of Lorentz' tenure, and this situation became painfully acute after the Foundation experienced a variety of financial setbacks. Fourthly, this decline in the laboratory's fortunes was amplified by the rise of universities as research institutes, which was closely connected to the structural changes scientific research was experiencing. Research increasingly involved a team effort and sensitive, expensive measuring devices.

Thus far one could therefore conclude that hiring Lorentz was essentially a fateful attempt by Teylers Foundation to maintain the position it had held throughout most of the nineteenth century, as the centre of Dutch experimental science. Two other aspects besides the Foundation's financial woes and Haarlem's increasingly peripheral status without a university contributed to this ill fate, i.e. the Foundation's laboratory being taken less and less seriously

71 "Op zich zelven komende verklaart Prof. Lorentz dat zijn plannen van vroeger helaas niet in elken deele verwezenlijkd zijn. Spreker heeft niet voor Teyler gedaan, wat hij had willen doen. Tal van omstandigheden hebben hem dat belet. Leiden liet hem niet los, de oorlog vroeg veel van zijn tijd even als de vrede. Toen kwamen de berekeningen voor de Zuiderzee enz. enz. Toch is er ook reden tot tevredenheid." Directienotulen, 14 May 1926. ATS, inv. no. 15.

within the scientific community: first, even if the funds had been available it would have been almost impossible to expand the laboratory building to a scale comparable to university laboratories, simply because the laboratory was in the middle of Haarlem and adjacent to Teylers Museum. And secondly, despite his brilliance and immensely high reputation Lorentz did not possess the charisma needed to compensate for Haarlem's peripheral status. As a result, by the time Lorentz passed away the Foundation was selling off those prints and drawings from its collection that it owned in duplicate in order to raise money.

This is remarkable, not so much because of what was being sold, but because of what the Foundation was *not* selling: the trustees did not decide to sell any of the historical scientific instruments – which fell under Lorentz' purview as curator – from their collection in order to gain some fresh income. This is even more noteworthy than the fact that they didn't sell any of the books from the library, coins from the numismatic collection, or fossils and minerals from the geological collection, because all of these were either unique or could still serve for research purposes. The instruments were all unique too of course, whereas the duplicate drawings and prints were obviously not – but many of the instruments were no longer of any use in the laboratory, or would clearly never be again; they no longer had any practical value. Given that the Foundation's predominant aim was clearly to support research, it is therefore easily imaginable that someone at the time might have come up with the argument that the old and useless instruments should be the first to be sold from any of the collections in order to guarantee that further research could be funded by the Foundation.

The prioritisation in selling off duplicate works of art before historical scientific instruments is therefore strongly indicative of some sense of the historical value of these artefacts of former science. This in turn chimes with broader developments that took place over the period of Lorentz' tenure in Haarlem, or rather two developments in particular: first, an increase in the historical appreciation of both past science and the artefacts of past science – the first edition of the journal *ISIS* for instance was published in 1912, the Dutch association for the history of science *Gewina* was founded in 1913, and its American counterpart, the *History of Science Society*, was set up in 1924; secondly, the establishment of numerous science museums at the beginning of the twentieth century – examples include the Deutsches Museum in Munich, conceived in 1903 and moved to its own premises in 1925; the National Museum of Science and Industry in London, which was separated from the South Kensington Museum in 1909; and the Technisches Museum für Industrie und Gewerbe in Vienna, which was built between 1909 and 1918. Although many of these science museums already incorporated historical instrument collections, as if in a sort of symbiosis of these two developments, by the 1930s museums for the history of science had emerged as a distinct class of museum.[72] The Museum of the History of Science opened to the public in Oxford in 1924; in Florence the permanent exhibition of the Istituto delle Istoria delle Scienze (predecessor of today's Museo Galileo) was completed by 1930, and in Leiden the Netherlands Historical Science Museum (precursor of Museum Boerhaave) was founded in 1931.

Teylers Museum, as both a museum and the home to a vast collection of historical scientific instruments, of course had enormous potential under these new circumstances, and this potential did not go unnoticed. As early as 1876, when the very first large-scale exhibition of historical instruments was held in London in South Kensington, Teylers Foundation sent a range of historical instruments across the Channel to be put on display. Some thirty years later, not only individual items from the collection, but both the entire collection and the museum had become noteworthy for their historical value. As much can be inferred from the title of a keynote speech the Dutch Nobel prize laureate Jacobus van 't Hoff held during the festivities surrounding the second annual meeting of the trustees of the newly founded Deutsches Museum in 1905 – the trustees were in fact still looking for both funding and inspiration for their institution, and one of its patrons, the Bavarian king, was amongst the audience in 1905. The title of Van 't Hoff's speech was "The Teylers Museum in Haarlem," and its subtitle "the significance of historical collections for science and technology."[73]

Yet despite this potential role perhaps even as figurehead of a new class of institution, one thing that is sure is that Lorentz, as curator of Teylers Museum's instrument collection, never made an active effort of highlighting it for its historical value; his focal point was the laboratory, not the museum. On the contrary, from the few references he made to the museum or the collection at all, two points clearly transpire: firstly, that in his eyes the instrument collection served as a sort of reservoir of instruments which could be used for laboratory

72 On this development see Jim Bennett, "European science museums and the Museum Boerhaave," in *75 jaar Museum Boerhaave*, ed. M. Fournier (Leiden: Museum Boerhaave, 2006), 73–78.

73 Jacobus van 't Hoff, *Das Teyler-Museum in Haarlem und die Bedeutung historischer Sammlungen für Naturwissenschaft und Technik* (Munich: Deutsches Museum, 1908). Van 't Hoff had previously published his speech in Dutch: Jacobus van 't Hoff, "Teyler's Museum en de beteekenis van geschiedkundige verzamelingen voor natuurwetenschap en industrie," in *De Gids* 70 (1906): 338–348. See also Martin P.M. Weiss, "Hoe Haarlem bijna München dwarsboomde. Van 't Hoffs lezing over Teylers Museum bij het Deutsches Museum," in: *'Dingen die ergens toe dienen.' Verhalen over Materiele cultuur van wetenschap*, eds. Esther van Gelder, Eric Jorink, Ilja Nieuwland, Marlise Rijks, Alice Spruit (Hilversum: Verloren, 2017), 182–185.

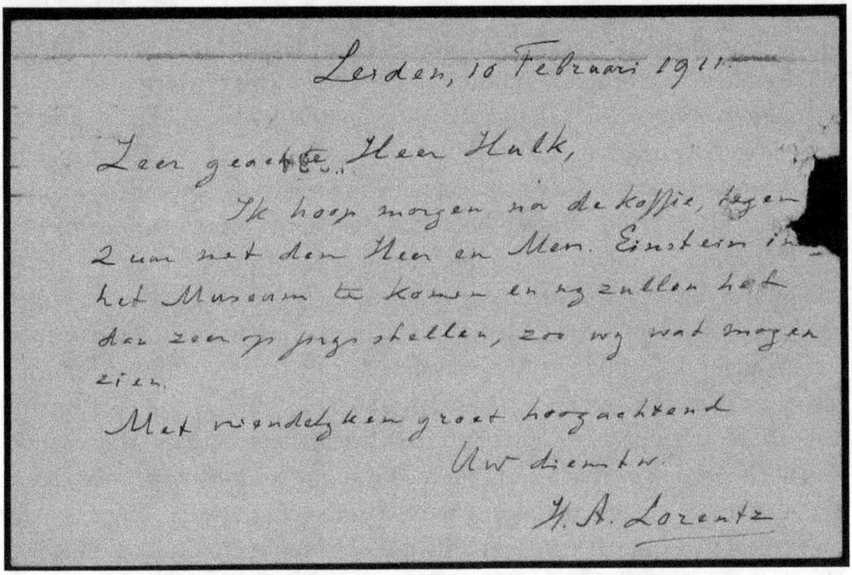

Leiden, 10 Februari 1911.

Zeer geachte Heer Hulk,

Ik hoop morgen na de koffie, tegen 2 uur met den Heer en Mevr. Einstein in het Museum te komen en wij zullen het dan zeer op prijs stellen, zoo wij wat mogen zien.

Met vriendelijken groet hoogachtend

Uw dienstw.

H.A. Lorentz

FIGURE 10.2 Lorentz' note to the curator of the art collections at Teylers Museum, Johannes Frederik Hulk, asking whether he can visit the museum with "Mr. and Mrs. Einstein"
COURTESY OF TEYLERS MUSEUM

work, should they prove to be of any assistance; and secondly, that to him the museum and laboratory were separate entities.

This second point is most clearly illustrated by a card he sent to the curator of the art department, Johannes Frederik Hulk, when he was expecting a visit from Einstein and his wife. It read:

> Dear Mr. Hulk, I hope to come to the museum tomorrow after coffee at about 2 o'clock with Mr. and Mrs. Einstein and we would greatly appreciate it, if we could see some things. Yours respectfully, H.A. Lorentz.[74]

In other words, "after coffee," Einstein and Lorentz – curator of the physics department – wanted to go on what amounted to a guided tour of the adjacent Teylers Museum. In his annual reports, Lorentz repeatedly stated how he resorted to instruments from the museum's collection for use in the

74 "Zeer geachte heer Hulk, Ik hoop morgen na de koffie, tegen 2 uur met den Heer en Mevr. Einstein in het Museum te komen en wij zullen het dan zeer op prijs stellen, zoo wij wat mogen zien. Met vriendelijken groet hoogachtend Uw dienst. H.A. Lorentz": Lorentz to Hulk, 10 February 1911. ATS, inv. no. 124. The author is grateful to Marijn van Hoorn for having drawn his attention to this letter.

laboratory. The most prominent example was the conversion of a Repsold universal instrument that had been acquired by Van der Ven's predecessor Van der Willigen in 1878, into a multipurpose spectrometer. This conversion amounted to a total cost of *f* 6,316.[75] In 1912, Lorentz remarked that "a lamp of Duboscq, present in the museum, provided good service" during some of Elias's spectrographic experiments.[76] Two years later, he reported that equipment required by Mrs. Bakker "could be assembled with the help of some instruments from the collection."[77] And in 1923, when money was getting tight, Lorentz wrote that only a few of the instruments required by Coster had had to be bought new because it had turned out that the laboratory was actually already well equipped for the purposes of his research, adding that some instruments had also been converted by Van Waveren.[78] At the same time, instruments from the collection were repeatedly lent to external researchers for their experiments.

This is not to say that Lorentz looted the museum or didn't take care of the instruments that were housed there. On the contrary, his annual reports also reveal that Van Waveren and his apprentice son spent time cleaning both the instruments and the showcases they were stored in. In 1912, for instance, Lorentz reported that "a number of old instruments, such as an electromagnet, a set of scales and a hot-air machine, were completely cleaned, checked up and revarnished."[79] In 1917, he wrote that "much care was also devoted to the cleaning of cupboards and instruments in the Museum and the restoration [*herstellen*] of some old apparatus."[80] One year later, Lorentz reported again that "the first and second amanuensis [Van Waveren and his apprentice] spent a lot of time checking and restoring instruments from the Museum."[81] It says a lot though that these thorough cleaning and repair jobs were undertaken towards the end of World War I, when no new instruments were being bought and research and other activities had been reduced to a minimum because of the fuel

---

75 Van Hoorn, "The Physics Laboratory of the Teyler Foundation," 16.

76 "Eene in het museum aanwezige lamp van Duborcq bewees [...] goede diensten": Jaarverslag 1911–1912, 6 April 1912. ATS, inv. no. 191.

77 "kon met behulp van eenige in de verzameling aanwezige instrumenten worden samengesteld": Jaarverslag 1913–1914, 8 April 1914. ATS, inv. no. 191.

78 Jaarverslag 1923–1924, 31 March 1924. ATS, inv. no. 191.

79 "Enkele oude instrumenten, als electromagneet, balans en heetelucht-machine, werden geheel gereinigd, nagezien en opnieuw gevernist." Jaarverslag 1911–1912, 6 April 1912. ATS, inv. no. 191.

80 "veel zorg werd ook besteed aan het schoonmaken van kasten en instrumenten in het Museum en het herstellen van eenige oude apparaten." Jaarverslag 1916–1917, 7 April 1917. ATS, inv. no. 191.

81 "de eerste en tweede amanuensis [hebben] veel tijd besteed aan het nazien en herstellen van instrumenten uit het Museum": Jaarverslag 1917–1918, 7 April 1918. ATS, inv. no. 191.

shortage. Again, it becomes clear that Lorentz' focal point at Teylers was the laboratory, and that he possibly didn't even recognise the museum's potential as a showcase for the history of science, alongside all the other collections that were being rediscovered throughout Europe precisely during Lorentz' tenure at Teylers.

At the same time, however, this also meant that, in principle, the original museum remained untouched and unchanged for another two decades. By the time Lorentz passed away, the Oval Room in particular clearly belonged to another era entirely, and the 1885 extension was more than 40 years old, too. Just five years after Lorentz' death, Fokker published a new guidebook to the instrument collection, with a strong emphasis on the rich history it illustrated. And in this sense it is also not surprising and maybe even symbolic that when Fokker's successor Jacob Kistemaker arrived in Haarlem in 1955 and the trustees took the decision no longer to continue funding research at Teylers, he entirely dismantled Lorentz' laboratory and sold all the instruments that it contained – but didn't touch any part of the museum.

One could therefore conclude that on the one hand Lorentz' tenure as curator at Teylers marked the end of an era: the era of Teylers Foundation's role as one of the most important, perhaps even the most important, patron of experimental research in the Netherlands. But on the other hand, because of events surrounding his tenureship and the fact that he took little active interest in Teylers Museum, the Foundation and its museum were soon able to take on a new role as beacons of the history of science.

CHAPTER 11

Scientists in Cowsheds: Disputes over Hygienic Milk Production in the Netherlands, 1918–1928

Floor Haalboom

In the history of science and medicine, locations of knowledge have received extensive programmatic attention by scholars of the 'practice turn.'[1] Initially, they focussed on places in which scientists are at home, most notably laboratories.[2] More recently, historians of science developed interest in places not solely used for scientific activity, such as locations of fieldwork.[3] These sites where scientists are often visitors can reveal interesting relations between scientists and other social groups. Among such unconventional locations are farms and cowsheds, which function primarily as the workplace of farmers, but which have also been the site of knowledge production.

In the Netherlands of the 1920s, cowsheds became the focal point of a dispute about national regulations on milk quality between farmers, milk traders, and a group of physicians, veterinarians and chemists who called themselves "hygienists." The debate found a temporary ending with the milk decree (*melkbesluit*), implemented in 1925, and the model milk decree (*modelmelkbesluit*), rejected in 1928. Moreover, hygienists were involved in farming practices themselves on a local level. In the first decades of the twentieth century, several paediatricians (child physicians) managed model farms where milk that met medical standards was produced for infants and children. During the 1920s, veterinarian Cornelis Folkert van Oijen and physician Evert Cornelis van Leersum challenged these model farms after having studied the cowsheds managed by the British physician Robert Stenhouse Williams. My aim here is to investigate

1 See for instance *The Place of Knowledge. The Spatial Setting and its Relation to the Production of Knowledge*, eds. Adi Ophir, Steven Shapin and Simon Schaffer. Special issue of *Science in Context* 4 (1991).

2 Bruno Latour even puts all workplaces of scientists under the header of 'laboratory' in his older work: Bruno Latour, *Science in Action. How to follow Scientists and Engineers through Society* (Cambridge, MA: Harvard University Press, 1987), 64.

3 Examples are: David N. Livingstone, *Putting Science in its Place. Geographies of Scientific Knowledge* (Chicago: University of Chicago Press, 2007); Robert E. Kohler, *Landscapes and Labscapes. Exploring the Lab-Field Border in Biology* (Chicago: University of Chicago Press, 2002).

 | DOI:10.1163/9789004264885_012

the role of cowsheds as locations of knowledge production and as centres of scientific disputes among Dutch hygienists in the 1920s.

1 Locating Cowsheds: The Countryside

The scientific interest in milk and cowsheds was part of the hygienist movement of the second half of the nineteenth century. Around 1840, a group of liberal Dutch physicians started to relate public hygiene to the health status of different social classes by means of statistics, roughly twenty years later than their British and French colleagues. Hygienists turned public health into a matter of political importance that could be improved by public hygiene interventions such as waste processing, better housing, and the production and distribution of safe drinking water and food. Apart from its medical initiators, the hygienist movement of the 1870s included such professions as lawyers, officials, engineers, teachers, chemists, physicists, and veterinarians. These professionals were very active in local 'health committees' (*gezondheidscommissies*), while the medically trained hygienists tried to achieve their goals at a national level through the Medical State Inspectorate (*Geneeskundig Staatstoezicht*) founded in 1865. In the late nineteenth century, a new generation of hygienists reframed the hygienist agenda in bacteriological terms and organized themselves into the Congress for Public Health (*Congres voor Openbare Gezondheidsregeling*). This Congress played an important role in the preparation of many national social and public health acts written in the new political situation of the early twentieth century, such as the Public Health Act (*Gezondheidswet*) of 1901, and, important for milk hygiene, the Meat Inspection Act (*Vleeskeuringswet*) and the Commodities Act (*Warenwet*) of 1919.[4]

As is currently the case, cowsheds in the 1920s were located primarily in geographical areas called 'the countryside' (*platteland*), rather than in urbanized areas. A negative image of farmers and other rural inhabitants prevailed among Dutch hygienists. For instance, the expert members of the State Committee for the Control of Tuberculosis (*Staatscommissie inzake de bestrijding der Tuberculose*, 1918–1921), who met regularly in the city centres of The Hague and Amsterdam, pictured the countryside as a place swarming with flies and bacteria, lacking modern sewage, water and electricity, having a high

4 Eddy Houwaart, *De Hygiënisten. Artsen, Staat en Volksgezondheid in Nederland 1840–1890* (Groningen: Historische Uitgeverij, 1991); P.A. Koolmees, *Symbolen van Openbare Hygiëne. Gemeentelijke Slachthuizen in Nederland 1795–1940* (Rotterdam: Erasmus, 1997) 54–56, 109–113.

incidence of tuberculosis, and whose inhabitants had "a distinct psychology" and were less civilized than their urban counterparts.[5]

Hygienists saw this poor rural hygiene as a dangerous source of milk contamination. The prominent veterinarian Henri Kroon, for instance, argued in 1920 that usually milk left the udder free of germs, but became contaminated in the cowsheds (*see figure 11.1*). Kroon listed numerous opportunities for contamination: cows and farm workers suffering from infectious diseases such as tuberculosis, dust, urine, and rinse water infected with typhoid bacteria. One such source of contagion was cow manure: "Mockingly, although not entirely incorrect, it has been argued: cow milk is almost always a more or less diluted cow manure solution."[6] Kroon compared the lack of hygiene in cowsheds to the rising levels of hygiene in human society thanks to the construction of a sewage system, water works, and women's education in home economics. These developments contrasted sharply with the situation in cowsheds. Kroon exclaimed: "The proverbial Dutch cleanliness is nowhere to be found in the cowsheds and in the way animals are cared for."[7] He listed several "primitive conditions," such as the soaking of dirty hands in the milk in order to make the milk "smooth," and concluded that "An Augean stable can be cleaned here!"[8]

In general, milk was thought to travel from a polluted rural environment, whose inhabitants paid little attention to hygiene, to a modern, urban one, where the state of public facilities and scientific knowledge was far more advanced. Hygienists praised especially the larger urban dairies (*melkinrichtingen*) that used modern technology to process milk.[9] For example, at a meeting of the Dutch Society for the Protection of Infants (*Nederlandschen Bond tot Bescherming van Zuigelingen*) in 1916, paediatrician Gabriël Hermanus Moll

---

5 Nationaal Archief, The Hague, (hereafter NA), 2.27.14 Staatscommissie inzake de bestrijding der Tuberculose 1917–1920 (later SC), inv. nr. 33 Notulen van de Subcommissie voor de directe bestrijding 1918–1920, 1 February 1919, 97, 102; 30 May 1919, 215; 13 June 1919, 260; 21 November 1919, 349–350, 35; inv. nr. 43 Notulen van de Subcommissie voor de behandeling van de lijders 1918–1920, 23 October 1918, 5. Other examples are: J. Alma, "Tuberculose-Bestrijding op het Platteland," *Nederlandsch Tijdschrift voor Geneeskunde* (hereafter *NTvG*) 52 (1908): 1209; Home visitor (*huisbezoekster*) Limburg, "Hygiëne ten Plattelande," *Tuberculose* 15 (1919): 84.

6 H.M. Kroon, "De Hygiëne der Melkproductie," *NTvG* 64 (1920): 204–214, there: 207. Dutch quote: "Spottend, doch niet geheel onjuist, is gezegd: de koemelk is bijna steeds een meer of minder verdunde koemestoplossing." All Dutch quotes have been translated by the author.

7 Kroon, "De Hygiëne," 207. Dutch quote: "De spreekwoordelijke Hollandsche zindelijkheid is in de stallen en de verzorging der dieren in den regel verre te zoeken."

8 Ibid. Dutch quote: "Hier valt een Augias-stal te reinigen!"

9 See for instance: Cornelia de Lange, *De Geestelijke en Lichamelijke Opvoeding van het Kind* (Amsterdam: Meulenhoff, 1908; ed. 1921), 142–143; NA, 2.15.33 Gezondheidsraad 1920–1956 (later GHR), inv. nr. 436 Stukken betreffende het adviseren aan de minister inzake hygiënische melkwinning 1925–1929, Minutes, 21 March 1925, 8.

FIGURE 11.1 Postcard titled *In den Potstal* ("In the Deep Litter House"), sent to veterinarian H.M. Kroon in 1916 by an unknown person. According to Kroon this type of shed (common in the eastern parts of the Netherlands) was "the worst" in matters of hygiene.
COLLECTION VETERINARY MEDICINE, UNIVERSITY MUSEUM UTRECHT

van Charante extensively discussed the unhygienic circumstances at dairy farms and the unhygienic attitude of farmers, noting the almost impossible task of urban dairies to turn this filthy rural product into something tolerable. He stated that the polluted milk entered the city where "it is partially taken over by a large dairy, which does what can be done to improve the contaminated product."[10]

Despite the dominant negative conception of rural hygiene, hygienists were not always negative about the countryside itself. It provided possibilities for realising an ideal of the 'natural life' as opposed to the 'artificial life' in cities. Alcoholism, prostitution and moral corruption in general were thought to be the particular problems of urban areas. This paradoxical relation between city and countryside reveals the ambivalent sentiments regarding industrialised, 'modern' society. While the modernisation of society was propagated, signs of a

10 G.H. Moll van Charante, "De Zorg voor het Verkrijgen en Beschikbaar Stellen van Modelmelk," *NTvG* 61 (1917): 787–802, there: 789. Dutch quote: "Zoo komt dan de melk in de stad. Daar wordt zij gedeeltelijk overgenoomen door een groote melkinrichting, die er dan nog aan doet, wat er aan te doen is, om het vervuilde product te verbeteren."

countermovement advocating natural life became visible, in particular among the urban elite. In his hygienist textbook of 1913, the physician R.H. Saltet noted that the rural environment was healthier than the city because farm labourers unintendedly ingested more of the necessary proteins than labourers in urban areas. When such a farm labourer moved to the city, the shortage of proteins would give him an "unpleasant feeling" and drive him to the pub.[11] The Dutch physician and novelist Frederik van Eeden established a rural colony, which he named *Walden* after the American transcendentalist Henry David Thoreau's *Walden; or, Life in the Woods* (1854). Van Eeden's *Walden* was populated by a mix of idealistic city elites – who were little successful in farming the land and milking the cows – and a few farmers. Another example of this back-to-nature-movement was the popular Society for the Preservation of Nature Monuments (*Vereeniging tot Behoud van Natuurmonumenten*), founded in 1905.[12]

The rural origin of milk was also pictured in positive terms. Milk was regarded as a 'natural' product because it was produced by living animals. Consequently, the urban environment was seen as a bad environment for milk production.[13] This positive image of the countryside clashed with its negative image as a dangerous pool of germs in discussions on milk production. Indeed, hygienists often distinguished the rural environment from its inhabitants. The rural environment carried the positive connotations of naturalness, fresh air and light, while farmers could easily spoil the beneficial effects of the rural environment with their lack of civilisation and education. They were in need of scientific supervision to make sure that the milk that ended up in the cities met scientific standards. In this way, the negative image of farmers and the positive image of the 'natural' countryside were able to co-exist.

Controlling the milk production in rural areas through scientific and governmental supervision would ensure the optimal use of the benefits found in both 'natural' rural and 'modern' urban areas. Indeed, this call for scientific supervision of farms was one of the main incentives for the milk decree in the 1920s and would become the central topic in a fierce dispute between farmers and hygienists. The milk decree specified the demands on milk production

11 R.H. Saltet, *Voordrachten over Gezondheidsleer* (Haarlem: F. Bohn, 1913), 340.

12 Jan Fontijn, *Tweespalt. Het leven van Frederik van Eeden tot 1901* (Amsterdam: Querido, 1990) chapter 20; Henry David Thoreau, *Walden; or, Life in the Woods* (Boston, 1854); Anneke H. van Otterloo, *Eten en Eetlust in Nederland (1840–1990). Een Historisch-Sociologische Studie* (Amsterdam: Bert Bakker, 1990), 195, 201; Annemarie de Knecht-van Eekelen, *Naar een Rationele Zuigelingenvoeding. Voedingsleer en Kindergeneeskunde in Nederland 1840–1914* (Nijmegen: Thieme, 1984), 199; Auke van der Woud, *Een Nieuwe Wereld. Het Ontstaan van het Moderne Nederland* (Amsterdam: Prometheus, 2006), 188–194.

13 See for instance: NA, 2.15.33 GHR, inv. nr. 436, Minutes, 21 March 1925, 7.

under the Commodities Act of 1919, an act regulating food inspection on a national level in the Netherlands. These national regulations on food inspection supplemented and replaced the local, urban inspection services already operating in many Dutch cities from the late nineteenth century onwards. As these urban inspection services had no power over rural areas, representatives of the Commodities Act Commission argued to Minister of Labour Piet J.M. Aalberse that it "was impossible to combat evil at the place of production, namely the farm."[14]

2 Milk as Living Environment

Hygienists did not just argue for supervision of milk farms on the grounds of their negative opinion of rural inhabitants and farm environments. Findings in milk science posed problems they argued could be solved only by increasing scientific control over farmers and cowsheds. Hygienists had found that bacteria could easily survive and multiply in milk, as it was an especially nutritious and digestible substance. The Congress for Public Health devoted the first edition of the *Codex Alimentarius* of 1907 to milk, illustrating the special position milk had in hygienist concerns about food adulteration and contamination.[15] Typhoid fever and tuberculosis were discussed as the main milk diseases.[16] Typhoid fever was predominantly regarded as an environmental problem because typhoid bacteria were spread by infected water or other infected liquids, especially milk. Hygienists argued that the spreading of typhoid fever via milk could occur through milk adulterated with infected water, through the use of infected milking equipment, and through people handling the milk who suffered from the disease.[17]

---

14 Moll van Charante, "De Zorg"; NA, 2.27.14 SC, inv. nr. 17; Nota's betreffende de bestrijding van tuberculose 1919; L. Heijermans, October 1919, 318; inv. nr. 33, Minutes, 21 November 1919, 349–50. The quote of the Commodities Act commission can be found in: NA, 2.15.37 Afdeling Volksgezondheid 1902–1950 (later VG), inv. nr. 329 Stukken betreffende de totstandkoming van het KB 23-06-1925 (no. 256) tot toepassing van de artikelen 14 en 15 van de Warenwet op melk (Melkbesluit) 1921–1926, Letter Commodities Act commission to Aalberse, 29 May 192[3], 1. Dutch quote: "was het niet mogelijk het kwaad te bestrijden op de plaats der productie, nl. op de boerderij."

15 De Knecht van Eekelen, *Zuigelingenvoeding*, 166–167.

16 *Codex Alimentarius No. 1 Melk* (Groningen: Noordhoff, 1920), 17–18.

17 M.D. Horst a.o., "Melkbesluit," *NTvG* 70 (1926): 492–500; Moll van Charante, "De Zorg," 793–797; Kroon, "De Hygiëne," 205–206; Chief inspector of public health (contagious diseases), "Bestrijding van Buiktyphus en Roodvonk," *Verslagen en Mededelingen betreffende de Volksgezondheid* (1925), 345–353.

During this period, tuberculosis in the Netherlands was called "enemy of the people number one," although reliable numbers on how many people suffered the disease are not available for the entire country.[18] The disease was related to milk in international scientific discussions on how tuberculosis in cows related to human tuberculosis: was bovine tuberculosis similar to human tuberculosis and thus a threat to public health? In 1901, the eminent German bacteriologist Robert Koch radically changed his original opinion that bovine and human tuberculosis bacteria were identical. In an address to the International Tuberculosis Congress in London, he argued that the tuberculosis bacteria in humans and cows were of a different kind and could not cross-infect. Koch's address caused a lot of unrest. Many bacteriologists did not agree with him, based on their own research, but also based on already far-reaching government policies to control bovine tuberculosis, in Great Britain for instance. In the following years, the issue was fiercely discussed among experts from different countries.[19]

Dirk Aart de Jong was the only veterinarian appointed as a member of the State Committee for the Control of Tuberculosis, being the Dutch expert on the implications of animal tuberculosis for human health. Influenced by De Jong's argument that animals could infect people with tuberculosis bacteria, the Committee noted in its report that "tubercle bacilli of different animal species that suffer from tuberculosis" should be considered dangerous for human health.[20] Tuberculosis bacteria stemming from cows were considered the most dangerous, especially because of the easy spread via milk and meat. De Jong thought tuberculosis-infected milk a particularly difficult problem. Milk from different cows was not traded separately, which made it impossible to prevent the consumption of infected milk through repressive inspection, as was the case with meat.[21]

---

18 Bartje Abbo-Tilstra, *Om de Sûnens fan it Fryske Folk. Tuberculose en haar Bestrijding bij Bevolking en Veestapel in Fryslân, 1890–1940* (Leeuwarden: Fryske Akademy, 2002), 2–3.

19 Abbo-Tilstra, *Om de Sûnens*, 52–57; Susan D. Jones, "Mapping a Zoonotic Disease: Anglo-American Efforts to control Bovine Tuberculosis before World War I," *Osiris* 19 (2004): 133–148, there: 133; Koolmees, *Symbolen*, 125–126; Keir Waddington, *The Bovine Scourge. Meat, Tuberculosis and Public Health, 1850–1914* (Woodbridge: Boydell Press, 2006), chapters 3, 7, 9 and 10.

20 W. Roëll a.o., *Verslag van de Staatscommissie ingesteld bij KB van 3 July 1918 no. 25 tot Voorlichting over Wettelijke Maatregelen tot Bestrijding van de Tuberculose en over de Beste Wijze van Bestrijding dier Ziekte* (Den Haag, 1922), 14.

21 D.A. de Jong, "Beteekenis van de Bestrijding der Diertuberculose en van Goede Voedselkeuring voor de Bestrijding van Tuberculose bij den Mensch," Roëll a.o., *Verslag*, 217–220.

In the late nineteenth century two solutions for milk bacteria were developed: sterilisation and pasteurisation. With the first technique, milk was boiled for a rather long period of time to sterilise it, or make it bacteria-free. Pasteurisation was done with lower temperatures around 60 to 70°C in order to kill most, but not all, bacteria. Sterilisation was generally thought to change the character of milk in such a way that it could hardly be called milk anymore. The colour of sterilised milk turned brown, for instance. According to De Jong, such changes took place when the milk was heated above 72°C. Pasteurisation was generally considered to be the better alternative. In the 1890s, the Amsterdam physician Josef Forster developed a method for heating milk for 25 minutes at a temperature of 65°C, which was generally used in dairies in Amsterdam. However, in the succeeding years pasteurisation was criticized, for instance by De Jong. He feared that the most virulent bacteria, especially tuberculosis bacteria, survived the treatment. I will refer to this as the first problem of pasteurisation.[22]

However, milk was not seen just as an ideal environment for germs, but it was also positively related to human health. It was thought to contain particularly nutritious elements that were easily digestible and contributed to the positive image of milk as a 'living' liquid. Due to experiences with food scarcity during the First World War, food scientists became increasingly interested in the quality rather than quantity of food. During the 1910s and 1920s, vitamins attracted much scientific attention, including those present in milk. Proteins, which had been known for some time already, were also abundantly found in milk. These and other nutritious components of milk contributed to its image as being a delicate and healthy foodstuff, especially for the 'weak' members of society, namely the sick, the young, the elderly, and the poor. During this period, milk was indeed mainly consumed by such 'weak' people and hardly by healthy adults, although the saying "milk is good for everyone" (*melk is goed voor elk*) was gaining popularity, boosted by advertisement campaigns of the dairy trade.[23]

---

22 De Knecht-van Eekelen, *Zuigelingenvoeding*, 61–62, 192–197; Van Otterloo, *Eten*, 73–75; Kroon, "De Hygiëne," 207–208; D.A. de Jong, "Pasteuriseeren van Melk en het Dooden van Tuberkelbacillen," *NTvG* 53 (1909): 147–154; NA, 2.27.14 SC, inv. nr. 33: 8 November 1918, 29; November 21, 1919, 349; De Jong, "Beteekenis," 219.

23 A.P. de Knecht van Eekelen and A. Albert de la Bruhèze, "De Witte Motor," in: *Techniek in Nederland in de Twintigste Eeuw. Landbouw en Voeding*, eds. J. Bieleman and A.H. van Otterloo, vol. 3 (Zutphen: Stichting Historie der Techniek, 2000), chapter 6 (II), 311–321; Pim Huijnen, *De Belofte van Vitamines. Voedingsonderzoek tussen Universiteit, Industrie en Overheid 1918–1945* (Hilversum: Verloren, 2011), 31–35.

In this context a second problem of pasteurisation was defined by hygienists: proteins and vitamins were found to be damaged during heating. The very procedure capable of eradicating milk's many bacteria destroyed its essential nutrients at the same time, turning it from a 'living' into a 'dead' liquid. As medical historian Annemarie de Knecht-van Eekelen has shown, these worries were already voiced in the 1890s, when new chemicals were discovered in raw milk. Also during the first decades of the twentieth century, the dairy industry's increasing use of pasteurisation to eradicate milk bacteria was cause for concern among hygienists – in the Netherlands as well as in an international context.[24] As a consequence, the need to supervise milk farms got another impetus: hygienists increasingly thought *preventing* bacteria from entering milk would be better than *killing* them through pasteurisation. As veterinarian Van Oijen noted: "Condemning good milk to pasteurisation is putting the cart before the horse."[25] Pasteurisation of milk was made compulsory in the Netherlands only in 1940.[26]

3 Model Farms

The ideal locations to produce milk hygienically, according to hygienists, were 'model farms,' built in the Netherlands after German examples. The first Dutch model farm was founded in Oosterbeek in 1905, after a prize-winning German model farm at the Hygienic Milk Exhibition in Hamburg in 1903. In the following years, several other Dutch model farms were founded (*see figure 11.2*). A major aim of model farms was to produce milk of exceptional hygienic quality in order to make raw consumption possible, or at least to limit heating to only once, namely in the consumer's home. In the US and the UK a similar rise in

---

24 De Knecht-van Eekelen, *Zuigelingenvoeding*, 196–197; Jones, "Mapping," 143–145; Peter J. Atkins, "The Pasteurisation of England: the Science, Culture and Health Implications of Milk Processing 1900–1950," in: *Food, Science, Policy and Regulation in the Twentieth Century. International and Comparative Perspectives*, eds. David F. Smith and Jim Philips (London & New York: Routledge, 2000), 37–51, esp. 37–48. Dutch concerns on pasteurisation during the 1920s can for instance be found in: Roëll a.o., *Verslag*, 53; NA, 2.15.37 VG, inv. nr. 329, Letter Health Council of The Hague to the local government, 26 October 1921, 2; De Lange, *Het Kind*, 143; E.C. van Leersum, "Over het Vitamine C-Gehalte van Rauwe en Gepasteuriseerde Melk," *NTvG* 70 (1926): 338–348, there: 348.

25 C.F. van Oijen, "Een Conferentie over het Pasteuriseeren van Melk," *Het Algemeen Zuivel- en Melkhygiënisch Weekblad* 20 (1924): 130–131, there 130: "Goede melk tot pasteurisatie te veroordeelen is het paard achter de wagen spannen."

26 A.C. Ruys, *Keerzijden van Vooruitgang* (Haarlem: Erven F. Bohn, 1959).

FIGURE 11.2 A model farm, probably 'Oud-Bussum' in Bussum, undated. Note the presence of light, the open windows (fresh air!), the high ceiling, white-washed walls and the tiles on the floor.
COLLECTION VETERINARY MEDICINE, UNIVERSITY MUSEUM UTRECHT

production of raw milk of special hygienic quality occurred, where it was referred to as certified milk.[27]

De Knecht-van Eekelen has argued that the popularity of model milk declined in the Netherlands after 1908, when a major advocate of pasteurisation, bacteriologist F. Basenau, declared unheated milk, even from model farms, dangerous.[28] However, my own findings show that model milk remained an appealing solution to milk-related problems for many physicians and veterinarians. This is supported by the fact that a 'model milk decree' was debated in the 1920s and implemented in 1940 where it remained in practice until 1950.[29] De Knecht-van Eekelen notes this model milk decree, but sees it more as a relic

---

27 De Knecht-van Eekelen, *Zuigelingenvoeding*, 198–204. For American and British certified milk, see: Susan D. Jones, *Valuing Animals. Veterinarians and their Patients in Modern America* (Baltimore & London: Johns Hopkins University Press, 2003), 68–69; Peter Atkins, *Liquid Materialities. A History of Milk, Science and the Law* (Farnham & Burlington: Ashgate, 2010), 247–276.

28 De Knecht-van Eekelen, *Zuigelingenvoeding*, 204.

29 NA, 2.15.37 VG, inv. nr. 366 Stukken betreffende de totstandkoming van besluiten inzake modelmelk (Modelmelkbesluit) 1928–1949 & inv. nr. 531 Modelmelkbesluit 1926–1950.

of old views than a sign of contemporary convictions of the need for model milk. However, the debate on the model milk decree in the 1920s was the result of contemporary rather than past debates on concerns regarding milk and its production places.

Attempts to control milk production at model farms were initiated by paediatricians. Their concerns about pasteurised or bacteria-infected cow milk being an unhealthy substitute for "the ideal infant food" – mothers milk – led them to start their own farms. A substantial number of model farms were initiated, owned, and managed by paediatricians like G.H. Moll van Charante, H.W. de Monchy, and E.G.A. ten Siethoff.[30]

Moll van Charante and several other physicians founded and partly owned the model farm *De Vaan* in Rotterdam.[31] The location of this farm was no coincidence: Rotterdam accommodated a large and active community of hygienists, among which the head of the State Serum Institute (*Rijksseruminstituut*), veterinarian Jan Poels, who conducted quality tests on milk. Moll van Charante spoke as an expert on model farming for the Society for the Protection of Infants in 1916. He referred to the agriculturalist J. Boersma and the director of the model farm A. Stehouwer who had given him advice on farming. Moll van Charante argued that the possibility of hygienic supervision of cowsheds was the most important reason for establishing model farms. Local inspection services in cities could only prohibit the selling of adulterated or contaminated milk, but had no power to *prevent* milk contamination, as they did not control the entire production process. Model farms under medical and veterinary supervision could and should do this, Moll van Charante argued.[32]

These model farms had to meet several physical criterions. Of great importance was the positioning of these farms near large cities, because "the transport of raw milk should be as short as possible."[33] In addition, water works and electricity, necessary for hygienic precautions such as the washing of hands and the inspection of cows in electric light, were available near the cities. The model farms required a lot of grazing fields for cows, sunlight and that fresh air should enter the cowsheds, analogue to ideals for healthy human living environments. Another important aspect of the farms was the separation of rooms for keeping, cleaning, and milking the cows, for cooling, storing and bottling

---

30 De Knecht-van Eekelen, *Zuigelingenvoeding*, 199, 204.

31 This model farm left archival documents (a rarity), which De Knecht-van Eekelen has studied: De Knecht-van Eekelen, *Zuigelingenvoeding*, 199–201.

32 *'Strenge wetenschappelijkheid en practische zin.' Een eeuw Nederlands centraal veterinair instituut 1904–2004*, ed. Peter Verhoef (Rotterdam: Erasmus, 2005), 31–3, 82; De Knecht-van Eekelen, *Zuigelingenvoeding*, 197–204; Moll van Charante, "De Zorg," 787, 790, 797.

33 Moll van Charante, "De Zorg," 790.

milk, and for cleaning dirty milk bottles. Only tap water should be used, unless water of very good quality from dune springs was available in the area. Like the milk, water had to be kept separate from the farming environments. When cows were grazing outside, the canals should be fenced off to prevent the cows from drinking water from canals. Several cowsheds were needed in order to be able to isolate suspect cows, sick cows, and calving cows.

The health of personnel and animals on the farms was assured by means of regular inspection by scientific experts. Personnel should be medically inspected twice a year and when the physician found them to suffer from tuberculosis of the lungs, typhoid fever, scarlet fever or diphtheria, they would be denied access to the cowsheds. The cows should be inspected thrice a year by a veterinarian and just as often milk should be tested for the presence of infectious bacteria through injection in guinea pigs. In order to prevent the cows from getting tuberculosis, Moll van Charante considered it best to breed cows on the model farm itself because resistance against contagious diseases – especially tuberculosis – was considered mainly a matter of heredity. Moreover, breeding at the farm prevented the introduction of alien animals, which could be unobserved carriers of disease. Every day (except on Sundays) samples of the milk from three cows should be sent to the State Serum Institute in Rotterdam for chemical and bacteriological research by Poels and his staff. These special and strictly supervised model farms should make it possible to produce germ-free milk that did not require pasteurisation.

The advocates of model milk, however, were faced with the problem that model milk was expensive. The price of model milk was around 20 cents per litre in the 1900s, which was twice as high as normal milk.[34] Because of its high price, model milk constituted only a tiny portion of the total amount of milk sold. For instance in Amsterdam, the Society for Dairy Industry and Milk Hygiene (*Vereeniging voor Zuivelindustrie en Melkhygiëne*) noted that "milk and buttermilk for infants, raw milk, [and] model milk" were products "of secondary importance."[35] In 1928, the consumption of model milk was estimated at 3000 to 4000 litres a day,[36] roughly 0.1% of the total national milk

34 De Knecht-van Eekelen, *Zuigelingenvoeding*, 199.

35 *De Melkvoorziening van Amsterdam. Uiteenzettingen van de Vereeniging voor Zuivelindustrie en Melkhygiëne naar Aanleiding van het Rapport van de Commissie van Onderzoek in zake het Instellen van een Gemeentelijk Melkbedrijf te Amsterdam*, ed. G.J. Blink (Den Haag: Vereeniging voor Zuivelindustrie en Melkhygiene, undated [1923]), 63.

36 NA, 2.15.33 GHR, inv. nr. 436: Letter commission Commodities Act to Minister of Labour, 29 May 1928, 4. Prof. W.C. de Graaff mentioned a trade of model milk of 2000 to 3000 liters per day in 1926: NA, 2.15.33 GHR, inv. nr. 436; Minutes, 19 November 1926, 3.

consumption per day.[37] Moll van Charante's model farm *De Vaan* provided lower class families with cheaper model milk thanks to a philanthropic 'milk fund' in Rotterdam. However, even richer people were unwilling to pay more for model milk, Moll van Charante noted, as everyone was used to paying low milk prices "[i]n our country abounding in milk, a country that counts 3 million cows compared to a population of more than 6 million people."[38] *De Vaan* worked at a loss, and was declared bankrupt in 1939.[39]

Nevertheless, for many physicians and veterinarians, the ideal of model milk was an appealing solution for the problems of pasteurisation, dangerous milk bacteria, and the dirty countryside. Farmers, however, were not so easily set aside. Milk farms became the centre of dispute during the preparation of national quality regulations on milk. For Dutch cowsheds, the 1920s were roaring indeed.

4 Disputes on Farm Authority

The cowshed as *the* place of milk production became the central point of disagreement in discussions between hygienists and farmers on the Commodities Act milk decree. During these discussions, taking place from 1920 until 1925, representatives of farmers consequently argued that the farmers did not want to lose authority over their own cattle, cowsheds and milking practices to learned 'theorists' from the cities. The discussion reached high-level Dutch politics: the successive ministers of agriculture, Hendrik A. van IJsselsteyn (1918–1922) and Charles J.M. Ruijs de Beerenbrouck (1922–1924), personally argued on the matter with minister of labour, Piet J.M. Aalberse (1918–1925).[40]

To their annoyance, representatives of large agricultural societies were involved in the decision-making only a couple of months after the first draft of

37 This total national milk consumption per day was estimated at three million litres in 1928: NA, 2.15.33 GHR, inv. nr. 436; Letter commission Commodities Act to Minister of Labour, 29 May 1928, 4. In Amsterdam on its own the entire milk consumption was already 350,000 litres per day in 1923, according to the local government as quoted in: *De Melkvoorziening*, ed. Blink, 70.

38 Moll van Charante, "De Zorg," 800. Dutch quote: "Men is bij ons, in het land overvloeiende van melk, het land dat bij een bevolking van ruim 6 millioen menschen meer dan 3 millioen koeien telt, zóó zeer gewend aan lage prijzen voor melk, dat ook de meer gegoeden [...] er tegen opzien om meer te betalen dan den laagsten prijs."

39 De Knecht-van Eekelen, *Zuigelingenvoeding*, 201; E. van Muijlwijk, "Hyg. melkstal 'De Vaan' wordt geliquideerd," *De Nieuwe Veldbode* 6 (22 September 1939), 11–12.

40 Archival papers on these debates can be found in: NA, 2.11.06 Directie van de Landbouw: Zuivelaangelegenheden 1914–1947 (later DLZ), inv. nr. 8 Melkkwesties 192[1]–1940.

the milk decree had been completed by the hygienist Commodities Act Milk Commission in January 1921.[41] One of the main problems farmers had with the first draft was the strong emphasis on the hygiene of the cowsheds and the milking process, topics that were thought vital by hygienists. An important argument to prevent inspection was to point out that small farmers had too little financial means to carry out all the required alterations of their cowsheds. According to the Dutch Farmers Federation (*Boerenbond*) "hundreds and hundreds" of small farmers existed.[42] Indeed, Dutch cattle farms were mainly small scale farms during the first half of the twentieth century. In 1910, on average 5.5 cows were held per farm and in 1953 this number had risen to only 7.4 cows per farm.[43]

In principle, farmers agreed with the claim that cowshed hygiene was important for good milk production. The Farmers Federation stated: "Good milk can only be obtained when it is produced well."[44] However, farmers opposed any interference by "government officials," referring to those hygienists who claimed to have superior knowledge on how such good milk should be produced. Farmers were familiar with the 'practice' of milking, while hygienists knew only the 'theory.' The representatives of the Royal Agricultural Committee agreed that the government could impose certain quality requirements on milk, but they should not "put the entire agriculture business under supervision."[45] Farmers should have the freedom to choose *how* to come up with clean milk, because "[t]he danger is not imaginary that the supervision

---

41 Representatives of the General Dutch Dairy Federation (*Algemeene Nederlandsche Zuivelbond*, F.N.Z.), the Royal Dutch Agricultural Committee (*Koninlijk Nederlandsch Landbouw Comité*), Dutch Farmers Federation (*Nederlandsche Boerenbond*), Christian Farmers and Market Gardeners Federation (*Christelijke Boeren- en Tuindersbond*). Other extraordinary members were: a representative of the Society of Dairy Industrialialists (*Vereeniging van Zuivelfabrikanten*), and veterinary professor Van Oijen. NA, 2.15.37 VG, inv. nr. 493: Stukken betreffende de instelling en benoeming van de Commissie ingevolge artikel 17 van de Warenwet 1920–1950; Letter Schoorl and Filippo to Minister of Labour, 14 December 1921; Official decree on extraordinary members of the Commodities Act Commission, 31 January 1922.

42 NA 2.11.06 DLZ, inv. nr. 8; Anonymous, "De uitvoering der Warenwet en de veehouders," *Nederlandsch Tijdschrift voor Melkhygiëne* 17 (1921): 208; NA, 2.15.37 VG, inv. nr. 329, H. Ruijter, "Rapport over het ontwerp-melkbesluit," 22 July 1922, 25–26, and attachment II.

43 J. Bieleman, "De georganiseerde rundveeverbetering," in: *Techniek in Nederland in de twintigste eeuw: landbouw en voeding*, eds. J. Bieleman and A.H. van Otterloo, vol. 3 (Zutphen: Walburg Pers, 2000), 150, table 10.6.

44 NA, 2.11.06 DLZ, inv. nr. 8; Minutes, 6 March 1925, 3. Dutch quote: "Goede melk kan alleen worden verkregen, wanneer ze goed is gewonnen."

45 NA, 2.11.06 DLZ, inv. nr. 8; Minutes, 6 March 1925, 2. Dutch quote: "Gaat men op dezen weg voort, dan kan men het heele landbouwbedrijf wel onder contrôle stellen."

will be entrusted to officials, who have too little knowledge about the circumstances in which the production takes place."[46] Thus, farmer representatives did not so much question the science of hygienists, but opposed the control hygienists claimed over the farms, the domain of farmers.

Farmer organisations argued that requirements to improve cowshed hygiene and the milk production process should *only* be adopted by farms that wanted to use the label 'improved milk production' for their products, like model farms. As such, farmers suspected scientists aimed to change all farms into model farms, a suspicion not entirely ungrounded.[47] Because of these concerns, requirements for milk bearing the label 'improved' were moved to a milk decree article that needed expansion. Eventually the discussion on improved milk production would be continued as the debate on the model milk decree in the late 1920s.[48]

The final milk decree required the inspection of milk's chemical composition because the permitted amount of contamination and number of bacteria were now regulated on a national level. Regarding cowshed hygiene and the milking process, however, the final milk decree contained little of the original Commodities Act Commission's draft. The dispute over cowshed authority was the major reason why the writing of the milk decree took five years and why many hygienists were disappointed with the eventual result. The Commodities Act Milk Commission referred to the improvement of the situation in the cowsheds as "the most difficult subject of the entire milk decree."[49] In medical circles, the final milk decree was generally seen as a failure, especially as medical supervision was not allowed on the milk farms. The Society for the Advancement of Medicine (*Nederlandsche Maatschappij tot Bevordering der Geneeskunst*) was indignant to find that the requirements for the hygiene of cowsheds were missing from the final decree: "It seems one has feared to make

---

46 NA, 2.11.06 DLZ, inv. nr. 8; Minutes, 6 March 1925, 2. Dutch quote: "Het gevaar is niet denkbeeldig, dat de contrôle in handen wordt gelegd van ambtenaren, die niet voldoende op de hoogte zijn met de omstandigheden waaronder de productie plaats vindt."

47 After the first presentation of the draft of the milk decree to the farmers on 16 June 1921 (which was meant to remain confidential), the daily *De Telegraaf* wrote that some of the farmers present feared the concept milk decree would mean their farms had to become a model farm. As quoted in: Anonymous, "De Uitvoering der Warenwet en de Veehouders," *Nederlandsch Tijdschrift voor Melkhygiëne* 17 (1921): 206–208, there: 207.

48 NA, 2.11.06 DLZ, inv. nr. 8; Letter Minister of Labour [to the Minister of Agriculture], 11 December 1922; Several letters between agricultural organisations and the Minister of Agriculture in 1923; Minutes of a meeting of the dairy commission of the Dutch Farmers Federation in Utrecht, published in: *Land en Vee* 17 (2 February 1923), 1–2; Concept milk decree, 14 June 1923.

49 NA, 2.15.37 VG, inv. nr. 329; Letter milk commission to Aalberse, 29 May 192[3], 3.

heavy demands upon practice, especially on cattle farmers."[50] The Netherlands Veterinary Association (*Maatschappij voor Diergeneeskunde*) commented that the requirements in the milk decree were "extremely modest" and regretted the fact that no opportunities for supervision of cowsheds were created.[51]

The large number of changes in the original hygienist milk decree due to agricultural lobbying against top-down inspection of cowsheds and the milking process did not improve the image of farmers in the eyes of hygienists. The latter continued their attempts to legalize requirements for the hygiene of cowsheds and the milking process in the discussions on the model milk decree in the second half of the 1920s. In these discussions, the distrust of hygienists towards farmers continued to play an important role. This is illustrated by worries about the practices of "fake model farms" (*schijnmodelboerderijen*) selling milk under the label "model milk" without scientific supervision.[52] The battle for scientific inspection of farms continued, as civilisation of and control over farmers had become even more pressing.

5 Model Farming in Barns

"For those who can afford it, model milk is an excellent product, for the masses it is unattainable."[53] This sentence written by Kroon reveals a major social concern related to the high price of model milk. Although milk was considered a cheap source of nutrients, model milk was too expensive for the lower classes, who were in particular need of its nutrients.[54]

Physician Van Leersum and veterinarian Van Oijen thought it possible to produce raw, hygienic milk under less costly conditions than those prescribed for model farms. Van Leersum was the founder of the Dutch Institute for Food for the People (*Nederlandsch Instituut voor Volksvoeding*) in Amsterdam in 1919.

---

50 The physicians compared the final decree to the draft of 1923, which had already been adjusted to the criticism of the farmers. Horst a.o., "Melkbesluit," 493. Dutch quote: "Het heeft de schijn, dat men bevreesd is geweest aan de praktijk, inzonderheid aan de melkveehouders, hooge eischen te stellen."

51 H.A. Vermeulen and J.J.F. Dhont, "Het Concept-melkbesluit," *Tijdschrift voor Diergeneeskunde* 52 (1925): 187.

52 NA, 2.15.33 GHR, inv. nr. 436; Letter Commodities Act commission to the Minister of Labour, 29 May 1928, 3.

53 Kroon, "De Hygiëne," 209. Dutch quote: "Voor wie het betalen kunnen, is de modelmelk een uitstekend product, voor de groote massa is zij onbereikbaar."

54 See also the concerns on the high price of model milk expressed by the Amsterdam socialist physician Louis Heijermans in: NA, 2.27.14 SC, inv. nr. 33; Minutes, 21 November 1919, 348.

He devoted the institute to the quality of the food of the Dutch population, especially vitamins. Van Leersum was especially interested in milk because of its meaning as a healthy, cheap food for the lower classes, at least when not devoid of 'living' nutrients like vitamins. Apart from pasteurisation, Van Leersum also criticized model farms: these produced an elite product in too luxurious circumstances. He thought "a beautiful design of the cowshed, with gleaming tiles, shining brass and so on" was neither necessary nor sufficient to produce bacteria-free, raw milk.[55] Van Leersum was even reluctant to use the phrase 'model milk' to describe his ideal milk, and preferred the label "pure, raw milk."[56]

Van Oijen was veterinary professor in the field of foods for human consumption with an animal origin. One of his aims was to apply the science of meat and milk hygiene in farming practice. He argued that necessary changes in milk production could only be achieved when farmers were involved. Strict laboratory control of all health risks associated with milk would result in a product meeting the highest scientific expectations, but too expensive for almost everyone. Van Oijen argued that problems of cowshed hygiene should be handled with the "most simple although sufficient means," making safe and raw milk available for the entire population.[57]

In their arguments for cheaply produced model milk, Van Oijen and Van Leersum were inspired by the work of the British physician Robert Stenhouse Williams (1871–1932), the director of the National Institute for Research in Dairying in Reading. According to historian Peter Atkins, this institute was "the intellectual focus of the clean milk movement between the wars and of dairy bacteriology generally" in Britain.[58] The ideal of Williams was to produce "clean milk" that did not need to be pasteurised, was cheap enough to become popular with consumers, and was produced by farmers who had sufficient knowledge of the dangers of bacterial contamination of milk. Williams argued for five basic principles of clean milk production, aiming to reduce and prevent milk contamination with bacteria. First, milk should be cooled

---

55 E.C. van Leersum, "Pure Melk," *Haagsch Maandblad* 4 N°6 (December 1925): 1–8, there: 5–6 (offprint). Dutch quote: "De productie van pure melk vereischt maatregelen van grooter beteekenis dan een fraaie inrichting van den stal, met glimmende tegels, blinkend koperwerk en wat dies meer zij."

56 Huijnen, *De Belofte*, 31–62; A. de Knecht-van Eekelen, "Het Vitaminen-Laboratorium," *Gewina* 19 (1996): 43–46; NA, 2.15.33 GHR, inv. nr. 436; Minutes, 23 March 1925, 6–7.

57 C.F. van Oijen, *Over de Grondslagen der Hygiëne ten opzichte van Vleesch en Melk* (Utrecht, 1919); C.F. van Oijen, "De Productie van Zuivere, Ziekte-kiemvrije, Versche Melk," *Het Algemeen Zuivel- en Melkhygiënisch Weekblad* 20 (1924): 155–158, 167–169, 179–182, there: 156 (quote).

58 Atkins, *Liquid Materialities*, 243.

within three hours after milking. Second, covered milk pails should prevent dust and dung from contaminating the milk. Third, steaming and sterilising equipment would reduce the growth of bacteria in the milk. Fourth, grooming and washing cows would reduce the risk of milk contamination. And, last, milking should be done by well-educated and motivated personnel. As many farmers lacked water works, the first principle of cooling proved to be problematic in practice, but Williams and his researchers concluded that sterilising milking equipment was a good alternative, as it was financially feasible for farmers. Atkins notes that Shinfield – where the old cowsheds were located that Williams used to demonstrate his principles – "gradually became a place of pilgrimage," referring to the visits of many British, ranging from milk dealers to public health officials.[59]

Some of Williams's pilgrims crossed national borders. Both Van Leersum and Van Oijen visited Reading and were personally informed on the principles of cheap hygienic milk production by Williams. Both became enthusiastic advocates of his method in the Netherlands.[60] According to Van Leersum, Williams succeeded in producing cheap 'pure milk' with extraordinary simple means like intensive cleaning of the milkers' hands, the cows and the milking equipment. He was "deeply impressed" by Williams's success in producing milk containing few bacteria in "an old-fashioned cowshed." In a similar manner, Van Oijen praised Williams's "practical mind" that prescribed simple requirements to improve existing farms that could be "followed everywhere."[61]

During his visit, Van Oijen researched the milk produced in the old shed at Williams's institute (probably Shinfield) on its bacteriological content and presented his findings in the Dutch Veterinary Journal (*Tijdschrift voor Diergeneeskunde*). The pure milk scored well on the number of bacteria found. Van Oijen argued that this was especially due to the use of sterilised milking equipment, as he showed with bacteriological results of milk produced with sterilised equipment and milk produced with non-sterilised equipment. Combined with bacteriological inspection for pathogenic germs, Van Oijen argued that this relatively simply produced milk could "compete with the best model milk."[62]

---

59 Ibid., 238–44.

60 Van Leersum, "Pure Melk"; Van Oijen, "De Productie"; C.F. van Oijen, "De Melkwinning en Melkcontrole onder Leiding van Dr. R. Stenhouse Williams, te Reading. Verslag van een Onderzoek ter Plaatse," *Tijdschrift voor Diergeneeskunde* 51 (1924): 297–306, 364–373.

61 Van Leersum, "Pure melk," 5; Van Oijen, "De Melkwinning," 299.

62 Van Oijen, "De Melkwinning," 368. Dutch quote: "Bij een goede contrôle op pathogene kiemen kan zij met de beste modelmelk concurreeren."

Williams's principle of well-trained and reliable personnel posed a problem for Van Leersum and Van Oijen in their argument that all farms could become model farms. According to Van Leersum, the contaminated environment of farms could be dealt with as long as the people handling the animals and the milk had excellent knowledge of bacteriology, analogous to the necessity of bacteriological training for wound treatment during war. Although Van Leersum thought model farms were too fancy, farmers did need to know some bacteriology to protect their milk, and as "laymen" they needed the supervision of academically trained professionals to overcome "their own notions on hygiene."[63] Van Oijen emphasized the need for inspection of hygienically produced milk in a similar way: "the appropriateness of this system stands or falls with the supervision."[64] In this sense, Van Leersum and Van Oijen differed little from the proponents of model milk: milk should be produced under strict scientific control.

Indeed, Van Oijen and Van Leersum used their argument for cheaply produced model milk to further the cause of medical and veterinary supervision of milk farms, which had failed during the writing of the Commodities Act milk decree. Inspection of milk and cows would be a rewarding new task for veterinarians who were in want of jobs, and inspection of farm personnel would ensure the authority of physicians in matters of epidemiology. Van Leersum and Van Oijen were both involved in the design of a model milk decree from 1925 to 1928. This model milk decree was not implemented, mainly on the grounds that the government could not guarantee safe consumption of raw milk stemming from an unreliable agricultural world, both because of its many bacteria and its stubborn inhabitants.[65]

6 A Case of Social Commerce: Puritas Milk

Model farming physicians like Moll van Charante, Van Leersum, and Van Oijen became involved in commercial activities because of their social ideals. Historian Pim Huijnen has shown how Van Leersum's Institute for Food for the People closely cooperated with the food industry from the 1920s onwards,

---

63 NA, 2.15.33 GHR, inv. nr. 436; Letter Van Leersum to Jitta, 29 January 1925, 2–3. Dutch quote: "melkveehouders en hunne arbeiders, die op dit gebied leeken zijn en er eigen begrippen omtrent zindelijkheid op na houden." See also: Van Leersum, "Pure Melk," 5.

64 Van Oijen, "De Melkwinning," 364, 372, Durch quote: "de doelmatigheid van dit systeem staat of valt met de gehouden contrôle."

65 Papers on the model milk decree can be found in: NA, 2.15.37 VG, inv. nr. 531; NA, 2.15.33 GHR, inv. nr. 436.

as Van Leersum hoped its industrially produced vitamins would benefit the food quality of the population at large.[66] Van Oijen became involved in a commercial farm during his quest for a cheaper alternative to model milk. In 1924, Van Oijen started experiments at the milk farm *Puritas* in Overschie (near Rotterdam). The aim was to investigate the economic possibilities of simply and thus relatively cheaply produced hygienic raw milk following the example of Williams's method. The farmers J. van der Kooij, W. and J. de Kroes had asked Van Oijen for assistance (probably via their local veterinarian M. Slager), inspired by Van Oijen's presentation of his experiences in Reading at the Dutch veterinary society (*Maatschappij voor Diergeneeskunde*).[67]

The farm in Overschie had to be adjusted to the new standards, as everything in the existing shed "mocked the most elementary notions of hygiene." "Nevertheless," Van Oijen noted, "a reasonable whole was composed with simple means."[68] This was exactly the aim of model farming with simple means: turn an old cowshed into a trustworthy one. The transition can be seen in figures 11.3 and 11.4. The shed in Overschie was provided with a new gutter and with litter. It was painted and electrical lighting was installed. As Van Oijen had found, sterilisation of the milking equipment was central to the success of Williams's method. For that reason, Van Oijen ordered sterilisation equipment to be installed in the room where the milk was cooled and bottled. Furthermore, Van Oijen ordered a machine that could cool the milk using spring water. From the cooler, the milk was poured into sterile bottles that were closed off with carton disks and a closing capsule.

Like all milk experts, Van Oijen pointed out that three types of periodically conducted research were of importance: veterinary inspection of cows, medical inspection of farm personnel, and bacteriological research of the milk. Van Oijen had two colleagues who took part in the testing and the management: local veterinarian Slager, who was responsible for the veterinary inspection of the cows,[69] and local physician Jacob Donkersloot, who was responsible for

---

66 This hope was hampered by the industry's interest in secrecy. Huijnen, *De Belofte*, 55–62.

67 Van Oijen discussed his research and involvement at the farm in Overschie (Delftweg 64) in: Van Oijen, "De Productie"; C.F. van Oijen, "Ervaringen met de Melkwinning volgens Dr. Stenhouse Williams bij Toepassing in Nederland," *Tijdschrift voor Diergeneeskunde* 52 (1925): 49–59. See also: Anonymous, "Belangrijke Vereenvoudiging in de Winning van z.g. Modelmelk," *Nieuwe Rotterdamsche Courant* (later NRC) morning (16 February 1924); Advertisement "Drinkt Zuivere Versche Melk," NRC evening (26 November 1923).

68 Van Oijen, "Ervaringen," 51. Dutch quote: "een echte spoelingstal, waar alles spotte met de meest elementaire begrippen van zindelijkheid. Toch gelukte het met eenvoudige hulpmiddelen een redelijk geheel samen te stellen."

69 Slager used the "ophtalmoreaction" to detect tuberculosis. This was done with a new kind of tuberculin produced at the State Serum Institute from 1924 onwards, which was put in

Vuile donkere spoeling-stal voor de verbouwing.

FIGURE 11.3 The cowshed of the *Puritas* farm in Overschie as veterinarian C.F. van Oijen found it in 1923. Van Oijen's caption reads: "Filthy dark cowshed before the alterations."
SOURCE: VAN OIJEN, "DE PRODUCTIE VAN ZUIVERE, ZIEKTE-KIEMVRIJE, VERSCHE MELK," 168

the medical inspection of the farm personnel. Van Oijen himself was responsible for the testing of the milk: he examined the presence of bacteria in the milk and carried out catalase tests to detect traces of udder infections and sediment tests to detect general filth. As the production at the farm in Overschie was too small to finance bacteriological and veterinary inspection, Van Oijen and his colleagues Slager and Donkersloot did the inspection work "disinterestedly."[70]

As specific disease-causing bacteria would often be present in small quantities only, Van Oijen argued for general bacteriological research on the number of coli bacteria, a bacterial species present in high numbers in the guts of humans and animals. If these bacteria could be found in milk, other specific

---

the eyes of animals with better diagnostic results than the traditional injection of tuberculin under the skin. This new product resulted in an important rise in the use of tuberculin. *"Strenge Wetenschappelijkheid,"* ed. Verhoef, 79, 81.

70 Van Oijen, "De Productie," 6–7, 22; Van Oijen, "Ervaringen," 53; Anonymous, "Belangrijke Vereenvoudiging."

Een gedeelte van den veestal na de verbouwing, Electrisch verlicht en met eenvoudige hulpmiddelen, volgens de eerste eischen der hygiëne, kraakzindelijk gehouden.

FIGURE 11.4 *Puritas* farm, Overschie, 1924. The caption reads: "A part of the cowshed after the alterations. Electrically lighted and with simple means, kept extraordinarily clean according to the first demands of hygiene."
SOURCE: VAN OIJEN, "DE PRODUCTIE VAN ZUIVERE, ZIEKTE-KIEMVRIJE, VERSCHE MELK," 169

disease-causing bacteria might be present as well. Van Oijen "researched the '*Puritas* Milk' for Bacteria coli daily" – in reality several times per month.[71]

Van Oijen's research of the *Puritas* milk was predominantly conducted in his own laboratory in Utrecht. As such, "The *Puritas*-milk had the disadvantage of a journey per mail," namely from Overschie to Utrecht.[72] The distance of approximately sixty kilometres was especially problematic because the transport resulted in a vast increase of bacteria and sending the samples on ice

71 Van Oijen, "De Productie," 12. Dutch quote: "Als scherpste reactief op de zindelijkheid en scherpste contrôle op de afwezigheid van ziektekiemen onderzoek ik dan ook dagelijks de 'Puritas Melk' op Bac. coli."

72 Van Oijen, "De Productie," 29. Dutch quote: "De Puritas-melk had nog het nadeel van een reis per post."

proved to be too expensive. Eventually it was decided that the milk was to be sent by means of express delivery, which brought Van Oijen the milk several hours after milking. The number of bacteria turned out to stay below 50,000 per cm^3 – the maximum number of bacteria allowed in model milk according to the *Codex Alimentarius*. The selection of milk was done by an anonymous person in Rotterdam, who bought a bottle of *Puritas* milk and a bottle of milk produced at the model farm *De Vaan* when ordered to by Van Oijen.[73]

Although Van Oijen predominantly worked in his own urban laboratory, he located his work on *Puritas* milk at the farm in Overschie, as part of a social project. Like Van Leersum, Van Oijen thought his involvement as a scientist in commercial activities would benefit the population at large, although such involvement was unusual. During a meeting on *Puritas* milk for Rotterdam housewives, Van Oijen was cited to have said:

> It is not his task to act in public like he did with the design of the farm in Overschie; his work is located in the laboratory. Nevertheless, a major public interest is related to this issue; and when the results that Van Oijen visions are reached, he will retreat into this workplace again.[74]

Such a study conducted outside the lab in the interest of the public had its disadvantages. Its "unforeseen factors" were incomparable to the controllable parameters of a laboratory experiment, especially "the mentality of the persons, who have to execute the necessary measures."[75]

Moreover, Van Oijen had to make concessions to laboratory practice, because the owners of the farm had to make a living with *Puritas* milk. The experiments were conducted from November 1923 to September 1924. Up until January, a period in which adjustments to the farm and cowshed were made, the milk of only eight cows was sold and the milk was hardly advertised. The official opening of the company by means of a local media campaign was in February. Van Ooijen played an important role in this campaign: he was a

---

73 Van Oijen, "Ervaringen," 55; Anonymous, "Belangrijke Vereenvoudiging"; Anonymous, "De Productie van Zuivere Ziektekiemvrije Melk," *NRC* morning (March 22, 1924); *Codex*, 20.

74 Anonymous, "De Productie"; Dutch quote: "Van Oijen besloot met te zeggen, dat het niet zijn taak is in 't openbaar op te treden als hij bij de inrichting der Overschiesche boerderij heeft gedaan; zijn werk ligt in het laboratorium. Maar er is met deze zaak een groot algemeen belang gemoeid; en zijn de resultaten, die professor van Oijen beoogt eenmaal bereikt, dan trekt hij zich weer in die werkplaats terug."

75 Van Oijen, "Ervaringen," 50. Dutch quote: "Tal van onvoorziene factoren spelen daarbij een rol, niet het minst de mentaliteit van de personen, die de te nemen maatregelen moeten uitvoeren."

prominent speaker at the press presentation of the *Puritas* farm as well as a meeting of the Rotterdam department of the Dutch Society of Housewives (*Nederlandsche Vereeniging van Huisvrouwen*). From this moment on, *Puritas* milk was regularly advertised in local newspapers. The eventual aim was to house a maximum of forty-five "carefully selected" cows producing an average of 700 litres per day, or ca. 250,000 litres per year. *Puritas* milk could be purchased at several milk sellers in Rotterdam. From October 1924 onwards, the distribution of the milk was farmed out to the large dairy Progress (*De Vooruitgang*) in Rotterdam. At this stage, a second *Puritas* farm was founded.[76]

Van Oijen saw his laboratory in Utrecht as a place where he could 'objectively' compare the merits of different locations of milk production. As Van Oijen had done with Williams's milk in Reading, he researched the bacteria content of *Puritas* milk produced in Overschie and compared it to the number of bacteria in model milk produced at Moll van Charante's model farm *De Vaan*. The latter turned out to contain more bacteria, as figure 11.5 illustrates. Van Oijen therefore thought it possible to provide a large percentage of the Dutch population with cheap model milk. Indeed, according to Van Oijen himself, *Puritas* milk was quite successful thanks to his public support. Nevertheless, his comments on its price reveal that it was hardly cheaper than original model milk.[77]

7 Facing Model Farmers

In arguing that *Puritas* milk did better than the model milk, Van Oijen walked on thin ice. A tight network of hygienists interested in milk existed in Rotterdam from the late nineteenth century onwards. They included founders of model farm *De Vaan* and physicians of the Society for the Protection of Infants (Moll van Charante, for instance, was involved in both). Van Oijen had to face these hygienists during his propaganda for model farming with simple means, using 'his' *Puritas* farm as shining example.

A direct confrontation between Van Oijen and representatives of *De Vaan* took place at a meeting of the Dutch Society of Housewives in Rotterdam, reported in detail in the local newspaper *Nieuwe Rotterdamsche Courant*. Hendrik Karel de Haas, a prominent Rotterdam physician and a member of

76 Van Oijen "Ervaringen," 50–51, 55, 58; Van Oijen, "De Productie," 22; Advertisement "Puritas melk," NRC morning (23 March 1924); Anonymous, "Belangrijke Vereenvoudiging"; Anonymous, "De Productie."

77 Van Oijen, "De Productie." Puritas milk cost 26 cents per 800 centiliter according to Van Oijen, "Ervaringen," 58.

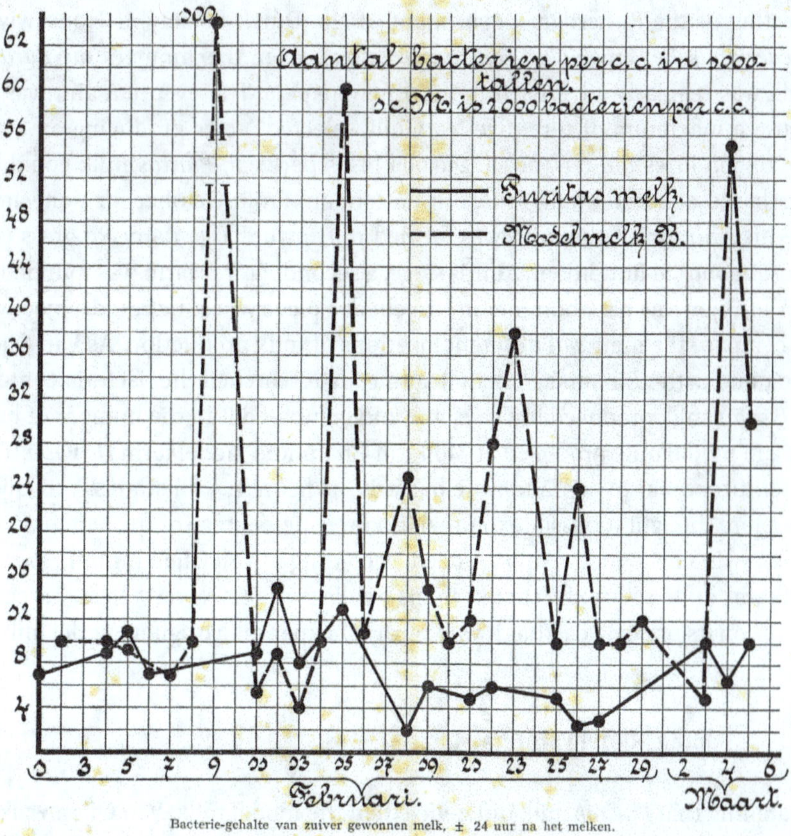

FIGURE 11.5 Van Oijen's graph of the number of bacteria present in *Puritas* milk and model milk from model farm *De Vaan*, February–March 1924. The caption reads: "Bacteria-content of purely produced milk, ca. 24 hours after milking."
SOURCE: VAN OIJEN, "DE PRODUCTIE VAN ZUIVERE, ZIEKTE-KIEMVRIJE, VERSCHE MELK," 181

the board of the Society for the Protection of Infants,[78] criticised Van Oijen for his counting of bacteria in *De Vaan*'s model milk, as food inspection services and the State Serum Institute in Rotterdam were equipped and officially authorised to do so. These institutes analysed model milk with much better results than Van Oijen did. Moreover, De Haas suspected Van Oijen's ties with the owners of the *Puritas* farm, leading De Haas to question Van Oijen's experiments. Van Oijen reacted furiously: "Does dr. De Haas doubt the objectivity of

78 H.J. Fliezinga [Flieringa, FH], "In Memoriam Dr. H.K. de Haas," *Rotterdams Jaarboekje* 2 (1954): 212–216.

my research?" Subsequently, Moll van Charante and manager A. Stehouwer of the model farm *De Vaan* criticized Van Oijen's farm for its lack of attention to the breeding of cows. Buying cows (as the *Puritas* farm did) was not without danger because these cows could be unobserved carriers of tuberculosis, which would result in severe losses for the model farm as well as endanger the health of the public. Only breeding on the farm itself could ensure the health of the animals and the superior quality of the milk. As such, the quality of the cows was very explicitly linked to their superior location at a model farm. Van Oijen could not simply overrule the experience of *De Vaan*. The housewives present at the meeting were also not amused by Van Oijen's supposed attempts to discredit *De Vaan*. Mrs Wenniger-Hulsebos argued that Van Oijen did not express "enough reverence ... for the model farms that have been practising for so long what he defends here." Physician De Haas's wife rhetorically asked Van Oijen whether he was still of the opinion that *Puritas* milk could be consumed unheated after all the critique, to which he answered: "Undoubtedly."[79]

A year later, Moll van Charante discussed Stenhouse Williams's principles in Reading and Van Oijen's *Puritas* farm in the same Rotterdam newspaper. He argued that although the results in Reading called for following, the Netherlands and the United Kingdom differed profoundly in matters of climate, soil composition, cow pasturing routines, disease incidence, farmers' hygiene,[80] and legal norms.[81] The article's conclusion, in short, was that Williams's principles were suitable for the United Kingdom, but not for the Netherlands. Especially the Dutch incidence of tuberculosis made it impossible to guarantee that milk produced by means of Williams's method was free from bacteria. The *Puritas* farm reacted to "the Dear Erudite Sir Dr. Moll van Charante" with an advertisement for one of Van Oijen's publications on *Puritas* milk (*figure 11.6*).[82]

Van Oijen triumphantly published the results of tests conducted by the Food Inspection Service in Rotterdam in 1928, which confirmed that *Puritas* milk contained less bacteria than milk produced at "two model farms" in Rotterdam. Van Oijen concluded that the results of the two model farms were

79 Anonymous, "De Productie." Dutch quotes: "Twijfelt dr. de Haas aan de objectiviteit van mijn onderzoek?" "voldoende piëteit ... tegenover de modelstallen, die al zoo lang in practijk hebben gebracht wat hij hier verdedigt." "Ongetwijfeld."

80 "... the average Englishman [is] much more hygienic ... than an inhabitant of the Continent" (Dutch quote: "... de gemiddelde Engelschman [is] veel reiner ... op zijn handen dan een bewoner van het Continent"). G.H. Moll van Charante, "Melk en de Melkvoorziening van een Groote Stad," NRC evening (16 May 1925).

81 British law recognised certified milk, Grade A tuberculin tested and Grade A veterinary inspected as different milk qualities. Atkins, *Liquid materialities*, 247–276.

82 Moll van Charante, "Melk"; Advertisement "Melkvoorziening eener Groote Stad," NRC evening (20 May 1925).

FIGURE 11.6
Advertisement of *Puritas* milk in response to physician G.H. Moll van Charante's criticism, *NRC*, 20 May 1925. The advertisement reads: "Milk Supply of a Large City. In response to the articles, in the last N. Rott. Courant by the Honerable Very Erudite Gentleman Dr. Moll van Charante, ask for our brochure On the Use of Pure Germ-free Milk by Prof. C.F. van Oijen, Professor of Hygiene of foods of animal origin, [...] which will be sent to you on request."

disappointing: "In my view, such milk does not deserve the name model milk anymore," while *Puritas* milk had been proven to be "*Model milk* in the true sense of the word."[83] When the model milk decree was implemented in 1940, the *Puritas* farm in Overschie was still operating as a model farm.[84]

8 Conclusion

This narrative about cowsheds and farming hygienists illustrates intricate ties between hygienist physicians and veterinarians, farmers and urban dairies. As hygienists framed cowsheds as the major source of milk contamination, linked

83 C.F. van Oijen, "Ervaringen met de winning van Modelmelk volgens de methode van Dr. Stenhouse Williams, in Nederland," *Tijdschrift voor Diergeneeskunde* 55 (1928): 686–8, there: 687, Dutch quote: "Zulke melk verdient m.i. den naam modelmelk niet meer"; 688, Dutch quote: "... de melk der boederij Puritas, volgens mijn aanwijzingen gewonnen overeenkomstig de in Engeland opgedane ervaring, is *Modelmelk* in den waren zin van het woord."

84 NA, 2.15.37 VG, inv. nr. 531; Letter Director-General of Food Supply to the Permanent Secretary of the Department of Social Affairs, 4 November 1941.

to hygienists' negative understandings of practices of rural inhabitants, they became central locations in debates on milk hygiene in the 1920s. As pasteurisation was increasingly criticised because of its unreliability in killing harmful bacteria and its damaging effects on milk's nutritious components, raw milk was framed as ideal milk, and the lack of scientific supervision over cowsheds became an even more pressing problem for hygienists. From 1900 onwards, such supervision was realised at model farms, commercial companies under medical and veterinary control with the aim of producing milk of exceptional hygienic quality, often started and owned by physicians. Hygienists tried to nationalise scientific supervision of milk farms via the milk decree of the food inspection act of 1919, the Commodities Act (*Warenwet*). However, they failed to achieve this goal, as farmers successfully opposed such interference in their domain of 'practice.'

A major problem of model milk was its high price, and, consequently, its low demand by consumers. As a solution, veterinarian C.F. van Oijen and physician E.C. van Leersum argued for model farming with simple means, in the fashion of the British physician R. Stenhouse Williams. Although Van Oijen's research into commercial *Puritas* milk was conducted in his usual laboratory in Utrecht, he pictured it as a social project, unusual for a scientist, but necessary to provide the population with healthy milk and to expand veterinary and medical control over cowsheds. In this project, the *Puritas* farm in Overschie served as an example of the possibility of turning every Dutch cowshed into a reliable milk farm, if only managed with expertise. By showing the positive results of bacteriological investigations of *Puritas* milk compared to model milk, Van Oijen argued that Dutch milk production as a whole could meet the demands of model milk without needing the unrealistic 'luxuries' of model farms. By getting involved in a commercial company that competed with model milk, Van Oijen became engaged in fierce debates with the physicians associated with the Rotterdam model farm *De Vaan*. As such, cowsheds were not merely locations where authority over milk was fought over by farmers and hygienists, but they became centres of fierce authority disputes on the knowledge of milking and related social concerns among hygienists as well.

CHAPTER 12

Blending Research: Linking Dutch Industrial and Academic Polymer Laboratories, 1940–1980

Marijn J. Hollestelle

1 Introduction

Plastics dominate our lives. Cars, packaging, clothing and electronics are all products that cannot exist without this material.[1] After the Second World War, plastics technology flourished in the Netherlands. Together with nuclear power, the recently discovered plastics were seen as a sign of progress. Until the oil crisis, which hit the Netherlands in 1974, the plastics industry was the fastest growing chemical sector. Between 1960 and 1973, the volume of production of the plastics branch grew by a factor of 13, whereas the entire Dutch chemical industry grew by a factor of 5.[2] Plastics – together with coatings, resins, adhesives and rubber part of a group of materials with long molecules called 'polymers' – became the number one product of the Dutch chemical industry. In order to be able to start up the manufacture of synthetic materials after the war, and to improve and develop new plastics, scientific knowledge of polymers became indispensable. At first, companies developed this knowledge themselves. They were pioneers in importing and developing new scientific insights concerning the behaviour and properties of large molecules. This imported knowledge was used by industry in manufacturing various types of plastic. But when developments started to speed up in the nineteen sixties, domestic research became indispensable for further improvement of products and for creating new types of plastic, in order to keep up with international competition. Just at this time, polymer science entered the Dutch *academia*. In the process of building this new academic field within the technical universities, the industry had a major input. Polymers proved to have a high practical

1 The author wrote this article in a personal capacity and wants to thank the Dutch Polymer Institute (DPI) and the Foundation for the History of Technology (SHT) for supporting the research that led to this article.

2 F. van der Most, E. Homburg, P. Hooghoff and A. van Selm, "Nieuwe synthetische producten: plastics en wasmiddelen na de Tweede Wereldoorlog," in: *Techniek in Nederland in de Twintigste Eeuw 2: delfstoffen, energie, chemie*, eds. J.W. Schot et al. (Zutphen: Stichting Historie der Techniek, 2000), 358–375, especially 363.

 | DOI:10.1163/9789004264885_013

utility, but also had a new and appealing scientific component. In the years to come, Dutch polymer science developed into a field with a high international standing, both at the industrial and the academic polymer laboratory. However, the relationship between academic and industrial laboratories is far from straightforward; these sites are two very different locations of knowledge, each with its own goal and distinct research culture.

During the Industrial Revolution, scientific knowledge was of secondary importance to technological developments. Inventors acquired knowledge at an *ad hoc* basis. Scientific institutions played a very small role in developing innovations or in stimulating industrial developments. This situation changed during the late nineteenth century, when strong ties with science and scientific institutions shaped the rise of the chemical, optical, electrical and refrigeration industries. Industrial research laboratories first emerged in Germany and the United States, and during the early twentieth century also in the Netherlands. This development is seen as the primary feature of the so-called Second Industrial Revolution. Dutch industry became more knowledge intensive, and employed academics for research and development. Contacts between industrial laboratories and universities were intensified. A number of professors visited industrial laboratories on a regular basis to give colloquia and advice, or were involved in industrial research.[3] Science began to play an important role for industry. Technology and industrial production became dependent on scientific knowledge and methods: industry became 'science based.' Some historians credit the universities for the rise of 'science based industry' and have suggested that these institutions were crucial in this process.[4] For instance,

---

3 See for instance: H. Lintsen et al., *Made in Holland. Een techniekgeschiedenis van Nederland (1800–2000)* (Zutphen: Walburg Pers, 2005), chapter 16; E.S.A. Bloemen, "Bezieling en 'espirit d'équipe.' Industriële research in Nederland in het interbellum," in: *Exercities in ons verleden. Twaalf opstellen over de economische en sociale geschiedenis van Nederland en koloniën 1800–1950*, eds. P. Boomgaard et al. (Assen: Van Gorcum, 1981), 153–167; H.G. Heijmans, *Wetenschap tussen universiteit en bedrijfsleven. De experimentele natuurkunde in Utrecht onder W.H. Julius en L.S. Ornstein 1896–1940* (Rotterdam: Erasmus, 1994); G.J. Somsen, *'Wetenschappelijk onderzoek en algemeen belang.' De chemie van H.R. Kruyt (1882–1959)* (Delft, 1998); M.J. Hollestelle, *Paul Ehrenfest. Worstelingen met de moderne wetenschap* (Leiden: Leiden University Press, 2011); P. Huijnen, *De belofte van vitamines. Voedingsonderzoek tussen universiteit, industrie en overheid 1918–1945* (Hilversum: Verloren, 2011).

4 See for instance: D.F. Noble, *America by Design: Science, Technology, and the Rise of Corporate Capitalism* (Oxford: Knopf, 1977), 3–19; *Science, Technology, and Everyday Life, 1870–1950*, ed. C. Chant (New York: Routledge/Open University, 1989), 320–321; B. Hunt, "'Practice vs. Theory': The British Electrical Debate, 1888–1891," *Isis* 74 (1983): 341–355, especially 354; L. Burchardt, "Wissenschaft und Wissenschaftswachstum: Industrielle Einflussnahmen auf die Wissenschaftspolitik im Wilhelminischen Deutschland," in: *Soziale Bewegung und politische Verfassung: Beiträge zur Geschichte der modernen Welt*, eds. U. Engelhardt, V. Sellin,

university researchers contributed directly to the German dyestuff industry.[5] Other historians oppose this view. Historian of chemistry Christophe Lécuyer, for instance, asks himself: "What do universities really owe to industry? Does *academia* always guide the industry on the scientific knowledge domain?" Lécuyer's answer is negative. He shows that the influencing can go the other way around. In his study he demonstrates that between 1955 and 1985, breakthroughs in industrial research laid the foundations for important scientific developments in micro-electronics and informatics at Stanford University.[6]

Likewise, in a leading article on electrical engineering before the First World War, Wolfgang König shows that the German *technische Hochschulen* benefited more from the knowledge developed in industry, than *vice versa*. König proposes to use a different terminology in this case, namely "industry-based science."[7] König and Lécuyer emphasize that industry can play a leading role in scientific research. So not only universities can affect industrial research; industry, in its turn, is capable of influencing universities and the research that is done there.

In this article, I will focus on how the interaction between universities and industry shaped Dutch polymer science and how this new field developed as a result of the interplay between these very different locations of knowledge. What exactly was the relation between polymer industry and the universities? How did these locations of knowledge shape each other, absorb each other's knowledge and affect one another in developing and strengthening polymer science as a new field of enquiry in the Netherlands?

H. Stuke (Stuttgart: Klett, 1976), 770–797, especially 772; P. Borscheid, *Naturwissenschaft, Staat und Industrie in Baden (1848–1914)* (Stuttgart: Klett, 1976), 172, 210; W. Rammert, *Sociale Dynamik der technischen Entwicklung: Theoretisch-analytische Überlegungen zu einer Soziologie der Technik am Beispiel der 'Science-Based Industry'* (Opladen: Westdeutscher Verlag, 1983); W. Wetzel, *Naturwissenschaften und chemische Industrie in Deutschland: Voraussetzungen und Mechanismen ihres Aufstiegs im 19. Jahrhundert* (Stuttgart: Klett, 1991), 96; P. Erker, "Die Verwissenschaftlichung der Industrie: Zur Geschichte der Industrieforschung in den europäischen und amerikanischen Elektrokonzernen, 1890–1930," *Zeitschrift für Unternehmensgeschichte* 35 (1990): 73–94, especially 73–74; P. Weingart, "Wissenschaftssoziologie und Technik: Die soziologische Analyse wissenschaftlicher und technischer Handlungsorientierung," in: *Zum Verhältnis von Wissenschaft und Technik, Erkenntnisziele und Erzeugungsregeln akademischen und technischen Wissens, Vortragstext einer Tagung*, ed. P. Lundgreen (Bielefeld: Kleine, 1981), 6–31, especially 7.

5 E. Homburg, "The emergence of research laboratories in the dyestuffs industry, 1870–1900," *The British Journal for the History of Science* 25 (1992): 91–111; Y. Furukawa, *Inventing polymer science. Staudinger, Carothers, and the emergence of macromolecular chemistry* (Philadelphia: University of Pennsylvania Press, 1998), 16.

6 C. Lécuyer, "What do universities really owe industry? The case of solid state electronics at Stanford," *Minerva* 43 (2005): 51–71.

7 W. König, "Science-based Industry or Industry-Based Science? Electrical Engineering in Germany before World War I," *Technology and Culture* 37 (1996): 70–101, especially 100.

2 The Dutch Knowledge Revolution

Companies nowadays are keen to advertise their products as the result of scientific research. The Dutch chemical company DSM uses the motto: "Bright science. Brighter living."[8] But it is a fact that historically viewed, Dutch companies turned to science rather late.

For a long period the Netherlands had a rather extensive industry that used chemical processes only to manufacture bulk materials, such as sulphuric acid. In 1864, a Polytechnic School had already been established at the Dutch city of Delft to train engineers who could be set to work in the manufacturing industry. But it took until around 1900 for Dutch companies to really show interest in scientific research. In neighbouring countries, such as Germany, industrial research laboratories had existed since the late nineteenth century. But at the turn of the century, the industrial research laboratory gained ground in the Netherlands, next to the already existing laboratories for technical support and quality control. Only then did Dutch industry start to employ chemical engineers trained at the Delft Polytechnic at a significant scale.[9] When Dutch parliament passed a patent law in 1910, this brought an end to the limitless use of foreign technology. This patent law appeared to be a tremendous stimulus for the rise of the Dutch industrial research laboratory.[10] Around 1910, the Philips company founded a chemical laboratory, and in 1914 the famous Philips *Natuurkundig Laboratorium* (Philips Physics Laboratory, better known as *NatLab*) was founded.[11] At the *NatLab*, the famous German physicist Gustav Ludwig Hertz (1887–1975), who in 1925 would receive the Nobel Prize in physics, did ground breaking research on gas discharges. Research in laboratories of similar industries, such as those of the American company General Electric, was interpreted as proof that conducting use-inspired basic research could be very useful to industries to provide them with commercial attractive

---

8 See DSM webpage: https://www.dsm.com/corporate/about.html (29 July 2019).

9 E. Homburg, A. Rip, J. Small, "Chemici, hun kennis en de industrie," in: *Techniek in Nederland in de twintigste eeuw* II, eds. J.W. Schot et al. (Zutphen: Stichting Historie der Techniek, 2000), 306.

10 The German industrial laboratories developed earlier then the Dutch, because Germany had adopted a patent law much sooner. See: W. Wimmer, *'Wir haben fast immer was Neues.' Gesundheitswesen und Innovationen der Pharma-Industrie in Deutschland, 1880–1935* (Berlin: Duncker & Humblot, 1994).

11 For the history of Philips NatLab, see: K. Boersma, *Inventing structures for industrial research. A history of the Philips Nat. Lab. 1914–1946* (Amsterdam: Aksant, 2002); M.J. de Vries and F.K. Boersma, *80 years of research at the Philips Natuurkundig Laboratorium 1914–1994* (Amsterdam: Amsterdam University Press, 2005).

technology.[12] New scientific breakthroughs were seen as direct sources of technological innovation.[13] In the Netherlands, the *NatLab approach* became the leading one, and spread during the *interbellum* years.

At this time, chemistry professors at the Dutch universities started to actively stimulate the deployment of academically trained chemists in industrial research.[14] They hoped to augment their discipline and to enlarge the job market for the rising number of science students. But also, they held a firm believe that science could benefit industry, and thus benefit society.[15] During the first fifteen years of the twentieth century, both the universities and the Delft Polytechnic actively tried to fortify ties with industry by employing *privaatdocenten* (university teachers) and extraordinary professors (professors who partly retained their jobs in industry), for instance in applied chemistry.[16]

Especially during the interbellum years, several large companies sought a closer cooperation with academic professors, who were recruited as industrial advisors and commissioners.[17] In 1911, Arnold Frederik Holleman (1859–1953), a professor at Amsterdam University, got involved as commissioner and advisor with a company that was started by one of his former PhD students, Jacques Coenraad Hartogs. The latter founded the *Eerste Nederlandse Kunstzijdefabriek Arnhem* (ENKA) after he had gained experience with the British artificial silk (rayon) industry.[18] In 1929, ENKA took over the German

---

12 Prominent research at General Electric was done for instance by Irving Langmuir (1881–1957). Cf. Ernst Homburg, *Speuren op de tast. Een historische kijk op industriële en universitaire research* (Inaugural lecture, Maastricht University, 2003), 31; M.J. Hollestelle, "Zuivere praktijk: Paul Ehrenfests opvattingen over toegepaste natuurkunde," in: *Onderzoek in opdracht. De publieke functie van het universitaire onderzoek in Nederland sedert 1876*, eds. L.J. Dorsman and P.J. Knegtmans (Hilversum: Verloren, 2007), 54–55.

13 Homburg, "Chemici, hun kennis en de industrie," 314–315.

14 G.J. Somsen, "Selling science: Dutch debates on the industrial significance of university chemistry, 1903–1932," in: *Determinants in the evolution of the European chemical industry, 1900–1939: new technologies, political frameworks, markets and companies*, eds. A.S. Travis et al. (Dordrecht, Boston and London: Springer, 1998), 143–168, especially 143–144.

15 Frans H. van Lunteren, *Uit de ivoren toren. Nederlandse natuurwetenschappers in het interbellum* (Inaugural lecture, VU University Amsterdam, 2003); F.H. van Lunteren and M.J. Hollestelle, "Paul Ehrenfest and the dilemmas of modernity," *Isis* 104 (2013): 504–536.

16 Homburg, "Chemici, hun kennis en de industrie," 309, 312.

17 Homburg, *Speuren op de tast*, 25; P. Baggen et al., "Opkomst van een kennismaatschappij," in: *Techniek in Nederland in de twintigste eeuw* VII, eds. J.W. Schot et al. (Zutphen: Stichting Historie der Techniek, 2003), 141–173, especially 160–165; D. Baneke, *De Groningse eeuw van de natuurwetenschappen* (Assen: In Boekvorm Uitgevers, 2005), 34–35; D. Baneke, *Synthetisch denken. Natuurwetenschappers over hun rol in een moderne maatschappij, 1900–1940* (Hilversum: Verloren, 2008), 57; Hollestelle, "Zuivere praktijk."

18 AKU-Research. *De NV Onderzoekingsinstituut "Research" van de Algemene Kunstzijde Unie NV* (1961), 18.

chemical company *Glanzstoff* and the *Hollandsche Kunstzijde Industrie* (HKI), after which the acronym ENKA changed into AKU (*Algemeene Kunstzijde Unie*). From 1929 onwards, Holleman was a member of the board of directors together with the Utrecht professor of colloid chemistry Hugo R. Kruyt.[19] In 1933, AKU centralized its research as a separate firm in a laboratory in Arnhem, called the *Onderzoekingsinstituut Research*.

Before the Second World War, Dutch companies, such as AKU, were involved in the production of artificial materials, based on natural products, for example casein (a milk protein), cellulose or rubber. These materials were used to manufacture buttons, rayon, and tires. One of the first fully artificial plastics to be mass produced in the Netherlands was Bakelite, a material that is a good electrical insulator.[20] Because of this property, electronics firm Philips started a state-of-the-art plant in 1925 that produced their own brand of Bakelite, called: *Philite*.

During these years, it became increasingly clear that all of these new materials were built up of macromolecules – molecules consisting of chains, made up of thousands of molecular parts, linked together. In 1912, the German chemist Hermann Staudinger had posed the existence of these large molecules. According to classical organic and physical chemistry, the existence of these extremely large molecular chains was not possible. Around the end of the 1930s, these macromolecules were proved to exist. It turned out that the properties of materials, such as cellulose, rubber, Bakelite and other plastics, could be explained by presuming that these materials were made up from macromolecules. The first large-scale test of the validity of macromolecular theory was conducted at the Du Pont laboratory in the United States.[21] In 1935, Wallace Carothers was able to develop nylon using the macromolecular theory.[22] The product instantly became a success. In 1939, the new nylon stockings flooded the marked. Within a year, 64 million pairs were sold. The nylon venture would become "a prototype for the science-based polymer industry."[23] It was clear that fully synthetic plastics had potential. During the economic crisis of the 1930s and the Second World War, they were used as alternative material. Chemical companies in the United States, England and Germany developed countless new products, such as synthetic fibres, polyethylene, polystyrene,

19 Homburg, "Chemici, hun kennis en de industrie," 308–309.

20 Bakelite was named after its inventor, the Flemish chemist Leo Hendrik Arthur Baekeland (1863–1944).

21 Furakawa, *Inventing polymer science*, chapter 3.

22 About Carothers, see: M.E. Hermes, *Enough for one lifetime. Wallace Carothers, inventor of nylon* (Washington: American Chemical Society / Chemical Heritage Foundation, 1996).

23 Furukawa, *Inventing polymer science*, 9.

polyvinylchloride (PVC) and artificial rubber. Examples of new applications were the perspex turrets mounted on bombers, nylon (and its German counterpart: perlon) parachutes; tires made from artificial rubber, and the newly discovered polyethylene played a crucial role in the development of the British radar system.

During this period, some Dutch companies carefully began to explore the possibilities of plastics. *Staatsmijnen* ('Dutch State Mines,' in 1973 renamed into the acronym DSM) was such a company. Its first experience with polymerizing new materials was part of a slow but steady diversification process, which changed the company from a local coal mining company into a large multinational chemical firm.[24] The German occupation of the Netherlands, which lasted from the tenth of May 1940 until the sixth of May 1945, accelerated research in the field of polymers as alternatives to oil and rubber, which had become scarce. In 1940, the Staatsmijnen opened a new *Centraal Laboratorium* (Central Laboratory), where, amongst other things, work was done on synthetic lubricating oil, synthetic diesel oil, and artificial rubber.[25]

The Dutch rubber industry was prospering before the war, because of the rubber that was imported from the colonial Dutch East Indies (now Indonesia). In 1936, the Dutch rubber industry founded the *Rubber-Stichting* (Rubber Foundation). This foundation conducted research and spread propaganda on rubber to stimulate its consumption. Because of the scarcity of rubber during the war, the research institute of the Rubber-Stichting redirected its activities and started research into synthetic plastics. The driving force of this development was the foundation's new director, Roel Houwink (1897–1988). Earlier, Houwink had been a chemical engineer, first at the Dutch tire factory *Vredestein* (1922) and later at the *Rijksrubberdienst*, a government-funded rubber laboratory. But Houwink also was the first technical director of Philips's famous *Philite* factory (1925–1939), the most modern Bakelite factory in Europe. Because of this background, he was acquainted with leading rubber and plastic experts worldwide, such as Leo Hendrik Arthur Baekeland (1863–1944) – the inventor of Bakelite – and the Austrian-American chemist Herman Francis Mark (1895–1992). In the midst of the war, Houwink founded the Plastics Institute (*Kunststoffen-Instituut*), which started research on synthetic plastics. In this way, Houwink prevented the seizure of laboratory equipment by the

24 For the history of Staatsmijnen and her R&D, see: A. van Rooij, *The company that changed itself. R&D and the transformations of DSM* (Amsterdam: AUP, 2007).

25 *Chemistry in wartime in the Netherlands. A review of the scientific work done by Dutch chemists in the years 1940–1945* (Amsterdam: Nederlandsche Chemische Vereeniging / D. B. Centen, 1947), chapter VIII: 10–12, chapter IX: 16–17. The Central Laboratory had started already in 1928 within an existing laboratory for quality checking.

German occupation authorities and saved his personnel from forced labour in Germany.[26] Earlier, in 1939, he had written a well-known book on plastics, to be followed in 1943 by a second one.[27] In 1942, Houwink was appointed as a *privaatdocent* (lecturer) at the Delft polytechnic, lecturing in "Chemistry and Technology of Macromolecular Materials." Although formally no university, the Delft polytechnic had the same status and rights as a regular university. So, Houwink can be seen as the first Dutch academic lecturer in what would become polymer science after the war.[28] According to Houwink, plastic was a material of limitless possibilities.[29]

Within industrial chemical networks, researchers and managers got enthusiastic about the post-war possibilities of plastics. Several academic researchers came into contact with the industrial plastics research when during the war they went into hiding at the laboratories of Philips and AKU.[30] As we will see, they contributed tremendously to making Dutch plastics research a successful discipline.

3 The Industrial Origin of Dutch Polymer Science

In the years immediately after the war, Dutch industry was greatly hampered in its plans to grow. The chemical industry could not wait to start the production of plastics. From America, exiting news about new plastics reached the Netherlands. Soon, Dutch companies, such as AKU and Shell, started their own production of nylon and PVC.[31] Simultaneously, the Dutch plastics processing industry made a jump start. Immediately after the war, existing companies started with the makeshift production of the first plastic consumer products, often on a solitary machine in the corner of the factory hall. Because of the low oil price, these new plastics were very cheap. This made it very difficult for companies to choose expensive alternatives, such as wood, leather, metals

26 For other examples, see: D. van Delft, "Preventing theft: The Kamerlingh Onnes Laboratory in wartime," in: *Scientific Research in World War II. What scientists did in the war*, eds. A. Maas and H. Hooijmaijers, (London and New York: Routledge, 2009), 62–76.

27 R. Houwink, *Chemie und Technologie der Kunststoffe* (Leipzig: Becker & Erler, 1939); Ibid., *Grundriss der Kunststofftechnologie* (Leipzig: Becker & Erler, 1943).

28 H.A.M. Snelders, *De geschiedenis van de scheikunde in Nederland 2. De ontwikkeling van chemie en chemische technologie, 1900–1950* (Delft: Delftse Universitaire Pers, 1997), 184.

29 Van der Most, "Nieuwe synthetische producten," 361.

30 The Section for Industrial Chemistry of the NCV organised a symposium about synthetic plastics during the war, see: *Chemisch Weekblad* 41 (1945): 112.

31 S. Howarth and J. Jonker, *Geschiedenis van de Koninklijke Shell. Deel 2. Stuwmotor van de koolwaterstofrevolutie, 1939–1973* (Amsterdam: Boom, 2007), 385.

and cotton.[32] A major problem hampering industry in these developments was the lack of skilled employees to fill the many emerging or expanding industrial laboratories. This triggered a run on graduate scientists and engineers.[33] Dutch universities simply lacked the capacity to train enough scientists. Another problem was that universities were running behind at some points, so they were not entirely up to date. In these years, the major new developments in chemistry took place not at the universities, but in industry. In some branches, industrial laboratories were the Dutch centres of excellence. In the 1950s, industrial research was ahead with respect to the university, especially in solid-state physics, solid-state chemistry, process technology, catalysis, and the physics and chemistry of polymers.[34] In essence, this remained so up to the 1960s and 1970s, when academic polymer science really took off. The companies that worked on polymers during the war played an important role in the emergence of polymer science as a field in the Netherlands, even if there was no polymer research done at the universities. This situation contributed significantly to the establishment of polymer science as a scientific discipline in the Netherlands.

Just as before the war, it was a firm belief in scientific circles that 'pure science' could benefit 'practical science,' and thus could contribute to the development of useful applications. People in government circles were convinced of the usefulness of scientific research for the rebuilding of the country.[35] The book *Science, the Endless Frontier* (1945) by the American presidential advisor Vannevar Bush (1890–1974) was seen as a new bible. Stimulation of so-called 'basic research,' together with unlimited freedom and resources for researchers would lead to an endless stream of innovations that would herald an era of unprecedented prosperity.[36] To make this dream come true, the Dutch

---

32 Van der Most, "Nieuwe synthetische producten," 363.

33 W. Hutter, "Chemie, chemici en wetenschapsbeleid," in: *De geschiedenis van de scheikunde in Nederland* 3, E. Homburg and L. Palm (Delft: Delftse Universitaire Pers, 2004), 19–35, especially 22–24; Staatsmijnen could not fill her job openings in research, see: Centraal Laboratorium Staatsmijnen, *Jaarverslag* 1948 t/m 1953. Also TNO had troubles filling positions, see: *Jaarverslag TNO* (1948): 72.

34 Homburg and Palm, *De geschiedenis van de scheikunde in Nederland,* 3, 13–15. Also for physics research this was the case; for instance Philips NatLab employed much better researchers during the first post-war years, and Dutch industry was then much better aware of the developments in international research than the Dutch universities. See: H.B.G. Casimir, *Het toeval van de werkelijkheid. Een halve eeuw natuurkunde* (Amsterdam: Meulenhoff, 1983), 277, 279.

35 Hutter, "Chemie, chemici en wetenschapsbeleid," 20–21.

36 H. Lintsen and A. van Rooij, "Een geschiedenis van TNO Industrie en Techniek," in: *Techniek in actie. Een geschiedenis van TNO Industrie en Techniek*, ed. H. Lintsen (n.p.: Stichting Historie der Techniek, 2007), 16–39, especially 22–24.

government stimulated the universities by doubling the budget for higher education between 1947–1954. In 1950 the government founded the Dutch Organisation for Pure Scientific Research (*Nederlandse Organisatie voor Zuiver Wetenschappelijk Onderzoek*, ZWO) to distribute the increasing research funds. Academic scientists made up the bulk of the ZWO board.[37] The Foundation for Chemical Research in the Netherlands (*Stichting Scheikundig Onderzoek in Nederland*, SON) was founded in 1956. Like ZWO, the goal of SON was "the promotion of basic research."[38]

According to Dutch industry, these measures were not adequate by far. Dutch industrial companies decided to donate unprecedented amounts of money to fortify the Dutch universities. In 1946, the *Bataafsche Petroleum Maatschappij* (BPM) donated three million guilders (comparable to fifteen million euro today) to the universities of Leiden and Delft.[39] As Delft harboured the only Dutch university of technology, Philips NatLab-director Gilles Holst (1886–1968), chairman of a government committee to reorganize the system of higher education in the 1940s and 1950s, called for a second university of technology in the Netherlands. After Philips had donated one million guilders (comparable to three million euro today), this second university of technology started in 1956 in Eindhoven – the very place where the Philips company was based (*figure 12.1*).[40]

To the taste of the Dutch industry, progress at the universities was still going too slow. In 1958 a committee, chaired by the new *NatLab* director and member of the Philips board Hendrik B.G. Casimir (1909–2000), called for a tenfold increase in the growth rate of the universities. The number of full professors should be tripled. The Dutch Ministers of Education carried out the recommendations of the Casimir report almost in its entirety. As a result, the resources for academic teaching and research grew dramatically between 1958 and 1970.[41] In 1964, a third university of technology opened at the town of Enschede.[42] Dutch chemistry was also on the rise in this period.[43] In 1960, it became possible for students to graduate in chemical technology at the University of Groningen – a

37 Hutter, "Chemie, chemici en wetenschapsbeleid," 20–21.

38 Ibidem, 22–24. In 1998 SON was integrated with the NWO Chemical Sciences division.

39 Homburg, *Speuren op de tast*, 41.

40 Ibidem, 42–43.

41 Ibidem.

42 Homburg and Palm, *Geschiedenis van de scheikunde*, 8. Chemist G. Berkhoff from Staatsmijnen became the first director of the university.

43 Ibidem, 5.

FIGURE 12.1 Equipment to make polymer fibers at the chemical technology laboratory at the Eindhoven polytechnic (later Eindhoven University of Technology)
PHOTO BY D. HEIKENS

technological study at a regular university. In 1962, the University of Nijmegen started a major in chemistry.[44]

An important contributor to the rise of post-war polymer research was the Rubber-Foundation and the associated Plastics Institute. Houwink, responsible for the foundation of the former institute during the war, was an important pioneer and promoter of polymer research in the Netherlands.[45] In 1946, he was one of the founding fathers of one of the first and most important international magazines in polymer science, the *Journal of Polymer Science*. He started this journal together with the Austrian-American chemist and dean of polymer science Herman Mark, who had immigrated to the United States during the war.[46] The AKU chemist Petrus Hendrik Hermans (1898–1979) became one of the advisory editors. The first year of its existence, Dutch polymer

44 N. Velthorst, "Een voortdurende uitdaging: het scheikundeonderwijs," in: Homburg and Palm, *Geschiedenis van de scheikunde*, 37–59, especially 45; Hutter, "Chemie, chemici en wetenschapsbeleid," 25.

45 *Chemisch Weekblad* 41 (1945): 14.

46 C.D. Andriesse, *Dutch Messengers. A History of Science Publishing, 1930–1980* (Leiden, Boston: Brill, 2008), 74–75, 84–86.

chemists – all of whom were working in industry – wrote 18 % of all articles in this journal.[47]

Although the Rubber-Foundation played an important role in establishing Dutch polymer science, it was short-lived. In the same year as the *Journal of Polymer Science* was founded, the Rubber-Foundation abandoned plastics research and continued with rubber research. The Plastics Institute was handed over to the *Nederlandse Organisatie voor Toegepast-Natuurwetenschappelijk Onderzoek* (TNO), the Dutch Organisation for Applied Scientific Research. This organisation was founded in 1932 to bridge the gap between academic science and industry.[48] Although the Rubber-Foundation continued with polymer (rubber) research, the laboratory was dismantled in the 1950s. Dutch rubber was no longer an important product because of the loss of the Dutch East Indies after the war.

But an important start was made, and the Plastics Institute gained national and international fame with its expertise in the field of plastics technology. The institute grew rapidly, from two employees in 1946, to 50 in 1947, 144 in 1951 and 200 in 1956.[49] In 1947, the physical chemist Albert Jan Staverman (1912–1993) came in charge of exploratory plastics research. He was the leader of the Central Laboratory of TNO between 1954 and 1970.[50] The Plastics Institute conducted research to service the entire plastic-producing industry (collective research), but also engaged in research contracts with individual companies.

In the establishment of polymer science as a scientific discipline, it was especially the company AKU that – next to TNO – played a role of importance which cannot be overstated. In 1941, in the midst of war, AKU founded an Institute for Cellulose Research in Utrecht. Until 1966, this was the place where basic physical chemical research took place.[51] AKU had a central laboratory at Arnhem, and Utrecht was sixty kilometres away. But the *Instituut voor Cellulose-Onderzoek* (or shortly the *Cellulose-Instituut*), as it was called, was based in Utrecht for a good reason. The institute's researchers did not wish to be bothered with all kinds of "practical questions that did not belong in its atmosphere." The choice of Utrecht was "partly made because of the vicinity of the university, which brought with it many advantages for this type of work

47 See: *Journal of Polymer Science* 1 (1946).

48 *Chemisch Weekblad* 41 (1945): 105.

49 *Plastica* 4 (1951): 269; A. van Rooij and H. Lintsen, "Materiaal voor onderzoek," in: Lintsen, *Techniek in actie*, 88–109, especially 90.

50 G. Challa, "Polymeren," in: Homburg and Palm, *Geschiedenis van de scheikunde*, 231–241, especially 231; Homburg, *Speuren op de tast*, 33–34.

51 D. Vermaas, *Research bij AKU* (1968), 95.

FIGURE 12.2 The 'Instituut voor Cellulose-Onderzoek' of AKU, in the centre of Utrecht, early 1950s. From left to right: D. Vermaas, D. Heikens, P.H. Hermans, P.F. van Velden. PHOTO BY D. HEIKENS

[basic research]."[52] In Utrecht, the institute benefited from the close relations with the Utrecht professor in colloid chemistry Hugo R. Kruyt, who had been a member of AKU's board of directors since 1933. The *Cellulose-Instituut* was under the inspirational guidance of the physical chemist P.H. Hermans.[53] As is illustrated by his position as advisory editor of *Journal of Polymer Science*, Hermans was well acquainted with the important players in the polymer field, and was a highly respected scientist and polymer expert (*see figure 12.2*).

The war provided Dutch industry with unexpected opportunities. When the war was over, German patents were confiscated by the allied powers and distributed amongst the different war-struck countries. The Dutch State came into possession of the German patent for the production of the chemical called 'caprolactam,' a raw material used for the production of a type of nylon. This gave an alternate route for producing a somewhat different type of nylon, bypassing the DuPont patents regarding its nylon production. The German patent enabled the Staatsmijnen to start producing caprolactam in 1952.[54] AKU bought this caprolactam and synthesized its own nylon. This was spun into

52 AKU-Research. De NV Onderzoekingsinstituut "Research" van de Algemene Kunstzijde Unie NV (1960?), 73.

53 Homburg, *Speuren op de tast*, 35.

54 E. Homburg and A. van Rooij, "Die Vor- und Nachteile enger Nachbarschaft. Der Transfer deutscher chemischer Technologie in die Niederlande bis 1952," in: *Technologietransfer aus der deutschen Chemieindustrie (1925–1960)*, ed. R. Petri (Berlin: Duncker & Humblot, 2004), 201–251, especially 240–241.

fibres, but was also made into granules, with which a rapidly emerging plastics processing industry shaped all kinds of plastic consumer products. AKU's nylon was one of the most important Dutch plastics. AKU conducted its own plastics research at its central laboratory research-based in Arnhem, to improve the quality and endurance of its products, but also to develop new kinds and types of plastic. To augment this, the *Cellulose-Instituut* stopped doing research on cellulose and focussed entirely on researching synthetic polymers after 1952. It became the Dutch centre of excellence for research on polymerization reactions.[55] Nevertheless, cellulose remained an important material for AKU, especially because of its use in tire yarn.[56]

The research that was performed at AKU's *Cellulose-Instituut* closely resembled the type of research done at universities. Researchers who were active in this type of research had a great deal of freedom, published in international scientific journals, and had little relation to the difficulties of the pilot plants and the factories (also think about the geographical divide between AKU's central laboratory and the *Cellulose-Instituut*), and were usually headed by a charismatic and leading scientist. Not without reason DSM employees called the division Basic Polymer Research "our university." Its research director Ronald Koningsveld was called 'the professor,' although Koningsveld never held an academic position.[57]

The alleged benefits of quite free and exploratory research were not up for debate at this time. According to AKU's annual reports, it was even a necessity: 'Basic research significantly deepened the understanding of the mechanism of the caprolactam polymerization; the obtained insight is used for technological research to improve the conditions for preparing and spinning of the polymer.'[58] Next to nylon, the reactions to synthesize polyester were researched: "Great attention was paid to research of a fundamental nature, as we believe that this kind of research work is essential not only to progress in the long term, but also because the knowledge of the backgrounds is of the utmost importance both for solving the everyday issues and for better management of production

55 Snelders, *Geschiedenis van de scheikunde*, 184–185.

56 It remained so up to 1996, when AKZO (a company that came into life after a merger between AKU and Koninklijke Zout) discontinued the production of rayon tire yarn in the Netherlands, despite the superior properties of this material with respect to nylon and polyester; the latter two are much cheaper, however. See: Koos Schwartz ,"De kwijnende markt van het rayon-bandengaren," *Trouw*, 19 October 1996.

57 H. Lintsen, J. Schueler, F. Veraart, "Naar een heroriëntatie van de research, 1970–1985," in: *Research tussen vetkool en zoetstof. Zestig jaar DSM Research 1940–2000*, ed. H. Lintsen (Zutphen: Walburg Pers, 2000), 82–103, especially 98.

58 AKU-Research, 41.

processes."[59] The report mentioned that the strength of tire yarn had increased during the last ten years by 50 to 60%.[60] Quality improvement and production control still were important roles of the laboratory, but the demand for basic research grew fast because it was seen as "necessary to keep up with the increasing competition."[61] Because of this, a new laboratory with a volume of 100,000 cubic meters was opened in 1959.[62]

4 Academic Polymer Science and Its Relation to Industry

Not only industrial laboratories were expanding. Around 1960, the measures to upgrade the universities started to bear fruit.[63] The number of full professors increased explosively, and a lot of new professorships were created – many of them in new subjects such as polymer chemistry. Several of the new positions – including those in polymer chemistry – were taken up by scientists previously working on these subjects in industry.[64]

This was not surprising because industry was leading with respect to polymer science. All polymer professors appointed before 1970 came from institutions or companies like TNO, DSM, DuPont, NYMA (another Dutch rayon producing company) and especially from AKU. Pioneer of the new Dutch academic polymer science was Jan Josef Hermans (1909–1997). In 1937 he defended his PhD thesis in Leiden, and in 1942 he went into hiding at AKU's *Cellulose-Instituut*, where Petrus Hendrik Hermans (not related) introduced J.J. Hermans into the physical chemistry of cellulose.[65] In 1946, J.J. Hermans was appointed a professor in physical chemistry at the University of Groningen. There he spent his time almost exclusively on the physical chemistry of polymers.[66] In 1953 he left Groningen for a similar position at the more prestigious University of Leiden. When he moved to the United States in 1958, he was succeeded by the physical chemist Staverman.

---

59 Ibidem.

60 *Jaarverslag* AKU (1956): 11–12.

61 *Jaarverslag* AKU (1958): 14.

62 *Jaarverslag* AKU (1959): 12.

63 Hutter, "Chemie, chemici en wetenschapsbeleid," 26.

64 Homburg, *Speuren op de tast*, 42–43; Challa, "Polymeren," 231; Homburg and Palm, *Geschiedenis van de scheikunde*, 13.

65 J.Th.G. Overbeek and J.H. van der Waals, "Levensbericht J.J. Hermans," in: *Levensberichten en herdenkingen* KNAW (Amsterdam, 1998), 41–46, especially 42.

66 J.J. Hermans, *Vormeigenschappen van grote molekulen* (Inaugural lecture, Groningen University, Groningen / Batavia: Wolters, 1947), 15, 17.

Like many researchers, during the war Staverman had to flee the university. He was sheltered by the Philips NatLab for the rest of the war.[67] There he came into contact with plastics research.[68] In 1947 he was put in charge of TNO's plastics research, and became director of the central laboratory of TNO from 1954 up to 1970.[69] Staverman became a professor in Leiden in 1958 and like his predecessor he conducted research on the physical chemistry of polymers.

In Delft, Houwink had briefly held a position in macromolecular chemistry, but this was a modest position and certainly not a professorship. The person who started to cultivate the polymer field in Delft was Dirk Willem van Krevelen (1914–2001), who already held an extraordinary professorship in energy generation since 1952. Van Krevelen was offered this professorship because of his expertise in coal science – for years he was research director at the Central Laboratory of DSM. In 1959 he made a career change and became research director and member of the board of AKU. A year later, he had his extraordinary professorship changed into 'process engineering of polymers.'[70] In 1962, Willem Prins (1929–1974) was appointed professor in physics and chemistry of macromolecular materials. He kept this position in Delft until he moved to the United States in 1968.

When J.J. Hermans accepted a position at Leiden University in 1953, no polymer expert was attracted to replace him at Groningen University – after all, he was a professor of physical chemistry, not a formal polymer chemist. In 1963 Derk Heikens (1921–2014) was appointed at Groningen University as an extraordinary professor in technical polymer chemistry.[71] Heikens had studied with J.J. Hermans when the latter was already a professor there. In 1950 Heikens started working as a chemist at J.J. Hermans's former employer, AKU's *Cellulose-Instituut*.[72] Heikens learned the tricks of the polymer-trade from J.J. Hermans's former boss and teacher, the institute's charismatic and creative professor and director P.H. Hermans. This other Hermans, who was a physical chemist by training, conducted groundbreaking work on x-ray analyses of polymers. In 1952, he even published his own textbook – not on polymer chemistry, but an advanced text on theoretical organic chemistry, *Inleiding tot de theoretische organische chemie*, in which he advocated the use of quantum

67 Homburg, *Speuren op de tast*, 37.
68 As follows from the credit-lines of some of his articles that appeared in 1947.
69 Homburg and Palm, *Geschiedenis van de scheikunde*, 293–321.
70 *Plastica* (1959): 347. In Dutch "Energie opwekking" and "Proceskunde der Polymeren," respectively.
71 D. Heikens, *Apologie van D. Heikens* (typescript), 1.
72 For a short biography and birds' eye view of his work, see: F. Valkema, *Professor Derk Heikens. Polymerenman in hart en nieren* (Eindhoven, 2011).

mechanics in organic chemistry.[73] In 1959, Heikens left AKU to become a member of the scientific staff of the brand-new Eindhoven Polytechnic (Technische Hoogeschool Eindhoven, later Eindhoven University of Technology). He started together with Jhr. Emile J.J.M. de Roy van Zuydewijn (1908–1963), who came from the rayon producing company *Kunstzijdespinnerij* NYMA.[74] After a short period of combining his position at Eindhoven with his extraordinary professorship in Groningen, Heikens discontinued his work in Groningen in 1963 and became a full professor in Eindhoven. De Roy van Zuydewijn had become a professor in plastics technology the same year, but died shortly thereafter. Heikens was succeeded in Groningen in 1965 by Ger Challa (*1928), his former colleague at the *Cellulose-Instituut*. Challa had written his dissertation on work conducted there. He became director of the Cellulose Institute after P.H. Hermans retired.

In 1964, Twente University of Technology was founded, and immediately started to attract polymer professors from industry. Jan Schuijer (1964, from the Centraal Laboratorium of DSM), Adriaan Bantjes (1968, from the American firm DuPont) and Cees A. Smolders (1969, again from AKU). At the same time as the number of polymer professors started to rise significantly during the 1960s, industrial polymer research was intensified. At the laboratory of DSM, exploratory research was enhanced (*figure 12.3*).

In 1962, a separate department for basic polymer research and creating new polymers was established.[75] Also AKU's central laboratory increasingly focused on basic research regarding new polymers and polymerisation techniques.[76] This kind of research accelerated so fast that it was again necessary to expand the laboratory, which had been finished only four years before.[77] In 1963 it was decided to erect a thirteen-story building, which was ready by 1967.[78] Looking at the numbers, we can indeed tell that 'research of a basic character' started to blossom around 1963. From 1962 on, the number of scientific papers written by AKU employees was on the rise, and quadrupled within five years; it increased from 5 papers in 1963 to 20 papers in 1967.[79]

At the universities, the new professors continued the work they started at their former industrial employer. Heikens had a background in doing basic

---

73 Snelders, *Geschiedenis van de scheikunde*, 144.

74 Heikens, *Apologie*, 1. The name 'NYMA' was a contraction of the places Nijmegen and Maastricht.

75 *Jaarverslag CL Staatsmijnen* 1962.

76 *Jaarverslag* AKU (1960): 30.

77 *Jaarverslag* AKU (1962): 15.

78 *Jaarverslag* AKU (1963): 13.

79 Also see the list of AKU publications in: Vermaas, *Research bij* AKU, 97–116.

FIGURE 12.3 Hall for plastics processing at DSM, 1960
PHOTO: WWW.DEMIJNEN.NL, DSM

polymer research at the AKU *Cellulose-Instituut*, but De Roy van Zuydewijn – who came from the rayon firm *Kunstzijdespinnerij NYMA* – had a more practical background in making fibres from polymers. Soon the Eindhoven polymer laboratory was filled with rolls, calenders, extruders, spinning machines, etcetera.[80] In a laboratory setting, only small amounts of polymer molecules can be synthesized. To be able to determine fundamental properties, such as the melting point of a polymer, the simplest way was to make a very thin wire from the material, and for instance to use a known method to determine the melting point of a polymer wire.

In Delft, Van Krevelen's 1972 book *Properties of Polymers* was an achievement beyond description.[81] In this internationally renowned text, he showed the connection between chemical structure and properties of a polymer. This method became known as the *Van Krevelen Method for Computational Modelling of Polymers*.[82] But Van Krevelen's academic work cannot be separated from the industrial atmosphere. The Delft laboratory was still small; when

---

80 D. Heikens, personal communication to the author, 31 March & 12 April 2011.

81 D.W. van Krevelen, *Properties of Polymers: Their Correlation with their Chemical Structure and their Estimation* (*and Prediction from Additive Molar Group Contributions* (Amsterdam: Elsevier, 1972).

82 J.A. Moulijn & H. van Bekkum, "De alleskunner. Van Krevelen: recordhouder chemische vernoemingen," in: *Chemie achter de dijken. Uitvindingen en uitvinders in de eeuw na Van 't Hoff*, eds. H. van Bekkum, J. Reedijk, S. Rozendaal (Amsterdam: Edita, 2001), 46–47.

Prins was appointed as a polymer professor in 1962, he described the Delft laboratory as being "in the foetal stadium."[83] At the same time, during the 1960s, the AKU central laboratory was well staffed and equipped and full of activities. Almost every week a new polymer was made, and it was examined whether the properties of the polymer were useful for new or improved applications. While working on this topic, the AKU researchers gained good insights in the relationship between physical properties and chemical structure. They kept record of all their findings in countless binders, which served as a basis for Van Krevelen's book.[84]

It is safe to assert that Dutch academic polymer research had a strong relationship with industrial practice. When he was appointed as dean of the chemical technology section in Eindhoven, Heikens invited four research leaders of big companies for advice. His plan was to match the profile of academic research with those of the industrial research. Every two years he asked industrial research leaders for input and feedback on the academic program.[85]

In many cases, large companies hired polymer professors as advisors. Heikens was an advisor for AKU, DSM and GE Plastics for many years.[86] The Groningen professor Challa was active as advisor for AKU, TNO and Philips NatLab. He also conducted some mutual research with his former company AKU, for shorter or longer periods of time.[87] Industrial laboratories were much better equipped for polymer research. They were constantly updated with new and more expensive machinery. Universities did not have the money to always buy state-of-the-art machinery. Until the nineteen eighties universities only got limited funding for polymer research from the Dutch science council ZWO (renamed NWO in 1988). Sometimes Eindhoven University would get some funds or equipment from GE Plastics in Bergen op Zoom. Only TNO got substantial funding from industry up to the 1970s, because of its expertise in fundamentals of polymerisation processes and the mechanical properties of polymers. In 1967, DSM spent 30% of its total research budget on polymers, from which 70% was spent on external research. TNO got an annual fee from

83 W. Prins, *Over elasticiteit en zwelling* (Inaugural lecture, Delft University, 1962), 18.

84 Interview with J. Bussink, 7 October 2011. It is said that the AKU employee Hoftijzer had practically written the entire book, but neither he, nor other scientific AKU employees shared in the credits; L. Struik, personal communication to the author, 16 December 2011.

85 Heikens, *Apologie*, 2. When his term as dean ended, so did this initiative.

86 Ibidem, 10.

87 G. Challa, "Veel meer polymer," in: *Werken aan scheikunde. 24 memoires van hen die de Nederlandse Chemie deze eeuw groot hebben gemaakt* (Delft: Delftse Universitaire Pers, 1993), 465–482, especially 468, 481.

DSM of 340,000 guilders (corresponding to 760,000 euro today), and consultant Heikens a "meagre" 7,000 guilders (corresponding to 15,500 euro today).[88]

Mutual research often took the shape of a academic dissertation. Companies sponsored PhD students in two ways. When they believed a certain topic could be relevant for them, companies supplied the student with funding to conduct research and to compose a dissertation. Also, industrial companies were quite liberal in making their cutting-edge laboratory equipment available to PhD students who had to work with outdated equipment at the universities. It was no secret that up to the 1970s, the real chemical expertise was found in the industrial polymer laboratories, which were better equipped and better staffed then their academic counterparts. Polymer professors benefited from the lasting connection with their former employees. As a result, for the 90 dissertations that were written on polymers between 1945 and 1970, half of the research involved was sponsored by and/or took place at industrial laboratories.[89] When the AKU research started to expand in the 1960s, so did the number of AKU-sponsored dissertations. From 1945 to 1962, only four AKU-sponsored dissertations appeared. From 1963 to 1970, the number of dissertations suddenly increased by ten – all of them had former AKU researchers as supervisor. Nine of them graduated from Delft, where they profited from the presence of research director and AKU board member Van Krevelen. One graduated in Groningen, with Challa. In total, between 1945 and 1970 AKU sponsored fourteen, TNO eleven and BPM/SHELL eight PhD dissertations.

A quarter of a century after the Second World War, there were a Dutch co-founded international journal on polymers, new professorships in polymer chemistry and technology, and a steadily rising number of PhD students. Polymer science had spread from industry to *academia*, and from physical chemistry to chemical technology and polymer chemistry. But until the 1960s, students could not graduate in polymer chemistry. This changed in 1967, when at the University of Groningen the first independent polymer laboratory (all others where affiliated with chemical technological groups) was founded.

Did this mean that now polymer science had finally become a separate academic discipline? Not according to a lot of other academic chemists. They still believed polymers belonged to the fields of colloid chemistry, physical chemistry and organic chemistry. A battle of many years still had to be fought to gain

88 *Jaarverslag Centraal Laboratorium van de Staatsmijnen* 1967, Central Archive DSM, Heerlen.

89 The primary source to track all Dutch PhD theses in polymers for the period 1946–1979 is *Catalogus van academische geschriften in Nederland verschenen* (Utrecht: Bibliotheek der Rijksuniversiteit, 1952–1979). After that, I went through all theses to see whether the credits, introduction or post scriptum contained indications of company involvement.

full entrance in the academic world. A still strong belief in the fundamentally different nature of so-called basic and applied research, meant that polymer scientists were excluded from ZWO and SON funding and were denied access to critical resources.[90] Only as late as 1973, the SON-board agreed to a modest start of an informal *study group* Macromolecules. Such a study group could organize meetings and conferences, but did not have access to SON funding. Polymer chemists first had to prove themselves as academic researchers. In 1979 the new SON-section Macromolecules finally started its work.[91] Until that time, the number of dissertations slowly grew. Other new disciplines also settled at the universities, all of them demanding new and expensive equipment. At the Eindhoven University of Technology, Heikens was challenged to use all his creative skill to get as much as he could from the scarce recourses. Everything had to be paid for from the tight university budget. Supportive staff grew slow, and the number of dissertations was low.

5 Industry and University: Blending Knowledge between Locations

It is clear that industry had good contacts with the academic polymer groups. Dutch industry kept abreast with academic publications, and modestly sponsored their former employees by paying for almost half of all PhD students during the period 1945–1970. This required a relatively small investment, because these students almost always proved to be an asset to the company's polymer laboratory after they had finished their studies. When students conducted industry-sponsored research, they were left alone and were free to do their research and publish on it as they pleased. Polymer professors regularly visited industrial research laboratories as advisors, although real research collaboration was rare. Patents where academic polymer scientists were mentioned as inventor did not exist. The proverbial exception to the rule was a patent of the Eindhoven professor Heikens that was assigned to his former employer, AKU. It was largely based on Heikens' already existing knowledge of polyester, a material he had worked with at AKU. His knowledge on crystallization of the material enabled him to come up with an idea that solved the problem of processing

90 In the interwar period, it happened that scientists whose work was too 'technical' and flirted with industry were denied professorships and positions in the Netherlands Academy of Arts and Sciences (KNAW). Examples are the physicists Michels and Ornstein, see: Hollestelle, *Paul Ehrenfest*, 194, 198–199.

91 D. Heikens, personal communication to the author, 31 March & 12 April 2011; G. Challa, "Staverman en de macromoleculen: een revue anno 1981," in: *Liber Amoricum A.J. Staverman* (n.pl. [Leiden], 1981), 19–24, especially 23–24; Challa, "Polymeren," 241.

the material into something other than yarn. After considerable further research and patents on processing and additives for the material, Heikens's patent enabled AKU to start with the production of a new type of polyester called Arnite, that was suitable for injection moulding of polyester objects.[92]

So direct spill-over of knowledge was rare and when it happened, it was because Heikens had mobilized the knowledge he obtained before he started his university career. AKU still asked for his regular advice, for instance because he knew a lot of the polymerisation mechanism of caprolactam. Much knowledge was transferred only through personal contact. The case of AKU's work on a new plastic called polyphenylenether or PPE provides us with a more telling example. In 1960, General Electric (GE) had started a small pilot plant in the USA for commercialising the production of this new plastic PPE.[93] At the same time, AKU started research on PPE as well; it was looking for a material that could be used as insulation for high voltage power cables. At his own initiative, Jan Bussink, a young chemist working at AKU's Central Laboratory, discovered the original 1959 publication on PPE by A.S. Hay. As it turned out, PPE was heat-resistant, flame-resistant and had a low electrical conductivity. In principle, it had better qualities than the widely used cable sheathing made from oil-drenched paper used on low voltage cables and hence was an attractive new polymer.

The mechanism that produced PPE, the so-called oxidative coupling, was completely new and not understood. After some months of research, Bussink had gained a lot of knowledge on the way the mechanism functioned. At a conference in the United States, he commented on a lecture of a GE chemist on the matter. The present GE personnel at the conference panicked; Bussink had undermined the basic patents of GE. Because AKU had built up advanced knowledge on the mechanism so quickly, and because GE already had their hands full on developing polycarbonate, GE sought rapprochement with AKU. Both companies joined their research to get rid of the difficulties with the production of PPE, and in 1964 they started the joint venture *Polychemie AKU-GE* in Arnhem.[94] AKU researched the possibilities for making PPE film, and

---

92 *Jaarverslag AKU* (1966), 12; R. de Jong, "Grootse moleculen," in: *Chemisch Weekblad* (26 March 2011): 30–31. D. Heikens and J.C.F. Kessler, NL pat. 131857: "Werkwijze voor het spuitgieten van voorwerpen, alsmede deze voorwerpen." Source: http://nl.espacenet.com/ (30 January 2012).

93 For reconstructing the history of PPE in the Netherlands, I strongly lean on two interviews I held with Prof. Drs. Jan Bussink, at 7 October 2011 and 16 January 2012. PPE was also called PPO, poly-phenylene-oxide.

94 *Jaarverslag AKU* (1964): 13–14. Vermaas, *Research bij AKU*, 40.

obtained the first results of its research on PPE fibres in 1966.[95] Commercially, film and yarn proved to be not feasible. The paper industry saw the threat and developed a better kind of paper that was well capable of being used in high voltage 500 kV cables as well.[96]

But because it was such a promising material, further use of PPE was explored, along with the possibilities to produce the material commercially. This proved to be quite a challenge, to put it mildly. The problems with PPE were numerous and concerned virtually all essential properties of a practical and useful polymer. To name a few: the exact chemistry of the polymerisation was still unclear. The reaction produced water, and water killed the polymerisation process. The molecular weight (length of the molecular chains, which determined the properties of the material) was hard to control; there were no chemical reactions that stopped the growth of the polymer. Polymerisation produced a massive amount of heat, and the reaction had to take place at a specific temperature, otherwise the material discoloured and also the wrong chemical elements would enter the polymer chain. Synthesis of the monomer that built the polymer chain was difficult. To make the oxidative coupling work, pure oxygen was bubbled through the mixture. In the first PPE plant in Pittsfield, GE found out the hard way how dangerously fast the reaction was. In no time all the monomer had polymerised and the excess oxygen built up at the top of the hot reaction vessel. This resulted in a tremendous explosion that blew the lid of the vessel clear off and through the factory roof. The ongoing problems caused Bussink to adorn PPE with the nickname "Permanent Pain Everywhere."[97]

A rather essential problem of PPE was that it was difficult to process. To make a plastic product out of it, it had to be heated and injected into a mould. But PPE could be injected only at a very high temperature. This meant that the material started to decompose before it could be processed. GE tried to come up with a practical solution.

In 1966, GE discovered that PPE was miscible with polystyrene (PS). This was a remarkable find, since most polymers are incompatible with one another and tend to produce separate phases when mixed. Mixing PPE with PS yielded 'Noryl': a mixture that became viscous at a much lower temperature and was suitable for injection moulding, retaining all the benefits of PPE. Suddenly, a whole new field of possibilities emerged.

---

95 *Jaarverslag AKU* (1966): 12.

96 H. Buiter, *The history of the power cable in the Netherlands* (Eindhoven: Stichting Historie der Techniek, 1997), 107–120, especially 113; Interview with Jan Bussink, 16 January 2012.

97 Interview with Jan Bussink, 16 January 2012.

Firms started mixing all kinds of polymers by trial and error. At Eindhoven, Heikens became interested in polymer blends and presumed that polymers – such as metals – could be blended into alloys, thus combining properties of the separate constituents. The first results of the investigation became visible after six to seven years.[98] Eindhoven became the first university laboratory to conduct systematic research on the influence of the mixing ratio on properties and on the adhesion of the different constituents within the blend. Heikens's work on polymer blends that had started in 1966 was inspired by the blending of PPE with PS. According to Challa, "Heikens's entire research was brought up in this atmosphere. When you read Heikens's first publications, he always mentions this [the PPE-PS blend, MH] as an example...."[99]

In the meantime, AKU pulled out of polychemistry in the late 1960s. GE took over, and established a large branch at Bergen op Zoom. They became the largest manufacturer of blends and composite plastics.[100] Bussink became research manager, and developed GE Plastics into a "fantastic institute."[101] He kept in touch with his former AKU colleagues Heikens and Challa. Heikens was an advisor to GE for a long time. The resources in Eindhoven were scarce, and Heikens and his crew had to do with a tight university budget. Bussink helped out Heikens by sometimes donating an amount of money to buy equipment, and on one occasion he even surprised Heikens with a truck load of new machines.[102] But Heikens had his own vision on research, conducting basic science with a view to potential applications but free from industrial interference.[103] It seems that the research at Eindhoven was inspired by industry, but in its turn did not affect it. In his 1993 overview of the developments in PPE and its blends in the *Kunststoff-Handbuch* for industrial use, Bussink never cites Heikens and his students.[104] GE Plastics developed a lot of practical knowledge on mixing polymers. Nevertheless, Heikens and Bussink regularly discussed what did or did not work and why.[105] Importantly, Heikens's PhD student Marijn Dekkers was employed by GE, and here he noticed that the most important thing that Eindhoven brought to GE was a model of how polymers mix, how

---

98 Heikens, *Apologie*, 8; Ibid., *Vrijheid in onderzoek en onderwijs* (Farewell lecture, Eindhoven University, 1986), 8.

99 Interview with G. Challa, 16 September 2011.

100 Valkema, *Professor Derk Heikens*, 34.

101 Interview with G. Challa, 16 September 2011.

102 Interview with Jan Bussink, 7 October 2011.

103 D. Heikens, personal communication to the author, 12 April 2011.

104 J. Bussink, W.L. Sederel, W. Minderhout, "Polyphenylenether-Blends," in: *Kunststoff-Handbuch*, eds. G.W. Becker and D. Braun (München: Hanser, 1993), 25–147.

105 Valkema, *Professor Derk Heikens*, 34.

morphology comes about and is reflected in properties and processing. "Because you understood the model, you understood how things went in the real world and why certain things worked."[106] So Eindhoven's know-why augmented GE's know-how on polymer blends. Bussink also confirms that Heikens had very original ideas that could put things into perspective. This is the kind of knowledge transfer between locations that does not take place in a published or codified shape. It is a tacit kind of knowledge, only transferred on a personal level. This is why Bussink and GE were interested in personal contacts with Heikens and his students. Heikens proved to be an original discussion partner, and Bussink attracted the best students from Eindhoven to be deployed in the polymer laboratories.[107]

Bussink had good contacts with his former AKU colleague Challa as well. While Heikens was working on the mechanical properties of blends, Challa had specialized in the chemistry of polymers. Groningen was farther away, and Challa did not have the same kind of close personal contact with Bussink as Heikens did. But also in Groningen students were sponsored by AKU or GE, and Bussink and Challa stayed in touch. While working at AKU, Challa had already become acquainted with research on PPE; Heikens had left AKU in 1959 and Challa stayed until 1966. Hence, at the AKU laboratory Challa had already worked on crystallisation of PPE.[108] When the Groningen polymer laboratory was build up and polymer science got accepted by SON and ZWO in the mid-1970s, Challa started research on topics that were connected with the work of the GE laboratory: blending of polymers, and related topics such as template polymerisation – in which two polymers do not mix, but one monomer grows along another polymer chain – and polymer-bound copper catalysts to be used in the production of PPE.[109] "All of this is due to Bussink's activities. Otherwise we would never have come up with this in the Netherlands."[110] People from the laboratory of GE Plastics were sometimes asked to provide Groningen with their expertise on a certain topic, and co-authored papers together with the Groningen polymer scientists. GE provided Groningen with occasional financial donations to buy a piece of equipment.

When asked, Heikens and Challa claim they did "basic research." But in fact, they almost always kept a keen eye on the practical use of their work. Heikens stated that with the results of his research in hand, industry was able

106 Ibidem.

107 Interview with Jan Bussink, 7 October 2011; 16 January 2012.

108 Interview with G. Challa, 16 September 2011; Challa, "Veel meer polymeer," 467.

109 Ibidem.

110 Interview with G. Challa, 16 September 2011.

to investigate the possibilities of every new polymer blend.[111] Challa puts it even more directly: "We *had* [my italics, MH] to start working on new polymers in the Netherlands, because GE had developed PPE, and this became a topic."[112] According to Bussink: "The technological development of PPE and all its blends has played an important part in the total development of polymer technology in the last forty years. The cause of this can be found in the extensive problems that were experienced while developing this new polymer."[113] The making of PPE with a new polymerisation principle, the oxidative coupling, and the application of PPE-PS mixtures "caused major scientific and commercial interest worldwide."[114] In the construction of PPE and its blends, scientific and commercial developments went hand in hand.

6 Conclusion

So how was the emerging Dutch polymer science shaped by the different locations where it developed? Dutch industry played a leading role in developing new synthetic materials in the Netherlands, and in the process it strongly contributed in constructing the science to go with it. To develop the knowledge for making and testing new plastics, the companies not only conducted their own research, but also stimulated the development of the academic polymer field. The basic research that was done at the universities augmented the basic research done at the industrial research laboratory. From scratch, Dutch industry and academia had to build a knowledge base for a new science. Dutch scientists such as P.H. and J.J. Hermans, Staverman and Houwink built up the foundations of polymer science at industry and *academia*. When the foundations were there, and companies such as Staatsmijnen and AKU were making their first polymers based on German technology, this was a first step in learning how to make plastics at a large scale. This required a lot of experimenting and learning by trial-and-error. When some of the first industrial pioneers in the Dutch polymer field – like Heikens and Challa – moved to the universities, they took with them a toolkit of basic and practical polymer knowledge acquired by years of pioneering research. Some of this knowledge was codified in articles, but the universities needed polymer specialists who knew far more

111 D. Heikens, personal communication to the author, 12 April 2011; Interview with G. Challa, 16 September 2011; Heikens, *Apologie*, 8.

112 Cited in: Valkema, *Professor Derk Heikens*, 28.

113 Interview with Jan Bussink, 16 January 2012. A similar statement can be found in: Bussink, "Polyphenylenether-Blends," 25–26.

114 Bussink, "Polyphenylenether-Blends," 25.

than what was codified; a new field of inquiry cannot be learned just by looking at articles and patents. Students had to be brought up in this discipline, and they had to start working on the pressing problems in the field. Because their way of doing research resembled the research they had done at the basic polymer laboratory departments in industry, and because in those days it was the dictum, they declared that they conducted "basic research." However, they acknowledged a fine and almost non-existing line between so-called 'basic' and 'applied' research. Their contacts with former colleagues in industry remained strong at a personal level. Because of this, there was a fairly strong interaction between industrial and academic research. For most common R&D questions, industry was serviced by TNO; academic research, by contrast was not done until the nineteen eighties. But although explorative academic research was not tailored to the exact immediate needs of the industrial research laboratory, it could provide a knowledge base from which industry could move forward on unfamiliar territory. Industry contributed in some way to the work of half of all PhD students, mainly because of the possible use of their work for industry, and because companies wanted to employ these students and acquire their knowledge and skill.

If we look Dutch polymer papers and patents up to the nineteen eighties, industry and academia move in different circles. Industry patents its findings, and tends not to publish its essential findings. Academics publish but do not patent their research. But when we look at a level deeper, we see lots of connections and a lot of mutual communication going on. Industry and academia were not that separated after all. The essential kind of knowledge that travels between locations is personal – students are employed, professors move to and from industry or visit as consultants to discuss pressing matters. It is by these connections that essential information is shared and ideas emerge, giving birth to new ideas and new innovations.[115] Dutch industry and universities therefore are quite like the different constituents of a polymeric blend: they seem incompatible, but in a certain mixing ratio they complement each other.

---

115 Even in a science that is highly codified like theoretical physics, discussion and personal contact is the most effective way of knowledge transfer and formation, especially at times of frantic scientific activity on a certain topic, for instance during the development of quantum mechanics during the nineteen twenties. See: Hollestelle, *Paul Ehrenfest*, 97–152, 284–310.

Index

www.ingramcontent.com/pod-product-compliance
Lightning Source LLC
LaVergne TN
LVHW010640300926
849161LV00005B/1000
* 9 7 8 9 0 0 4 2 6 4 8 7 8 *